Atomic & Molecular Symmetry Groups and Chemistry

ATOMIC & MOLECULAR SYMMETRY GROUPS AND CHEMISTRY

S. C. Rakshit, Ph.D.
Formerly Professor of Chemistry,
Burdwan University

CRC Press
Taylor & Francis Group
Boca Raton London New York

CRC Press is an imprint of the
Taylor & Francis Group, an **informa** business

LEVANT
Levant Books
Kolkata, India

First published 2022
by CRC Press
2 Park Square, Milton Park, Abingdon, Oxon, OX14 4RN

and by CRC Press
6000 Broken Sound Parkway NW, Suite 300, Boca Raton, FL 33487-2742

CRC Press is an imprint of Informa UK Limited

British Library Cataloguing-in-Publication Data
A catalogue record for this book is available from the British Library

Library of Congress Cataloging-in-Publication Data
A catalog record has been requested

ISBN: 9781032075310 (hbk)
ISBN: 9781003208266 (ebk)

LEVANT

To My Teachers

Professor B. N. Ghosh
who led me through the puzzling areas of research into the inner recess of chemistry

Professor P. C. Rakshit
whose brilliant art of teaching was ineffable and inimitable

Professor S. K. Siddhanta
whose scintillating lectures on Group Theory opened the door

Preface

The book scripted by me namely, *Molecular Symmetry Groups and Chemistry* had a welcome response from the readers. This emboldened me to have its present incarnation, ***Atomic & Molecular Symmetry Groups and Chemistry.*** This is a trimly fattened edition of the old one strengthened by the incorporation of the different aspects of atomic symmetry.

Atomic Symmetry groups, being continuous groups, are just a fallout of the Lie Groups and Lie Algebras. The description of Lie Groups, for reasons of complexities, has been confined to the barest minimum so as not to produce much of indentations in the smooth description of atomic symmetry. Atoms are structurally simpler than molecules, but atomic symmetry is more complex than molecular symmetry. In quantum mechanics, we study atoms first and then the molecules. In symmetry studies, we just do the reverse.

Apart from the theories, the descriptions of both the symmetry groups are attended with adequate applications.

S. C. Rakshit

Contents

Chapter 1

Symmetry Elements And Symmetry Operations

When we look at certain objects or figures, we intuitively feel the presence or the lack of symmetry in them without being fully conscious of what is really meant by symmetry . A ball, a circle, a parallelogram, a formate ion–each of these possesses some kind of symmetry while the latter is lacking in a plant, in a triclinic crystal or in the structure of acetic acid molecule. A body is said to possess symmetry if certain operations performed on the body throw the latter into an indistinguishable equivalent or identical configuration. The concept inherent in this definition is linked with the ideas of a couple of terms, viz., (i) symmetry elements and (ii) symmetry operations.

1.1 Symmetry Elements, Symmetry Operations And Symbols

Objects, geometrical figures, crystal lattices, molecules or any other thing may have one, more or none of the following four types of symmetry elements.

(i) Proper axis of symmetry (Rotational axis)
(ii) Plane of symmetry (Reflection plane or mirror plane)
(iii) Centre of symmetry (Inversion centre)
(iv) Roto reflection axis of symmetry (Alternating or Improper axis)

To develop the ideas of symmetry elements, we proceed to illustrations before coining the definitions. We shall heavily draw on molecular illustrations and refer to others only sparingly.

Consider the nuclear frameworks of planar H_2O, triangular pyramid NH_3, planar formate ion $[H—C{\leq}^{O}_{O}]^-$ eclipsed and staggered C_2H_6 (Fig. 1.1a-e). In H_2O molecule the imaginary line, passing through the O atom and bisecting the $\angle$HOH, is a 2-fold rotational axis (a symmetry element) since taking advantage of it one can rotate the molecule through specified amounts, viz., $\frac{2\pi}{2}$ or $\frac{2\pi}{2} \times 2$ causing the H atoms to trade places once or twice respectively. The H_2O molecule assumes equivalent configuration in the first case and identical in the second case, but each is indistinguishable from the original starting configuration. These rotations through specified amounts are the symmetry operations. We thus realise that symmetry operations are possible only because of the existence of a symmetry element, viz., a 2-fold rotational axis in H_2O molecule.

In pyramidal NH_3 (Fig. 1.1 b), the three H-atoms constitute an equilateral triangle and the N-atom lies at the apex of the pyramid. It is easy to imagine a line passing through the N-atom and the centroid of the equilateral triangle. This imaginary line serves as a 3-fold rotational axis (a symmetry element) permitting symmetry operations of rotations through $\frac{2\pi}{3}$, $\frac{2\pi}{3} \times 2$ and $\frac{2\pi}{3} \times 3$. While the last operation will result in identical configuration, the first two will end in equivalent configurations in which the three H-atoms merely permute their places. All these resulting configurations are indistinguishable from the initial configuration of NH_3. These three rotations about the axis are symmetry operations.

Before considering further examples, the situation is ripe now for defining an axis of symmetry formally.

A body is said to possess an n-fold axis of symmetry if a rotation of $\frac{2\pi}{n} \times p$, *where p ranges from 1 to n, of the body about the supposed axis causes the body to assume equivalent or identical configurations indistinguishable from the original.*

This n-fold symmetry axis bears the symbol C_n and the permissible rotational operations, as indicated in the definition, are the symmetry operations. Thus the symmetry element C_n permits of the symmetry operations of rotations through $\frac{2\pi}{n}$, $\frac{2\pi}{n} \times 2$, $\frac{2\pi}{n} \times 3 \cdots \frac{2\pi}{n} \times n$ which bear the symbols C_n, C_n^2, $C_n^3 \cdots C_n^n$ respectively. Viewed in this light NH_3 molecule has a C_3 symmetry axis (symmetry element) and the rotational symmetry operations are C_3, C_3^2 and C_3^3=identity operation, symbolically represented by E. Here p ranges from 1 to 3. A value of $p > 3$ for NH_3 will yield for a rotation, $\frac{2\pi}{3} \times p$, an equivalent configuration similar to either that produced by C_3 or by C_3^2 or C_3^3 and hence does not constitute a new operation. Integral values of p greater than n are thus redundant.

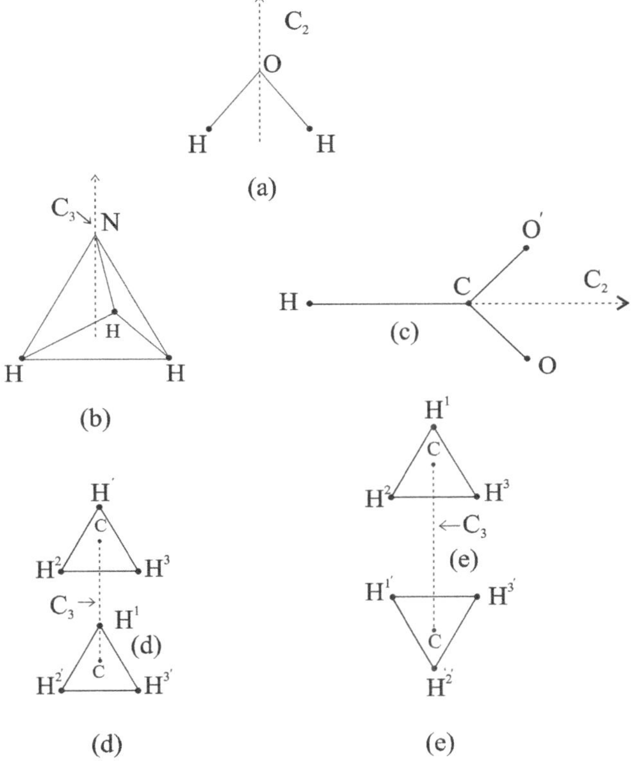

Fig. 1.1: (a) H_2O (b) NH_3 (c) Formate ion (d) Eclipsed C_2H_6 (e) Staggered C_2H_6

Reverting to instances once again, it is thus easy to see that the formate ion (fig. l.lc) has a symmetry axis C_2, collinear with H-C bond and the two symmetry operations C_2 and C_2^2=E are performable on it. Figs. 1.1d and 1.1e show the eclipsed and the staggered ethane molecules respectively. In staggered ethane, the lower CH_3- moiety is juxtaposed at a deviation of 60^0 relative to the upper half H-atoms. It is easy to discern that both the eclipsed and the staggered species have C_3 symmetry axis and hence the possible operations are apparent. Sometimes, a geometrical figure as simple as a parallelogram tends to be deceptive. A parallelogram has only one C_2 axis normal to its plane and passing through its center. The lines bisecting the opposite parallel sides of a parallelogram are not axes of symmetry.

So long we have been on to systems with one axis of symmetry (except ethane). Molecules with more than one symmetry axis are abundant. Take the examples of planar BCl_3, trigonal bipyramid PCl_5, benzene molecule and the square planar $PtCl_4^{2-}$ion. All these are characterised by possession of plurality of symmetry axes, one of a higher-fold order and the rest are each of 2-fold order. BCl_3 molecule has a C_3 axis normal to the molecular plane and three 2-fold (i.e., C_2) axes in the molecular plane (Fig. 1.2). These three 2-fold axes (C_2) are all perpendicular to the highest fold axis, namely, C_3 which is termed the principal axis. C_6H_6 has the principal axis C_6 besides six C_2 axes, all perpendicular to C_6 axis and all these six C_2's lie in the molecular plane. The symmetry axes in PCl_5 and $PtCl_4^{2-}$ are also apparent (Fig. 1.2).

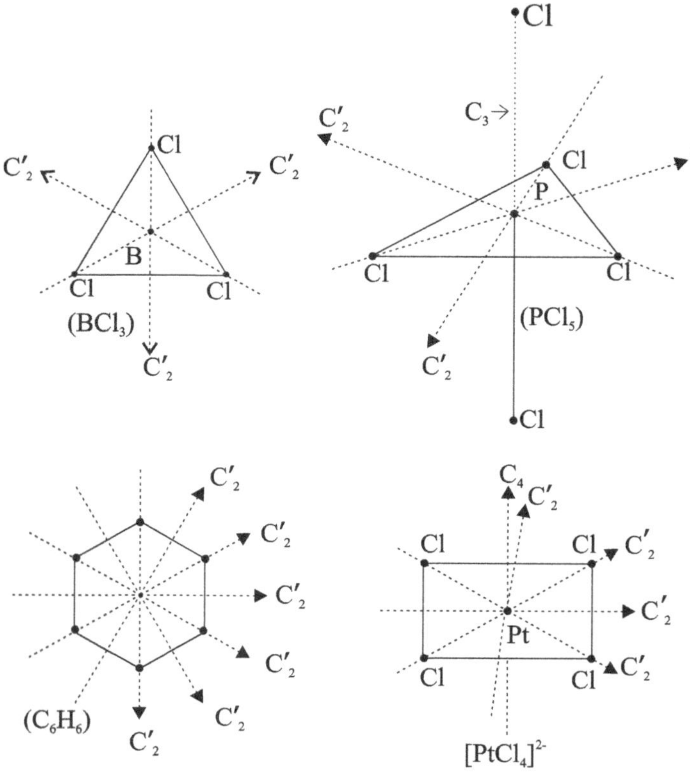

Fig. 1.2: Symmetry axes in different molecules and ions

One general conclusion emerging from such systems characterised by plurality of symmetry axes and having one highest fold axis C_n is that *there would be a total of n number of* C_2 *axes lying in a plane and to all of which* C_n *is normal.* One can now easily decide as to the type and number of symmetry axes in eclipsed C_2H_6 molecule mentioned earlier. For an orthorhombic crystal ($a \neq b \neq c$; $\alpha = \beta = \gamma = 90^0$) there are three mutually perpendicular C_2 axes; so if one of these be arbitrarily honoured as the principal axis, there would remain two other 2-fold axes conforming to the general rule.

Complications arise when one encounters objects or molecules having a number of the same higher-fold symmetry axes accompanied by a number of lower-fold ones. Highly symmetrical objects or molecules of the cubic systems such as the CH_4 molecule or $[CoF_6]^{3-}$ ion do not just have one highest-fold or a unique principal axis. There are *four* 3-fold, *three* 2-fold proper symmetry axes in tetrahedral methane molecule and in the latter case, $[CoF_6]^{3-}$, there occur *three* 4-fold, *four* 3-fold axes besides another bunch of *six* C_2 axes. A proper accounting of the symmetry axes in such tetrahedral and octahedral molecules will be taken up later (Sec. 2.3.4).

What symmetry axes are present in linear molecules? For any diatomic molecule such as N_2, CO or linear polyatomic molecule (e.g. CO_2, N_2O) the principal axis of symmetry coincides with the nuclear axis and it is of C_∞ type, since any infinitesimally small rotation about this axis results in equivalent configuration. Besides having C_∞, in N_2 or any other homonuclear diatomic molecule or in symmetrical linear polyatomic molecule (e.g. CO_2) there exist an infinite number of C_2 axes, to all of which C_∞ is normal. Heteronuclear diatomic molecules and unsymmetrical linear polyatomic molecule, N_2O, do not possess any symmetry axis other than C_∞.

All objects, figures or molecules, whether symmetrical or unsymmetrical, always possess an arbitary number of trivial symmetry axis C_1 and the consequential symmetry operation, a rotation of $\frac{2\pi}{1}$, results in identical configuration. This sort of operation is tantamount to a 'do-nothing-operation' and termed the indentity operation, E. In molecules or systems with C_n, the operation C_n^n=E, comprises the effect of C_1 and hence the existence of C_1 is not separately mentioned in such molecules or systems.

1.2 Symmetry Planes

We now turn to the second symmetry element, namely, symmetry plane and once again shall focus our attention mainly on molecules.

If in any molecule we can conceive of a plane bisecting the former such that each half is the mirror-image of the other, then this plane constitutes a symmetry plane or reflection plane. Commencing with H_2O we readily perceive the existence of two reflection planes, one bisecting the $\angle$HOH and the other being the molecular plane itself. Both these planes contain the C_2 axis of H_2O (Fig. 1.3). In NH_3, there are three

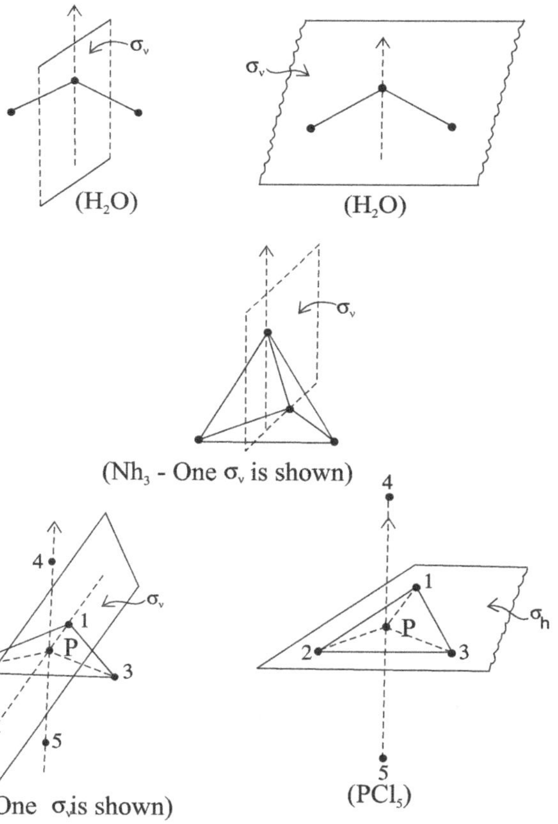

Fig. 1.3: Symmetry Planes in Molecules

such planes and each such reflection plane contains the apical N-atom, one of the three H-atoms and bisects the basal equilateral triangle. All these reflection planes contain the principal axis i.e., the C_3 axis. The symbol for these symmetry planes is σ_v; the significance of the subscript will be clarified shortly. Just like a 2-fold symmetry axis, the existence of any such reflection plane (a symmetry element) permits of only two symmetry operations of reflections, viz., σ_v. and σ_v^2 =E resulting in equivalent and identical configurations respectively. Fig. 1.3 shows one of the three σ_v elements in NH_3 molecule. In $PC1_5$ molecule (trigonal bipyramid) the equatorial Cl atoms are numbered 1,2,3 and the axial ones 4 and, 5 respectively (Fig. 1.3). Each of the planes containing 14P5, 24P5, 34P5, where P is the phosphorus atom, is a symmetry plane, σ_v and all these three σ_v's contain the principal C_3 axis. Further we detect another symmetry plane σ, passing through 123 (equatorial plane) to which C_3 is perpendicular. This latter symmetry plane is, therefore, called a horizontal symmetry plane, σ_h and allows two symmetry operations, symbolised as σ_h and σ_h^2=E, to yield equivalent and identical configurations respectively. The other three mentioned earlier, which contain the principal axis, are vertically poised with respect to σ_h. Therefore, these σ_v's are called vertical reflection planes making the significance of the subscript 'v' apparent. Arguing in the same vein, we can think of six σ_v's, all containing the principal C_6 axis in C_6H_6 molecule and one σ_h which is the molecular plane.

For certain mathematical subtleties that will be evident in the discussion of classes and characters (Chapter 5) vertical reflection planes occurring in even numbers (e.g., 4 in $PtCl_4^{2-}$, 6 in C_6H_6) are subdivided into two sets, half the number being termed σ_v's and the rest σ_d's, the dihedral reflection planes, i.e., symmetry planes bisecting the angles between two C_2 axes.

1.3 Centre of Symmetry

Molecules are said to possess a *centre of symmetry or inversion centre if a straight line joining any arbitrarily chosen atom to the supposed inversion centre when produced an equal distance beyond the centre meets a similar equivalent atom.* The existence of such an inversion centre (symmetry element) is symbolised by 'i' and the process of inverting the atoms constitutes the symmetry operations which are two in number, represented by i and i^2=E. Molecules or ions with inversion centre are many such as trans dichloroethylene, C_6H_6, $[PtCl_4]^{2-}$, CO_2, staggered ethane etc. Eclipsed ethane has no inversion centre.

1.4 Roto-reflection Axis of Symmetry

In some molecules or objects, a two-tier operation of suitable rotation about an axis followed by reflection across a plane normal to the rotating axis becomes a symmetry operation, although the axis of rotation and the reflection plane may or may not be a proper symmetry axis or a proper σ_h respectively. This *two-step operation of rotation and reflection (or vice versa, since the operations commute) giving rise to equivalent configuration is termed an "S_n" ($C_n\sigma$) operation, i.e., an improper rotation.* The corresponding symmetry element, indicated by S_n, is called a roto-reflection (alternating or improper) axis. Rather than delving into descriptions, a few examples will suit our purpose better. Consider the circle (Fig. 1.4a) containing four aerials pointing upwards and downwards alternately and located at the extremities of the X and Y axes respectively. Each aerial is twisted through $\frac{\pi}{2}$ relatively to its previous member.

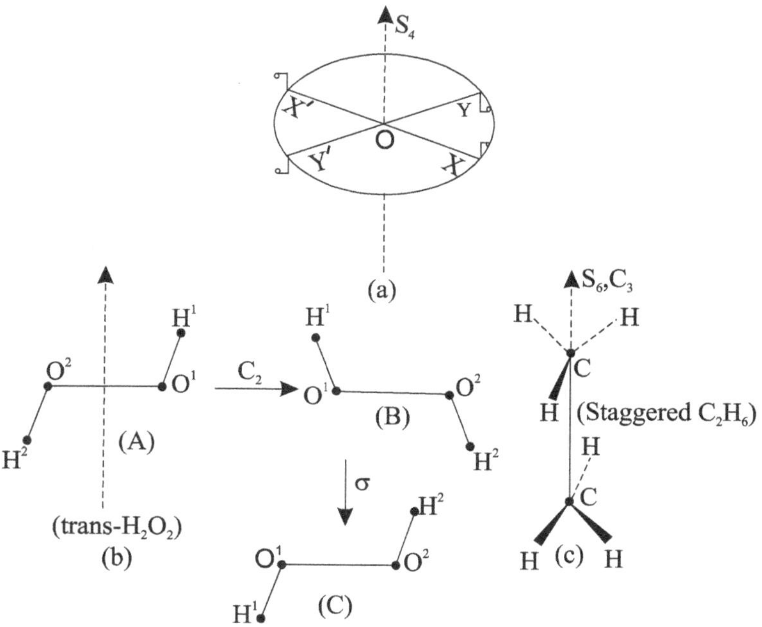

Fig. 1.4(a-b): Different S_n Axes

It is evident that the line normal to the plane and passing through the centre O is not a proper C_4 axis nor is the circular plane a σ_h. Yet an S_4 operation, i.e., a σC_4 operation (a rotation of $\frac{2\pi}{4}$ about the line passing through O and then σ across the circular plane) causes the figure to assume equivalent configuration and is, hence, a symmetry operation. The non-equilibrium planar trans form of H_2O_2 (Fig. 1.4b) has an improper axis S_2 in the plane of the molecule. A rotation of $\frac{\pi}{2}$ about the dotted line in the molecular plane produces the configuration (B) whereupon a σ operation across a plane normal to the improper axis and passing through O-O bond gives rise to configuration C indistinguishable from A. This coupled process of improper rotation and improper reflection yields the symmetry operation S_2. Classic examples of such roto-reflection axes are also provided by staggered ethane molecule (Fig. 1.4c) and ferrocene. If we consider the C-C bond of ethane molecule to lie along the Z-axis, then the three H-atoms attached to the upper carbon atom form an equilateral triangle and the lower H-atoms also constitute another such triangle of which the vertices are displaced relatively to those of the upper triangle through 60°. We now carry out an S_6 (i.e., $C_6\sigma$) on it. It is seen clearly that this staggered C_2H_6 molecule, although bereft of a proper C_6 and a proper σ_4, has nonetheless an improper S_6 axis collinear with the proper C_3. The fact that C_6 and σ commute can be verified by the reader by first applying σ on the molecule and then following it up with C_6. Ferrocene molecule (Fig. 2.7) provides an S_{10} axis collinear with proper symmetry axis C_5.

The number of distinct symmetry operations that are possible on a molecule having an S_n axis depends on whether n is an even integer or an odd one. With even n there are n operations possible including E, e.g., with symmetry element S_4, the operations are S_4, S_4^2, S_4^3 and S_4^4=E and with the symmetry element S_6, these are S_6, S_6^2, S_6^3, S_6^4, S_6^5 and S_6^6=E. Remembering that an S_n operation involves a couple of consecutive operations, C_n and σ, and further that C_n and σ operations commute it is easy to show that S_n^k operations become identical with pure rotational operations C_n^k when k is even, but are distinct (i.e., $C_n^k\sigma$) with odd values of k. A list of symmetry operations is provided here for the improper axis S_6 that occurs in staggered C_2H_6, C_6H_6 and others.

$$
\begin{aligned}
S_6 &\equiv C_6\sigma \\
S_6^2 &\equiv (C_6\sigma)(C_6\sigma) = (C_6)(C_6)\sigma^2 = C_6^2E = C_6^2 \ [\because C_6 \text{ and } \sigma \text{ commute}] \\
S_6^3 &\equiv (C_6\sigma)(C_6\sigma)(C_6\sigma) = C_6^3\sigma^3 = C_6^3\sigma E = C_6^3\sigma \\
S_6^4 &\equiv (C_6\sigma)(C_6\sigma)(C_6\sigma)(C_6\sigma) = C_6^4\sigma^4 = C_6^4 \\
S_6^5 &\equiv (C_6\sigma)^5 = C_6^5\sigma^5 = C_6^5\sigma \\
S_6^6 &\equiv (C_6\sigma)^6 = C_6^6\sigma^6 = EE = E
\end{aligned}
$$

Turning now to the case of S_n when n is *odd*, the total number of symmetry operations becomes $2n$ including E. These are S_n, S_n^2, $S_n^3 ... S_n^{2n}=E$. With S_5, the permissible operations are S_5, S_5^2, S_5^3, S_5^4, S_5^5, S_5^6, S_5^7, S_5^8, S_5^9 and $S_5^{10}=E$. Like the previous set of S_n^k, some of these ten latter symmetry operations are identifiable with pure rotations and the rest are distinct. Thus

$$
\begin{aligned}
S_5 &\equiv C_5\sigma \\
S_5^2 &\equiv (C_5\sigma)^2 = C_5^2 E = C_5^2 \\
S_5^3 &\equiv (C_5\sigma)^3 = C_5^3\sigma^3 = C_5^3 = C_5^3\sigma \\
S_5^4 &\equiv (C_5\sigma)^4 = C_5^4\sigma^4 = C_5^4 \\
S_5^5 &\equiv (C_5\sigma)^5 = C_5^5\sigma^5 = E\sigma\sigma^4 = \sigma_h \\
S_5^6 &\equiv (C_5\sigma)^6 C_5^6\sigma^6 = C_5 \\
S_5^7 &\equiv (C_5\sigma)^7 = C_5^2\sigma \\
S_5^8 &\equiv (C_5\sigma)^8 = C_5^3 \\
S_5^9 &\equiv (C_5\sigma)^9 = C_5^4\sigma \\
S_5^{10} &\equiv (C_5\sigma)^{10} = E.
\end{aligned}
$$

1.5 Multiple Symmetry Operations, Inverse Operations and Simplified Symbols for Symmetry Operations

Consider a molecule possessing a number of symmetry elements including a C_4, an i and symmetry planes σ_v's and σ_d's. If we start from the initial configuration of the molecule and carry out upon it successively two or more symmetry operations permitted by the individual symmetry elements, such as $i\sigma_v C_4^3$ (extreme right operation is to be applied first), the net result of these multiple symmetry operations (technically called a 'product') is an equivalent configuration. That is the product of two or multiple symmetry operations is also a symmetry operation. This can be easily tested by taking, say, square planar $[PtCl_4]^{2-}$ as an illustration (Fig. 1.5) where the chlorine atoms are numbered.

Now starting from some initial configuration of a molecule if we perform any permissible rotational symmetry operation (counterclockwise) C_n^k, we can always subsequently carry out an inverse operation C_n^{n-k} (still counterclockwise) to make the molecule resume its original identical configuration E. Thus C_n^{n-k} is the inverse of C_n^k. To wit, C_5^3 is

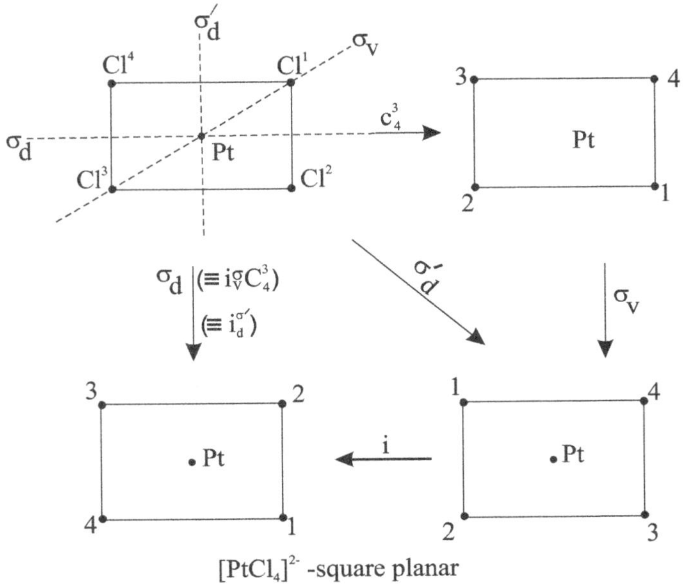

Fig. 1.5: Multiple Symmetry Operations

the inverse of C_5^2 and vice versa. This inverse C_5^3, a counterclockwise rotation, is equivalently sometimes expressed as $\overline{C}_5^2$ meaning a clockwise rotation $\frac{2\pi}{5} \times 2$. In other words, $\overline{C}_5^2$ is the inverse of C_5^2 and vice versa.

Each symmetry operation has its inverse, that is, when the symmetry operation is effected and the inverse next applied, the result (i.e., the product) is identical configuration E.

σ_v, σ_d, σ_h and i are each their own inverses. For improper rotations, S_n^k, the inverse operations are S_n^{n-k} for even n and S_n^{2n-k} for n *odd*, all rotations being in the same sense, e.g., counterclockwise. Even if one does not remember this rule, one may easily work out the inverse of any S_n^k by following the listing in Sec 1.4. Thus $S_5^4 \equiv C_5^4\sigma^4{=}C_5^4$. Hence its inverse is C_5 which is the same as S_5^6. An $S_3^5 \equiv (C_3\sigma)^5{=}C_3^2\sigma$. Hence its inverse is $C_3\sigma$, i.e., S_3.

It is a convention to simplify the symbols for proper and improper rotations by using the smallest possible numerals, wherever possible, without affecting the magnitude of the symmetry operations. For instance, the operations associated with C_6 axis are C_6, C_6^2, C_6^3, C_6^4, C_6^5 and $C_6^6{=}E$. It is easy to see that C_6^2, which is a rotation through 120°, can be expressed simply as C_3. Similarly C_6^3 and C_6^4 are equivalently and conventionally written as C_2 and C_3^2 respectively. Consequently, the set of operations with C_6 axis are, C_6, C_3, C_2, C_3^2, C_6^5 and E. Similar practice is followed in simplifying symbols of improper rotations wherever possible. Thus S_6^3 is replaced by S_2 which is incidentally equivalent to i. S_6^2 is written as C_3 and S_6^4 as C_3^2.

1.6 Choice of Origin and Axes

The choice of an origin and the setting up of x, y and z axes within a molecule not only make the description of symmetry operations more graphic but become imperative also in the mathematical follow-up of the symmetry transformation processes. Some convention is followed in regard to the choice of origin and x, y and z axes. If the molecule has an inversion centre, as in homoligand-octahedral complex or benzene, the origin is located at that point. If, however, it is lacking in inversion centre as in NH_3 molecule, the origin may still be chosen either at the centre of gravity or at some convenient point on the z-axis, the convention for the selection of the latter being the following.

Choice of z-axis:

(i) If the molecule has only one rotational symmetry axis, the latter is chosen to be the z-axis.
(ii) In case of several symmetry axes, the highest fold axis is regarded as the z-axis.
(iii) If there be several symmetry axes of the highest fold (e.g., three 4-fold axes in homoligand octahedral complexes, or three 2-fold axes in ethylene) the axis passing through the largest number of atoms constitutes the z-axis.

Choice of x-axis:

(i) If the molecule is planar and z-axis lies in the plane, positive x-direction is normal to the plane and points towards the reader.
(ii) If the molecule is planar and the z-axis is normal to it, then the x-axis, lying in the molecular plane, is so chosen as to pass through the largest number of atoms.
(iii) If the molecule is non-planar and z-axis has already been chosen, the x and y axes are chosen in such a plane normal to the z-axis as to contain a larger number of atoms than in any other plane.
In every case, choice of axes is so made as to form a right-handed reference frame.

1.7 Active and Passive Modes

While describing the symmetry transformations (operations) in the previous sections, it is implicitly assumed that the molecule is set in a fixed coordinate frame with the origin situated somewhere inside the molecule. It is the molecule which is moved, i.e., rotated (properly or improperly), reflected or inverted. This process is called an *active mode* of symmetry transformation. We can conversely think of symmetry transformation enabling a molecule achieve equivalent configuration as one in which the axial system is rotated, reflected or inverted with the molecule remaining fixed. This alternative way of performing symmetry transformations is termed passive mode of transformation. In the development of the subject we shall sometimes concern ourselves with active and sometimes with passive modes of operations depending on situations. The nature of the transformation relations against the background of assumed sense of rotation (counterclockwise) will tell the mode which is employed.

Finally, irrespective of the nature of the mode being employed in a particular instance, it is to be understood that a C_4 or C_5^3 or S_3 etc. would mean in this text a counterclockwise rotation while a bar sign, viz., $\overline{C}_4$, $\overline{C}_4^3$, $\overline{S}_3$ etc. would involve a clockwise sweep.

The definition of symmetry provided before is somewhat limited and covers systems, viz., atoms, molecules and material objects or geometrical figures or drawings. The actual definition has much wider implications and includes symmetry of phenomena, physical situation, physical theory, displacements in space and time. These demand operations that are not confined to merely fixed-origin coordinate transformations but are widely different depending on the systems concerned. The readers are asked to go through symmetry chapter of the book "*Six Not So Easy Pieces*" authored by the inimitable physicist, Richard P Feynman (*Basic Books* – Member of Perseus Publishing Group.)

Chapter 2

Groups And Molecular Point Groups

The aim of the present chapter is to acquaint ourselves with some basic concepts of the group theory and to show that the symmetry operations performable on the molecules can be woven into what is generally meant by a group. Group theory itself is an abstract and a very powerful topic of higher mathematics. But what chemists should know about it are some definitions and concepts of the fringe areas of group theory. Additionally one should have some knowledge of a number of theorems in the group representation theory at some level of sophistication. Chemists thus can avoid the inner complex fabric of the group theory and justifiably, perhaps, only nibble at it without doing the probing.

2.1 Groups, Definition, Elucidations And Multiplication Tables

A set of elements is said to constitute a group, if the following four conditions are satisfied by the member elements:

(i) The combination, technically called a 'product' (according to a defined law), of any two elements of the group is already included in the set.

(ii) The associative law holds good for the product of the elements.

(iii) The set must include an *identity element*, E, such that the product of any other element and E gives the element itself RE=ER=R, where R is any member of the group.

(iv) Each element in the set must have its *inverse* element included in the set so that the product of the element and its inverse gives the identity element E, i.e. $RR^{-1}=R^{-1}R=E$.

Conditions (iii) and (iv) demand that all the elements commute with E, the identity element, and with their own respective inverses. We now attempt to elucidate the definition and the content of the associated conditions with the help of a number of illustrations. Firstly it is to be noted that there is no restriction imposed upon the nature of the elements that form a group. The group elements may be pure numbers, matrices, geometrical operations on a body or even permutations of a set of things.

Illustrations

(a) Take the set of infinite number of integers both positive and negative including zero... -4, -3, -2, -1, 0, 1, 2, 3........ These elements constitute a group (an infinite group since the number is infinite) when the combination law is one of arithmetic addition. The "product" (here arithmetic addition only) of any two elements, say 18 and -6, which is 12 is already present as a member in the collection satisfying condition (i). This requirement is complied by other pairs of the group elements. The associative law (ii) is also obeyed for the product of any three elements, e.g., 2, 16 and -4 is $\{2+16\}+(-4) = 2+\{16+(-4)\} = 14$. Incidentally, it is to be noted that in this particular group since all the elements commute mutually, the sequence in association becomes immaterial. But commutation is not a general property for other defined laws of combinations of group elements of different natures. Hence the *sequence* in association should be preserved.

The third condition is satisfied by the group elements since one can easily pick 0(zero) as the identity element which yields the product $0 + i = i + 0 = i$, where i is any member of the group. Finally, condition (iv) is also obeyed by noting that any positive and the *corresponding* negative integers are the mutual inverses of each other. Hence this set of elements constitutes a group with respect to the combination law of addition.

(b) It is easily verified that a set of algebraic entities $\cdots \frac{1}{a^3}, \frac{1}{a^2}, \frac{1}{a}, a^0 = 1, a, a^2, a^3 \cdots$ infinite in number do form a group with the product law as one of algebraic multiplication. $a^0 = 1$ serves as the identity element and any quantity and its reciprocal serve as inverses of each other. Hence this set of elements satisfies all the formalities of a group.

(c) The infinite number of two-dimensional non-singular matrices with real elements constitute a group with respect to the combination law as one of simple matrix multiplication. The identity element in this group is $\begin{pmatrix} 1 & 0 \\ 0 & 1 \end{pmatrix}$ and since only the non-singular matrices are considered, each should have its inverse matrix already present in the set.

Apart from such infinite groups, we can also consider some examples of finite groups.

(d) Take the following six matrices

$$E=\begin{pmatrix} 1 & 0 & 0 \\ 0 & 1 & 0 \\ 0 & 0 & 1 \end{pmatrix} \qquad P=\begin{pmatrix} 1/2 & -\sqrt{3}/2 & 0 \\ \sqrt{3}/2 & 1/2 & 0 \\ 0 & 0 & 1 \end{pmatrix}$$

$$Q=\begin{pmatrix} -1/2 & -\sqrt{3}/2 & 0 \\ \sqrt{3}/2 & -1/2 & 0 \\ 0 & 0 & 1 \end{pmatrix} \qquad R=\begin{pmatrix} -1 & 0 & 0 \\ 0 & -1 & 0 \\ 0 & 0 & 1 \end{pmatrix}$$

$$S=\begin{pmatrix} -1/2 & \sqrt{3}/2 & 0 \\ -\sqrt{3}/2 & -1/2 & 0 \\ 0 & 0 & 1 \end{pmatrix} \text{ and } T=\begin{pmatrix} 1/2 & \sqrt{3}/2 & 0 \\ -\sqrt{3}/2 & 1/2 & 0 \\ 0 & 0 & 1 \end{pmatrix}$$

The set of six matrices forms a group with the product law as one of matrix multiplication. The identity element is E, T is the inverse of P and S of Q. E and R are self inverses. The group conditions are squarely fulfilled by these matrices.

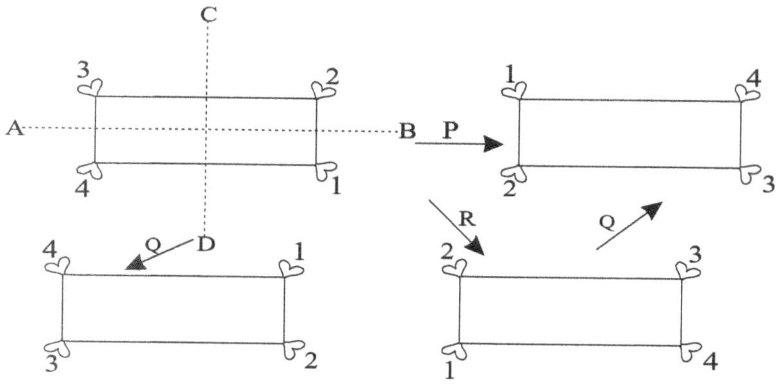

Fig. 2.1 Symmetry operations on a metallic strip. The product QR=P

(e) As one further example, let us consider the geometrical operations on a rectangular metallic strip with carvings at the corners which are numbered in order to keep tracks of these (Fig 2.1). The operations are

(i) E is a 'do-nothing operation'.
(ii) P=a counterclock rotation through π about an axis passing normally through the midpoint of the strip $\equiv C_2$.
(iii) Q=a counterclock rotation through π about the axis AB in the plain of the strip$\equiv C_2'(AB)$.
(iv) R=a counterclock rotation through π about CD axis bisecting the plain $\equiv C_2'(CD)$.

Assume the combination law to be one of execution of geometrical operations successively starting from the initial configuration. It is easily found on checking the effects of the corresponding operations on the Figures 2.1, that these four geometrical transformations do behave as the elements of a group. E serves as the identity operation and each element is its own inverse. Since each geometrical transformation here throws the figure into an equivalent nondistinguishable configuration, each such geometrical operation is a 'symmetry operation' and hence such a group may be termed a "*symmetry group*".

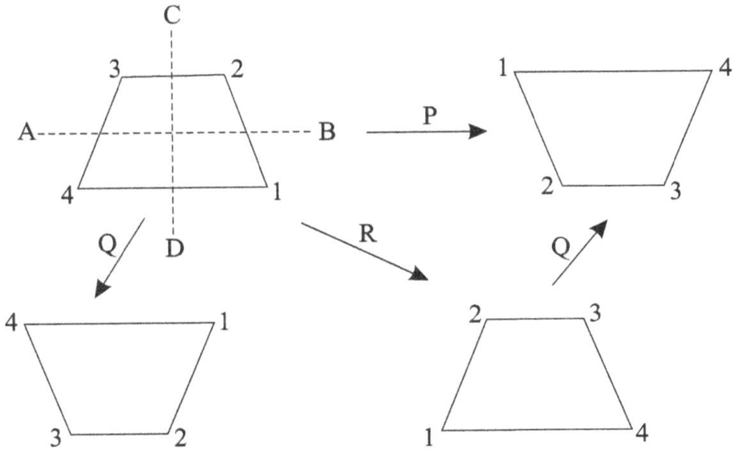

Fig. 2.2: Geometrical but not symmetry operations on a trapezium.

Had the strip been of the form as in Fig. 2.2 and the same operations E, P, Q, R been applied to it, these would still constitute a group notwithstanding the fact that P and Q would no longer be symmetry operations in this case. E, P, Q, R would, therefore, be proper elements of a group which, however, should not be termed a symmetry group and the effect need not be equivalent configuration.

Group Multiplication Table

The conditions attending the definition of a group are merely the laws which a product of the group elements is to obey. This whole set of product relations of a group can be given in a tabular form, called the group multiplication table, which is also a direct manifestation of the satisfaction of the conditions of the group. In fact each group is characterised by its group multiplication table. As an example we consider the group of six matrices of (d). To save space, we represent the matrices by E, P, Q, R, S and T and write these in order in a horizontal row at the top and in a vertical column on the left of the table. Each entry in the table is the product of the element labelling the row times the element labelling the column.

	E	P	Q	R	S	T
E	E	P	Q	R	S	T
P	P	Q	R	S	T	E
Q	Q	R	S	T	E	P
R	R	S	T	E	P	Q
S	S	T	E	P	Q	R
T	T	E	P	Q	R	S

The group elements (viz., the matrices) only appear as the products and each occurs once and only once in any particular row or column (proof Sec. 2.2.4). One can easily read the product of three or more elements from the group multiplication table. To illustrate, let us ask for the product QRP. It is seen from the table that QRP= Q(RP)=QS=E. The symmetry group of example (e) carries the following multiplication table:

	E	P	Q	R
E	E	P	Q	R
P	P	E	R	Q
Q	Q	R	E	P
R	R	Q	P	E

(f) A final illuminating illustration of a finite group is one in which the group elements are *permutations* of three electrons with artificial tags e_1, e_2, e_3 distributed at three locations with coordinates $(1)(x_1, y_1, z_1)$ $(2)(x_2, y_2, z_2)$ and $(3)(x_3, y_3, z_3)$ respectively. We start from the initial arrangement $\begin{pmatrix} 1 & 2 & 3 \\ x_1y_1z_1 & x_2y_2z_2 & x_3y_3z_3 \\ e_1 & e_2 & e_3 \end{pmatrix}$

The total number of permutation of the electrons at the given three locations is ${}^3P_3 = 6$. Starting from the initial arrangement, the six permutations are

Operations symbols	Nature of operation	Final Arrangement
P_0	Leave that arrangement unchanged	$(e_1\ e_2\ e_3)$
P_{123} (Anti-clock cyclic)	Change the electrons in positions 1, 2, 3 cyclically and anticlockwise	$(e_3\ e_1\ e_2)$
P'_{123} (clock wise cyclic)	Change the electrons in positions 1, 2, 3 clock wise cyclically	$(e_2\ e_3\ e_1)$
P_{12}	Exchange the electrons in positions 1 & 2	$(e_2\ e_1\ e_3)$
P_{23}	Exchange the electrons in positions 2 & 3	$(e_1\ e_3\ e_2)$
P_{13}	Exchange the electrons in positions 1 & 3	$(e_3\ e_2\ e_1)$

The six permutations P_0, P_{123}, P'_{123} etc. behave like the elements of a group. P_0 is the identity element E. Thus the product of any two elements, e.g., $P_{13}P_{23} = P_{13}(e_1, e_3, e_2) = (e_2e_3e_1) = P'_{123}$, which is an element of the group. Again, $P_{123}P'_{123} = P_{123}(e_2, e_3, e_1) = (e_1e_2e_3) = P_0 = E$. Thus P_{123} and P'_{123} are each other's inverses. The rest are each their own inverses. This group, having permutations as the elements is called a symmetric (not symmetry) group and is symbolized by S_n. In the present instance, the group is S_3 and is characterized by the following multiplication table.

	P_0	P_{123}	P'_{123}	P_{12}	P_{23}	P_{13}
P_0=E	P_0	P_{123}	P'_{123}	P_{12}	P_{23}	P_{13}
P_{123}	P_{123}	P'_{123}	P_0	P_{13}	P_{12}	P_{23}
P'_{123}	P'_{123}	P_0	P_{123}	P_{23}	P_{13}	P_{12}
P_{12}	P_{12}	P_{23}	P_{13}	P_0	P_{123}	P'_{123}
P_{23}	P_{23}	P_{13}	P_{12}	P'_{123}	P_0	P_{123}
P_{13}	P_{13}	P_{12}	P_{23}	P_{123}	P'_{123}	P_0

2.2 Basic Concepts and Some Theorems

It is time one gets familiar with some technical terms used in group theory. A group is finite or infinite depending on whether it comprises a finite or an infinite number of elements. A group is *Abelian* if all the elements (not just one or a few) commute with each other. When it is not so, it is *non Abelian.* Our group of six matrices (Ex. *d*), the number group of infinite elements (Ex. *a*) and the symmetry group (Ex. *e*) are all Abelian. The permutation group (Ex. *f*) is non Abelian as can be verified readily from the multiplication table. The number of elements in a group is called its *order.**

2.2.1 Generators

Generators of a group G (order g) are the minimum requisite number of elements (one or more, but less than g) the repeated self products and interproducts among which generate all the elements of the group G. The group $\{i, -1, -i, 1\}$ defined with respect to the combination law as one

*One should be conscious of the different meanings attributed to the symbols, such as P_{12}. by different authors. By P_{12} some authors convey the idea that electrons in the positions 1 and 2 change places. Accordingly $P_{12}(e_3e_1e_2)=(e_1e_3e_2)$ which is entirely consistent with the sense of permutation used by us. In another sense, $P_{12}(e_3e_2e_1)$, i.e., electrons e_1 and e_2 exchange places.

of multiplication has the single generator i, since repeated self products i, i^2, i^3, i^4 give rise to the entire group. Consider the illustrations of the groups in Sec. 2.1 Ex.(a) has a couple of generators, viz., a positive and a negative integers. In Ex.(d) the group is based on a single generator, viz., the matrix P and P^2, P^3, P^4, P^5 and P^6 generate the remaining matrices Q, R, S, T and E. The symmetry group (Ex. e) has two generators C_2 (about the vertical axis) and C_2' about AB. P_{12} and P_{23} can be chosen as the generators of the symmetric group, i.e., the permutation group (Ex. f). Of course, other alternatives for choice of generators also exist in such cases.

2.2.2 Subgroups

If within a group G we can find a set of elements which by themselves form a smaller group H with the same combination law as of G, then this H is a subgroup of G. Thus the subset of elements $\{E, P\}$ of the symmetry group (Ex. e) and also of the elements $\{P_0, P_{123}, P'_{123}\}$ of the permutation group (Ex. f) are their respective subgroups. The identity element E always constitutes a single element trivial subgroup in all groups.

2.2.3 Cosets

If within a group G we single out a subgroup H with elements H_1, H_2, ..., H_k, ..., H_h and if K_1, K_2 etc. be the other elements of G not belonging to H, then the set of elements K_1H, K_2H etc., which must all belong to G according to the product law, are called the left cosets of the subgroup H with respect to the elements K_1, K_2 etc. Like the left cosets, one can also form the right coset HK_1, i.e., H_1K_1, H_2K_1......, H_kK_1..., H_hK_1. It should be noted that the cosets are not subgroups of G since the identity element does not occur in them.

2.2.4 Some Finite Group Theorems

Theorem 2.1 *Every binary product of elements in a group (or a subgroup) occurs once and only once.*

Proof: Let $\{E,\ G_l\ G_2, ..., G_m, ..., G_{g-1}\} \in$ G (the symbol $\in$ is to be read as 'belong to'), which is a group of order g. Then the sequence of products EG_m, G_1G_m, G_2G_m, ...,G_mG_m, G_kG_m,....., $G_{g-1}G_m$, where $G_m \in G$, must also be present in G. Let us suppose, contrary to the theorem, a product occurs twice, i.e.,

$$G_kG_m = G_l = G_fG_m \qquad \text{where } G_f \in G.$$

$$G_k = G_fG_mG_m^{-1} = G_fE = G_f$$

This identity of G_k, G_f makes the group G of an order less than g, which is not true. Hence no product occurs more than once in the product sequence (cf. Sec. 2.1 multiplication table). Every binary product, therefore, occurs once and once only in a row or column.

Rearrangement Theorem

Considering the validities of the row related and the column related statements about the distribution of the group elements in the group multiplication table, it follows

'No two rows (columns) of a multiplication table can be identical. They must differ row wise (column-wise) from each other in respect of permutation of the group elements. In other words, the group elements rearrange (i.e., permute) themselves differently row wise and column wise in building up the multiplication table.

Theorem 2.2 *A subgroup H and any of its coset K_1H cannot have any element in common.*

Proof: Let $\{H_1, H_2, ..., H_k, H_l, ..., H_h\} \in H$ and K_1H be any coset. If, contrary to the theorem, coset K_1H and H possess any common element, let it be $K_1H_k = H_l$ where $H_l \in H$

$$\therefore K_1 = H_lH_k^{-1} = H_m \text{ (say)}$$

where $H_m \in H$ according to product law. Hence it follows $K_1 \in H$, which goes against the condition for the formation of coset. **[Coset condition is $K_1 \notin H$]**. K_1H should necessarily have all elements different from those in H.

Theorem 2.3 *The cosets K_1H, K_2H of a subgroup H with respect to the elements K_1, K_2 are either identical or completely different in having no element in common.*

Proof: Let us suppose that there is just one common element, $K_1H_k = K_2H_l$; where H_k, $H_l \in H$

$$K_2^{-1}K_1 = H_lH_k^{-1} = H_m \text{ where } H_m \in H$$

Postmultiplying by H

$$K_2^{-1}K_1H = H_mH = H,$$

for H_mH must generate all the elements of the subgroup H, may be, in different order.

$$\text{Coset } K_1H = \text{Coset } K_2H$$

That is, in the event of the occurrence of just one common element, all the elements in the Coset K_1H become identical with those in the Coset K_2H.

Theorem 2.4 *The order of a group is an integral multiple of the order of its subgroup (Lagrange's theorem)*

Let H of order h be a subgroup of G of order g. Now every element of G must occur either in H or in one of its distinct cosets which we suppose to be $(l-1)$ in number. Since no element in G can occur more than once and since all the elements in H and in its distinct cosets are different from each other (Theo. 2.2 and 2.3) we can write

$$G \equiv H + \text{Coset } K_1H + \text{Coset } K_2H + \ldots\ldots + \text{Coset } K_{l-1}H$$

Now since the number of elements in G is g and in H is h, it follows

$$\begin{aligned} g &= h + \{h + h + \ldots(l-1) \text{ terms}\} = h + (l-1)h \\ &= lh \text{ which satisfies the theorem.} \end{aligned}$$

2.2.5 Generators and Generation of Group Elements

The previous relation $G \equiv H + K_1H + K_2H + \ldots + K_{l-1}H$ is of prime importance in fabricating all the group elements. Suppose we desire to construct a new group starting from two generators P_1 and P_2. Using the repeated self products of P_1, viz. P_1 P_1^2, P_1^3, ...P_1^h=E we generate completely the subgroup H. The total group G≡ H+Coset P_2H. What is basically needed in this case is to start with a given set of generators. Adequate illustrations of the application of this technique will appear under molecular symmetry groups (Sec. 2.3 – 2.3.3) with which the chemists are mainly concerned. It is with this end in view that we have so long surveyed the ideas of cosets and established the theorems.

2.2.6 Conjugate Elements and Classes

Let {E, P, Q, ... Y, Z} form a group.

Two elements A, B of a group are said to be conjugate to each other if these are connected by a relation $B=RAR^{-1}$, where R is a group element. The inverse relation, $A=R^{-1}BR$, should, therefore, be automatically valid. B is the similarity transform of A and A, the similarity transform of B under R.

Applying this technique of premultiplying P by R and postmultiplying by R^{-1}, where R runs successively over the group elements, it is possible to collect such group elements as are conjugate to P. This collection, called a class, will include

(i) P, since $P=EPE^{-1}$

(ii) all such elements as are conjugate not only to P but are also mutually conjugate.

This second characteristic feature can be proved easily as follows. Suppose W and S are two members of this class and hence, conjugate to P.

$$\begin{aligned} XPX^{-1} &= W \qquad \text{or } P = X^{-1}WX \\ \text{and } YPY^{-1} &= S \qquad \text{or } P = Y^{-1}SY \end{aligned} \tag{2.1}$$

whence,

$$X^{-1}WX = Y^{-1}SY$$

$$\text{or,} \quad W = (XY^{-1})S(YX^{-1}) = TST^{-1} \tag{2.2}$$

where $T = (XY^{-1})$ must be another element of the group. This means W and S are mutually conjugate.

After finding the above class, we can find a second class containing group elements conjugate to an element other than P and not already included in the first class. In this way a complete group may be split into several classes that are mutually exclusive in view of the following theorem.

Theorem 2.5 *No two classes of a group can share a common element.*

Proof: If at all it be possible, let the element P be common to two classes and let W and U be two other elements, mutually non-conjugate, which are present in the two classes, one in each. Accordingly,

$$\begin{aligned} X^{-1}WX = P &= V^{-1}UV \\ W = (XV^{-1})U(VX^{-1}) &= ZUZ^{-1} \end{aligned} \tag{2.3}$$

Eq. (2.3) demands W and U be conjugate which is contrary to their real nature. Therefore, our assumption is wrong and no element can be common to two classes.

The element E of any group forms a single-element class in that E commutes with all other group elements. Besides, in Abelian groups since all the elements mutually commute, all the classes are single-element classes. The matrix group (Ex. *d*) decomposes into single-element classes. Discussion of further examples of classes is deferred since one will come across a plenty of such illustrations in the treatment of molecular symmetry groups (Sec. 2.3, 4.3). Finally it may be added that, like cosets, classes save the one containing E are not subgroups. The class with the element E alone is, however, a trivial subgroup.

Apart from the generalisation of some sort made in the foregoing para, the formal method of finding which of the symmetry operations of a group belong to a class of their own is obviously tedious and lengthy. These exist, however, several short cuts for ready recognition of mutually conjugate elements of a class.

(a) Reflection planes: If the body (molecule, crystal) or any geometrical figure having a set of reflection planes also possesses a suitable symmetry axis such that a definite rotation about it causes a mutual interchange of any two reflection planes or a trading of places amonst a number of them, then these reflection planes are mutually conjugate. Referring to Fig. 2.3, it is seen that a rotation of $\frac{2\pi}{3}$ about the 3- fold symmetry axis of the equilateral triangle causes the reflection planes σ_{v_1}, σ_{v_3} and σ_{v_2} to trade places. Hence σ_{v_1}, σ_{v_2} and σ_{v_3} belong to a class.

In the cases of H_2O an isosceles triangle or a rectangle (Fig 2.1 a) there are two reflection planes (σ_v's) together with one or more proper axes of rotations. But no proper rotation about the axis (H_2O, isosceles triangle) or the axes (rectangle) causes the reflection planes to exchange places. Hence, these reflection planes are not conjugate and belong to separate classes.

(b) Rotations about different axes: If a body, having proper rotational axes, also possesses reflection plane or planes, such the reflection across a suitable mirror plane makes two symmetry axes the mirror image of each other, then rotations of the same magnitude (clockwise or anticlockwise) about the two axes belong to a class. Thus the three C_2's in BCl_3 molecule are mutually conjugate. But in a square bipyraid, of the four C_2's, two belong to a conjugate class, the other two to another class

Readers may also reflect on the validity of the statement that in a sphere

(i) All possible rotations, through an angle α, about all axes passing through the center of the sphere belong to a class
(ii) All reflection planes, containing the center of the sphere, are mutually conjugate and such operations belong to a class.

2.2.7 Invariant Subgroup

Consider a subgroup $H = \{H_1, H_2, H_3, ..., H_h\}$ of the group G. Obtain all the elements conjugate to those in H with respect to R, i.e., RH_1R^{-1}, RH_2R^{-1}, ..., RH_hR^{-1}. If these conjugate elements are always the same set of group elements as are present in H for all R's in the group G then this subgroup H is termed as invariant subgroup. To illustrate, the group of symmetry operations executable on a triangular pyramid contains a subgroup E, C_3, $C_3^2{=}\overline{C}_3$ which is invariant.

2.2.8 Direct Product Group

Suppose F and G be two groups in which $(F_1, F_2, F_3, ..., F_f) \in F$ of order f and $(G_1, G_2, G_3, G_g) \in G$ of order g each with its own combination law. If all the elements in F commute with those in G, then a direct product group $A{=}F\otimes G$ may be defined which consist of the *fg* number of elements that are the interproducts of the elements in F and G. The direct product group of *fg* order follows a multiplication law which is decided by (i) the combination laws of F and G and (ii) commutation of F elements with those G. For example

$$A_pA_q = (F_iG_j)(F_kG_l) \quad = \quad (F_iF_k)(G_jG_l)$$

$$= F_m G_n$$
$$= \text{An element in A}$$

That is the product of two elements in the direct product group A is an element of the group. The principle of direct product group underlies the technique of building bigger group from the smaller ones (Sec. 2.3.8) when the group elements in both F and G are symmetry operations. In this instance,

$$A \equiv \sum_{i=1}^{f} \sum_{j=1}^{g} F_i G_j \text{ (where } E = F_1 = G_1 = \text{identity element)}$$

In other words a direct product group is the product of its subgroups (Sec. 2.3.8 for illustrations).

2.3 Molecular Symmetry Groups (Point Groups)

While discussing the illustrations of groups in Sec. 2.1, we have touched upon an example (Ex. e) in which the symmetry operations on a metallic strip serve as group elements. Naturally one then pounces upon the molecular targets as these are already known (Sec. 1.1) to be the storehouse of symmetry elements providing opportunities for symmetry operations. But the question arises 'do these symmetry operations on any particular molecule really constitute a group obeying its formal conditions'? If they do, this set of symmetry operations will constitute a *molecular symmetry group.* The answer to this query is in the affirmative. But how to prove it and classify the molecules groupwise?

The solution depends on two alternative approaches: (i) subjective and the other (ii) objective. In the subjective approach one makes a random selection of molecules or ions, completes a list of the entire set of symmetry elements and hence of the performable symmetry operations on-the selected molecule or ion and then tries to draw up a group multiplication table for it. Everytime he attempts it on any randomly selected molecule, he comes out successful and thus can really name the symmetry group (Sec. 2.3.1) to which the species belongs. But since the number of different chemical molecules is unbounded, one has to extrapolate one's conclusions on the molecules taken for trial and to brand the statement, namely, "with each molecule one can associate a suitable molecular symmetry group (point group)" as a general truth.

Let us exemplify this approach with two concrete simple cases
(a) Ammonia molecule, NH_3, with triangular pyramidal structure *and*
(b) cis dichloroethylene, 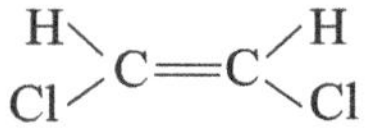 which is a planar configuration.

(a) *Ammonia molecule*: Let the equilateral triangle (Fig. 2.3) represent the trace of the base of the pyramid with the H atoms (marked

Symmetry Element	Symmetry operations	Remarks
C_3 collinear with Z-axis $\sigma\nu_1$, $\sigma\nu_2$, $\sigma\nu_3$, each being the normal bisectors of the sides opposite to the H atoms $(1, 2, 3)$ respectively and each passes through N-atom.	C_3, ${C_3}^2$, ${C_3}^3$=E $\sigma\nu_1$, $\sigma\nu_2$, $\sigma\nu_3$	The operations $\sigma\nu_1^2 = \sigma\nu_2^2 = \sigma\nu_3^2$ = E have been left out, since E operation is already included under C_3– operation set

1, 2 and 3) located at the vertices of the triangle. A cartesian coordinate frame is set up with the origin at the centroid of the equilateral triangle and the positive direction of Z-axix pointing upward towards the N atom. X and Y axes are as shown in the figure. The foregoing represents the table of symmetry elements and of the symmetry operations in NH_3 molecule.

The six operations (shown in previous page) constitute a group as is borne out by the following multiplication table:

	E	C_3	C_3^2	σ_{v1}	σ_{v2}	σ_{v3}
E	E	C_3	C_3^2	σ_{v1}	σ_{v2}	σ_{v3}
C_3	C_3	C_3^2	E	σ_{v2}	σ_{v3}	σ_{v1}
C_3^2	C_3^2	E	C_3	σ_{v3}	σ_{v1}	σ_{v2}
σ_{v1}	σ_{v1}	σ_{v3}	σ_{v2}	E	C_3^2	C_2
σ_{v2}	σ_{v2}	σ_{v1}	σ_{v3}	C_3	E	C_3^2
σ_{v3}	σ_{v3}	σ_{v2}	σ_{v1}	C_3^2	C_3	E

The entries in the table can be ascertained by checking the effects of the successive symmetry operations on the equilateral triangle in the manner as shown for the sample case $C_3^2\sigma\nu_1$ (Fig. 2.3).

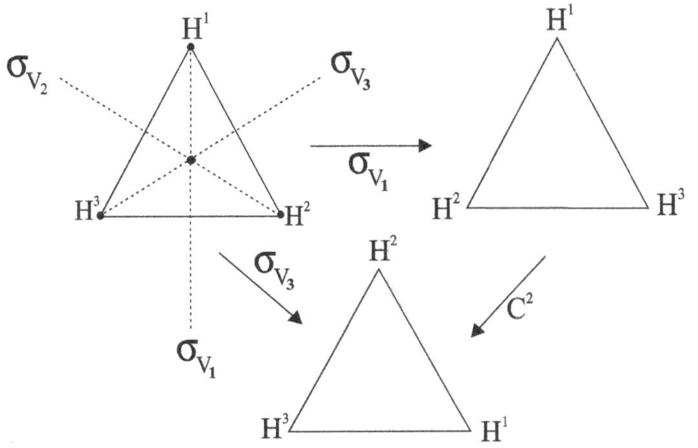

Fig. 2.3: $C_3^2\sigma_v1$ operation on ammonia molecule

(b) *Cis dichloro ethylene*: The C_2 axis is collinear with the Z-axis and lies in the plain of the molecule. There are two symmetry planes $\sigma\nu_1$, coplanar with XZ plane and $\sigma\nu_2$ with the molecular plane YZ (Fig 2.4). The distinct symmetry operations are E, C_2, σ_{v_1} and σ_{v_2}. The multiplication table appears below:

	E	C_2	σ_{v1}	σ_{v2}
E	E	C_2	σ_{v1}	σ_{v2}
C_2	C_2	E	σ_{v2}	σ_{v1}
σ_{v1}	σ_{v1}	σ_{v2}	E	C_2
σ_{v2}	σ_{v2}	σ_{v1}	C_2	E

To follow the specific entries, one may turn to fig. 2.4 where the steps leading to the final results are shown for $\sigma\nu_1 C_2$ and $\sigma\nu_1\sigma\nu_2$.

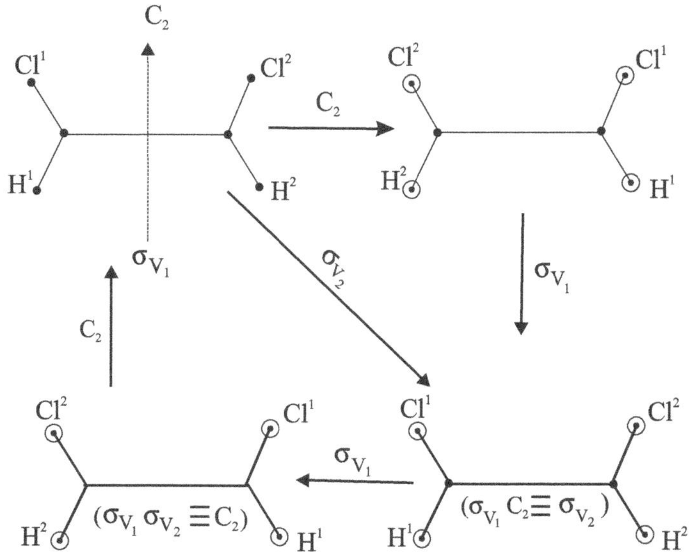

Fig. 2.4: Products of $\sigma v_1 C_2$ and $\sigma v_1 \sigma v_2$ in cis dichloroethylene; atoms within circles represent upfaces turned downward.

The molecules NH_3 and cis (HClC=CClH) treated above belong to groups C_{3v}, and C_{2v} respectively. All molecules, possessing just the set of symmetry elements as in NH_3 and no more or less, are classified under the symmetry group C_{3v}. In a similar manner, plenty of molecules can be placed under the group C_{2v}. Before taking up the alternative objective approach of forming symmetry groups, the conventions followed in the classification and naming of the symmetry groups are dealt with in the next section.

2.3.1 Classification of Point Symmetry Groups and Group Symbols

Just as symmetry elements and operations are specified by their corresponding symbols (Sec. 1.1, 1.2, 1.3, 1.4), so also are the various point symmetry groups of which examples are provided by geometrical figures, crystals and molecules. The aim of the present section is two fold. (A) a symmetry based classification of the point symmetry groups (also called point groups) and (B) an acquaintance with the conventions for attributing symbols to different point groups.

Most of the molecules and all crystals belong to one of the following four broad divisions into which the point symmetry groups can be classified.

(a) Non-axial groups, (b) Axial groups, (c) Cubic groups and (d) Special groups of linear molecules. Of course there are some molecules having symmetry higher than that of even a cube. Only a grazing reference to these will be made later.

(a) Non-axial point groups are those symmetry groups in which the set of group elements does not contain any rotational symmetry operation, C_n (except for n=l, which is trivial) but contains operations arising only from the presence of a σ or an i or even C_1. Group symbols for the non-axial groups are (i) C_1 with group element E, e.g. the molecule, CH_3COOH (ii) C_i with group elements {E, i} e.g., 1, 2-dichloro 1,2-difluoro ethane (anti) and (iii) C_s with group elements {E, σ_h}, the h subscript is a little arbitrary as there is no rotational axis normal to σ_h. Example: Pyrimidine, Quinoline.

(b) Axial point groups: These point groups comprise the following types. (i) Symmetry groups containing one n-fold symmetry axis with n greater than two. Other rotational axes, if present, should be less than n-fold. (ii) The groups where the highest fold rotational axis is C_2 itself. (iii) The groups with one improper axis, S_{2n}, and no 2nfold proper axis. The family tree for the axial group helps explain the involvement of generators in the branching of the axial groups.

The generation of the axial groups, their group elements, examples of molecules belonging to these groups and certain other aspects will be given in the sections (2.3.2—2.3.4). It will be noted, as indicated here, that all the axial point groups (C_n, D_n, C_{nv}, C_{nh}, D_{nd} and D_{nh}) are the handiworks of either one [viz., C_n], two [viz., (C_n+C_2'), (C_n+σ_v), (C_n+σ_h)] or at most three [(C_n+C_2'+ σ_d), (C_n+C_2' +σ_h)] generators. The group symbols of the axial groups are written by using the symbols C_n, D_n or S_{2n} (indicative of the axial nature of the group and also of the generators) together with the additional subscript (ν, h, d) depending on the presence of the additional generator, if any. For generation of groups like $C_{n\hbar}$, $D_{n\hbar}$ with even n, some authors use the inversion element *i* instead of $\sigma_{\hbar}$ as the generator. We have, however, used σ_h in preference to incorporation of another generator i for developing these groups.

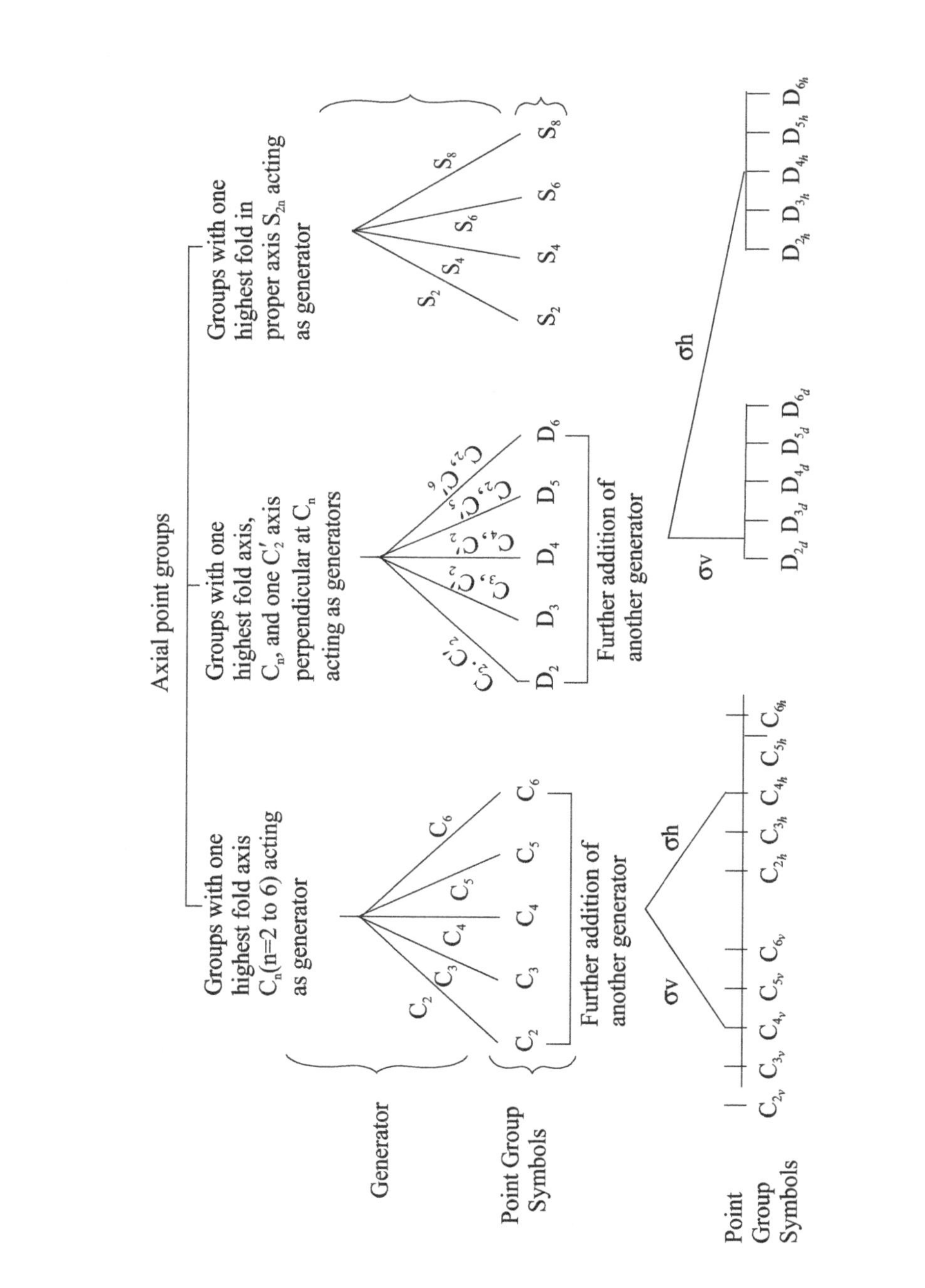

(c) The cubic point groups: These groups are characterised by the presence of several same highest-fold symmetry axes accompanied by some lower-fold ones. As the name suggests, the cubic groups include systems (crystals and molecules) ranging from the possession of a limited number of symmetry elements of a cube to those possessing the full symmetry of the cube. The cubic groups have five subdivisions (i) T (ii) $T_\hbar$ (iii) T_d (iv) O (v) $O_\hbar$. While the group T is itself formed from a few generators, the latter four are generated from T with additional generators. From the viewpoint of molecules, the tetrahedral, T_d, and the octahedral, O_h, groups are of prime importance in chemistry. In view of this a complete section (2.3.4) will be devoted to these two groups exclusively dealing with the detailed methods of the generation of these groups, the group elements and examples of molecules belonging to such groups.

(d) The special groups of linear molecules will be treated in Section (2.3.5).

2.3.2 Generation of Point Symmetry Groups: Axial Point Groups.

We now turn to the objective approach of forming point groups. The principle, already laid bare in the classification of axial groups (2.3.1), involves the following steps.

(a) Conceive of a number of simplest sets A_1, B_2, C_1....., of symmetry elements as generators. With the repeated self products of the generator and interproducts (if there be more than one generator in a set) point groups are obtained and named. Molecules (and crystals) that permit just the same symmetry operations as are obtainable from A_1 belong to the group G_{A_1}. Similarly the point groups G_{B_1}, G_{C_1} are generated from B_1 and C_1 sets respectively.

(b) To this set of generator(s) in A_1, a new generator is added producing a new generator set A_2 from which a new bigger group G_{A_2}, having G_{A_1} as its subgroup, is obtained. Equivalently, one can say that addition of a generator to the group G_{A_1} yields G_{A_2} with the help of the coset of G_{A_1} (2.2.5). That is $G_{A_2} \equiv G_{A_1} +$ Coset $K_1 G_{A_1}$, where K_1 is the generator.

Once again molecules (and crystals), carrying the same set of symmetry elements as the generators in A_2, belong to this newly framed and newly named point group. This continual addition of generators is stopped after one or two steps for one of the following reasons:

(i) want of compatibility of the added generator with the already existing symmetry elements,
(ii) scarcity of molecules answering to all the symmetry requirements demanded by the point group going to be formed,

(c) Once the exercise with A_1 is over, one takes up B_1, C_1 in turn. The process continues and terminates for each set.

We now summarily construct the axial point groups with no further fetish for explanation except for what is given under the 'Remarks' column. The nonaxial point groups C_1, C_s and $C_i(\equiv S_2)$ have already been given in (2.3.1).

Axial Point Groups:

Point group	Generators	Group Elements	Remarks
$\underline{C_n}$			
C_2	(C_2)	$\{E,\ C_2\}$	(a) No. of group elements$=n$
C_3	(C_3)	$\{E,\ C_3, C_3^2 = \overline{C}_3\}$	
C_4	(C_4)	$\{E,\ C_4,\ C_2, C_4^3 = \overline{C}_4\}$	(b) For simplified symbols used here see (Sec. 1.5)
C_5	(C_5)	$\{E,\ C_5,\ C_5^2,\ C_5^3,\ C_5^4 = \overline{C}_5\}$	
C_6	(C_6)	$\{E,\ C_6,\ C_3,\ C_2,\ C_3^2,\ C_6^5 = \overline{C}_6\}$	

Point group	Subgroup+ Generators	Group Elements ≡ Subgroup Elements+Coset	Remarks
$\underline{C_{nv}}$			
C_{2v}	$C_2 + (\sigma_v)$	$\{E, C_2, \sigma_v E,\ \sigma_v C_2\}$ $= \{E, C_2, \sigma_v, \sigma_v\}$	(a) No. of group elements=$2n$
C_{3v}	$C_3 + (\sigma_v)$	$\{E, C_3, C_3^2, \sigma_v E, \sigma_v C_3,$ $\sigma_v C_3^2\} = \{E, C_3, C_3^2, 3\sigma_v\}$	(b) Product of $\sigma_v C_n^k$ are a set
C_{4v}	$C_4 + (\sigma_v)$	$\{E, C_4, C_2, C_4^3$ $2\sigma_v, 2\sigma_d\}$	of σ_v's which form a class for
C_{5v}	$C_5 + (\sigma_v)$	$E, C_5, C_5^2,$ $C_5^3, C_5^4, 5\sigma_v\}$	n=odd and two classes for
C_{6v}	$C_6 + (\sigma_v)$	$\{E, C_6, C_3, C_2, C_3^2,$ $C_6^5, 3\sigma_v, 3\sigma_d\}$	n=even

Point group	Subgroup+ Generators	Group Elements ≡ Subgroup Elements+Coset	Remarks
$\underline{C_{nh}}$			
C_{2h}	$C_2 + (\sigma_h)$	$\{E, C_2, \sigma_h, i\}$	(a) No. of group elements=$2n$
C_{3h}	$C_3 + (\sigma_h)$	$\{E, C_3, C_3^2, \sigma_h, S_3, S_3^5\}$	(b) For products σ_h C_n^k see listings in sec. 1.4.
C_{4h}	$C_4 + (\sigma_h)$	$\{E, C_4, C_2, C_4^3, \sigma_h, S_4, i, S_4^3\}$	
C_{5h}	$C_5 + (\sigma_h)$	$\{E, C_5, C_5^2, C_5^3 C_5^4, \sigma_h, S_5, S_5^7 S_5^3 S_5^9\}$	
C_{6h}	$C_6 + (\sigma_h)$	$\{E, C_6, C_3, C_2, C_3^2, C_6^5, \sigma_h, S_6, S_3, i, S_3^5, S_6^5\}$	
$\underline{D_n}$			
D_2	$C_2 + (C_2')$	$\{E, C_2, C_2', C_2''\}$	(a) No. of group elements=$2n$
D_3	$C_3 + (C_2')$	$\{E, C_3, C_3^2, 3C_2'\}$	
D_4	$C_4 + (C_2')$	$\{E, C_4, C_2, C_4^3, 2C_2', 2C_2''\}$	(b) Products of C_2', C_n^k equal other $C_2'^{\,s}$ which are in one class when n=odd & in two classes for even n.
D_5	$C_5 + (C_2')$	$\{E, C_5, C_5^2, C_5^3, C_5^4, 5C_2'\}$	
D_6	$C_6 + (C_2')$	$\{E, C_6, C_3, C_2, C_3^2, C_6^5, 3C_2', 3C_2''\}$	
$\underline{D_{nh}}$			
D_{2h}	$D_2 + (\sigma_h)$	$\{E, C_2, C_2', C_2'', i\ \sigma_v', \sigma_v'', \sigma_v'''\}$	(a) No. of elements=$4n$
D_{3h}	$D_3 + (\sigma_h)$	$\{E, C_3 C_3^2, 3{C_2}', \sigma_h, S_3, S_3^5, 3\sigma_v\}$	(b) Products $\sigma_h C_2'$ = σ_v's which are distributed in one or two classes depending on odd or even values of n
D_{4h}	$D_4 + (\sigma_h)$	$\{E, C_4, C_2, C_4^3, 2C_2', 2C_2'', \sigma_h, S_4, i, S_4^3, 2\sigma_v, 2\sigma_d\}$	
D_{5h}	$D_5 + (\sigma_h)$	$\{E, C_5, C_5^2, C_5^3 C_5^4, 5C_2',, \sigma_h S_5 S_5^7, S_5^3, S_5^9, 5\sigma_v\}$	
D_{6h}	$D_6 + (\sigma_h)$	$\{E, C_6, C_3 C_2, C_3^2 C_6^5 3C_2', 3C_2'',, \sigma_h S_6, S_3, i, S_3^5, S_6^5, 3\sigma_v, 3\sigma_d\}$	

Point group	Subgroup+ Generators	Group Elements ≡ Subgroup Elements+Coset	Remarks
D_{nd}			
D_{2d}	$D_2 + (\sigma_d)$	$\{E, C_2, C_2', C_2'', 2\sigma_d, S_4, S_4^3\} = \{E, C_2, C_2', C_2'', 2\sigma_d, 2S_4\}$	(a) Elements=4n (b) $C_n^k\sigma_d = \sigma_d$'s all equivalent.
D_{3d}	$D_3 + (\sigma_d)$	$\{E, C_3, C_3^2, 3C_2', 3\sigma_d, S_6, S_6^3, S_6^5\} = \{E, C_3, C_3^2$ $3C_2', 3\sigma_d, 2S_6, i\}$	
D_{4d}	$D_4 + (\sigma_d)$	$\{E, C_4, C_2, C_4^3 4C_2', 4\sigma_d, S_8, S_8^3, S_8^5, S_8^7\}$	(c) All C_2' are equivalent.
D_{5d}	$D_5 + (\sigma_d)$	$\{E, C_5, C_5^2, C_5^3, C_5^4, 5C_2', 5\sigma_d, S_{10}^3, S_{10}^5 = i, S_{10}^7, S_{10}^9\}$	(d) Products of σ_d and set of C_2'/'s lead to series S_{2n},${S_{2n}}^3$... ${S_{2n}}^5$... which can then be put into classes
D_{6d}	$D_6 + (\sigma_d)$	$\{E, C_6, C_3, C_2, C_3^2, C_6^5, 6C_2', 6\sigma_d, S_{12}\ S_{12}^3 = S_4, S_{12}^5, S_{12}^7, S_{12}^9 = S_4^3, S_{12}^{11}\}$	

Point group	Generator	Group Elements	Remarks
S_{2n}			
S_2	(S_2)	$\{E, S_2 = i\}$	(a) Elements=2n
S_4	(S_4)	$\{E, S_4, C_2, S_4^3\}$	(b) $S_2 \equiv C_i$ already include in the nonaxial group.
S_6	(S_6)	$\{E, S_6, C_3, i, C_3^2, S_6^5\}$	(c) S_n group with n=odd are identical with C_{nh} point groups
S_8	(S_8)	$\{E, S_8, C_4, S_8^3, C_2, S_8^5, C_4^3 S_8^7\}$	

2.3.3 Features of Group ElementsClasses and Products

The group elements recorded in the foregoing section can be presented in somewhat more condensed form by making use of the concept of classes (Sec. 2.2.6). The class idea has been partially but not uniformly used throughout in the foregoing tables. It is useful to remember the following while writing the group elements classwise.

(i) The inversion i and horizontal reflection σ_h always form separate classes by themselves as also E.

(ii) The reflections (σ_v's) either belong to one class or split into two classes. If they are of one class, they are written as n σ_v's. If two classes are involved, these elements are grouped under two heads n σ_v's and n σ_d's.

(iii) The C_2' rotations follow the same type of groupings as for σ_v's. In D_{nd} point groups, all the C_2''s belong to a single class, so also do all the σ_d's.

(iv) Rotations of C_n point group and also the group elements of the higher groups C_{nh} and S_{2n} mutually commute. All the elements of these point groups form separate classes.

(v) In higher groups containing C_2"s and (or) σ_v's, as in D_n, D_{nh}, D_{nd}, pairs of proper rotations (C_n^k and C_n^{n-k}) form a class as can be verified by the similarity operation RAR^{-1}. Thus C_6 and C_6^5 belong to a single class and are written as $2C_6$. Similarly C_5^2 and C_5^3 of the groups D_5, D_{5h}, D_{5d} are expressed in a class as $2C_5^2$. In class notations, the lowest possible numerals are used.

The convention for grouping into classes the improper rotations of the higher point groups (D_{nh}, D_{nd}, O_h, T_d) is the same as for proper rotations of these groups. Thus S_6 and S_6^5 belong to a class ($2S_6$) and so do the group elements S_5^3 and S_5^7 to the class $2S_5^3$.

To illustrate the above classwise groupings, we rewrite the elements of some groups.

$$C_4 \rightarrow \{E,\ 2C_4,\ {C_4}^2(=C_2)\}$$
$$D_4 \rightarrow \{E,\ 2C_4,\ {C_4}^2(=C_2),\ 2{C_2}',\ 2{C_2}''\}$$
$$C_{5v} \rightarrow \{E,\ 2C_5,\ 2{C_5}^2,\ 5\sigma_v\}$$
$$D_{6h} \rightarrow \{E,\ 2C_6,\ {C_6}^3(=C_2),\ 2C_3,\ 3{C_2}',\ 3{C_2}'',\ \sigma_h,\ 2S_6,\ i,\ 2S_3,\ 3\sigma_d,\ 3\sigma_v\}$$
$$D_{6d} \rightarrow \{E,\ 2C_6,\ 2C_3,\ C_2,\ 6{C_2}',\ 2S_{12},\ 2S_4,\ 2{S_{12}}^5,\ 6\sigma_d\}$$

Attention may be drawn to the product relations under the 'Remarks' column of the group tables (Sec. 2.3.2). The products of C_2' C_n^k (with different k's) are nC_2's grouped into one or two classes and similarly for the generated $n\sigma_v$'s. Although these operations, C_2's and σ_v's are all associated with distinct symmetry axes and symmetry planes, the group elements as indicated above do not reveal their locations. This specification of locations, though often unnecessary from the viewpoint of application, can nevertheless be ascertained with a little bit of geometry and co-ordinate transformation. To acquaint the reader with the principle of specifying the locations of C_2's and σ_v's, etc., we take up the issue in the more difficult cases of the cubic groups discussed in Sec. 2.3.4.

2.3.4 Cubic Point Groups

As mentioned earlier (Sec. 2.3.1) there are five subdivisions T, $T_\hbar$, T_d, O and $O_\hbar$ under the cubic group. Of these different cubic systems the tetrahedral, T_d, and the octahedral, $O_\hbar$, groups are of extreme importance in chemistry and these two groups will mainly be discussed in this section. Unlike the axial group molecules, molecules of the cubic groups have several C_n axes with n>2. A tetrahedron can be inscribed in a cube (Fig. 2.5) with vertices positioned at the opposite corners of each face of the cube and this tetrahedron shares some of the symmetry elements (not all) of the cube.

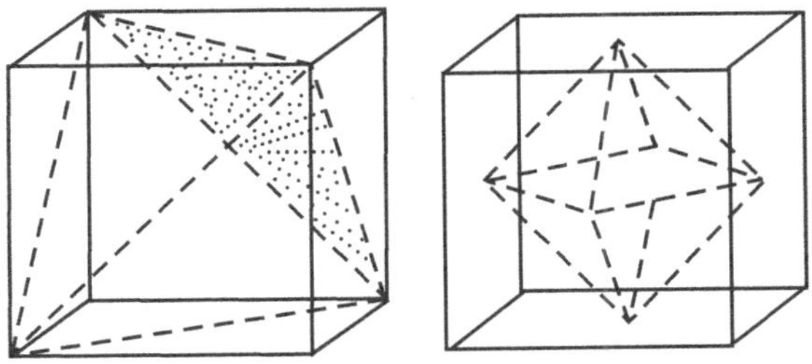

Fig. 2.5: (a) A tetrahedron inside a cube. (b) An octahedron inscribed in a cube.

On examining this tetrahedron we can make a list of the following symmetry elements and symmetry operations possible in a molecule with tetrahedral symmetry.

Table : Symmetry elements and their locations

Symmetry elements and their locations	Symmetry operations	No. of distinct operations	Remarks
(a) Four 3-fold axes collinear with the body diagonals of the cube.	4x(C_3, C_3^2, C_3^3=E)	9	All the E operations are equivalent to one operation.
(b) Three C_2-axes normal to the faces of the cube and incidentally passing through the mid points of the edges of the tetrahedron.	$3x(C_2, \{C_2^2 = E\})$	3	Operations within { } are redundant since E has already been considered under C_4 operations.
(c) Three S_4-axes collinear with C_2 axes	$3x(S_4, \{S_4^2 = C_2\}, S_4^3, \{S_4^4 = E\})$	6	Operations within { } are redundant.
(d) Six σ_d's passing through the diametrically opposite edges of the cube. Each such σ_d comprises a pair of vertices and accommodates C_3 and a C_2 axis.	$6x(\sigma_d, \{\sigma_d^2 = E\})$	6	Operations within { } are redundant

The total number of symmetry operations on a tetrahedral molecule is thus 24 and the group elements written classwise are {E, $8C_3$, $3C_2$, $6S_4$, $6\sigma_d$}. Examples of molecules falling under T_d are many of which CH_4,

CCl_4, tetrahedral homoligand complexes such as $[MnCl_4]^{2-}$, $[CoCl_4]^{2-}$ and some simple inorganic ions like SO_4'', ClO_4' are typical. As in the case of the axial groups, the generator method of drawing up the T_d group will be taken up along with that of the O_h group later in this section.

To consider the symmetry of an octahedron, we turn to Fig. 2.6 where we may suppose that the six ligands of the octahedral complex, ML_6, occupy the mid-points of the six faces of the cube. A cartesian frame of reference axes (right handed) is set up such that the x, y and z axes accommodate six apices of the octahedron. With the aid of a model, one can discern eight types of symmetry elements the nature and locations of which are listed below along with the consequential symmetry operations.

	Symmetry elements and their locations	Symmetry operations	No. of distinct operations	Remarks
(a)	Three 4-fold axes passing through the midpoints of the opposite faces of the cube.	3x(C_4, C_4^2=C_2, C_4^3,C_4^4=E)	10	Four E's are just equivalent to a single E.
(b)	Three S_4 axes collinear with C_4.	3x(S_4, {S_4^2= C_2 } S_4^3, {S_4^4=E})	6	Operations within { } are redundant
(c)	Six C_2' passing through the mid-points of the opposite edges of the cube and also through the center.	6x(C_2' { $C_2'^2$=E })	6	,,

	Symmetry elements and their locations	Symmetry operations	No. of distinct operations	Remarks
(d)	Six σ_d's each passing through a pair of diametrically opposite edges of the cube.	6x(σ_d, {σ_d^2=E})	6	,,
(e)	Three $\sigma_\hbar$ perpendicular to the C_4 axes.	3x($\sigma_\hbar$, {$\sigma_\hbar^2$=E})	3	,,
(f)	i, inversion center.	(i, $\{i^2 = E\}$	1	,,
(g)	Four 3-fold symmetry axes collinear with the body diagonals of the cube.	4x(C_3, C_3^2, $\{C_3^3 = E\}$)	8	,,
(h)	Four S_6 improper axes collinear with the body diagonals of the cube.	4x(S_6, $\{S_6^2 = C_3\}$),$\{S_6^3 = i\}$, $\{S_6^4 = C_3^2\}S_6^5$, $\{S_6^6 = E\}$)	8	,,
		Total number	48	

A perfectly octahedral molecule can, therefore, be subjected to forty-eight symmetry operations which, clustered classwise, are

$$\{E,\ 6C_4,\ 3C_2,\ 6C_2',\ 6\sigma_d,\ 3\sigma_\hbar,\ i,\ 8C_3,\ 6S_4,\ 8S_6\}$$

Examples of molecules and ions belonging to $O_\hbar$ group abound in the literature of coordination complexes. All undistorted unidentet homoligand octahedral complexes (molecules and ions) of transitional metals fall under the $O_\hbar$ group.

We now take up the construction of T_d and O_h groups using the method of generators and hence of subgroups and cosets. In order to understand the interrelations more deeply, we follow a detailed nomenclature as given below.

Proper Symmetry Axes

(i) C_4's and C_2's are expressed as C_4^z, C_2^x and so on indicating the axes collinear with the z and x directions respectively.

(ii) C_2' axes are written as C_2^{xy}, $C_2^{x\bar{y}}$ etc. The second one indicates that the line joining the origin and the point (x, -y, 0) of the cube serves at the C_2' axis.

(iii) C_3 axes are indicated by writing C_3^{xyz} or $C_3^{x\bar{y}\bar{z}}$, etc., The latter one indicates the C_3 axis collinear with the line joining the origin and the point (x, -y, -z).

Improper Symmetry Axes

S_4's are expressed in the manner of C_4's and C_2's, and S_6's in the manner of C_3's.

Reflection Planes σ_d**'s**

A reflection plane $\sigma^{\bar{y}\bar{z}}$ will indicate that symmetry plane to which the line joining the origin and the point $(o, -y, -z)$, is normal. Similar meaning is to be ascribed to the other reflection planes. Reflection planes σ^x, σ^y, σ^z will indicate the symmetry planes to which x, y and z axes are normals respectively.

The symmetry operations involving these symmetry elements will also bear similar symbols.

The simplest group of cubic symmetry is the T group having the group elements

$$\left\{E, C_2^x, C_{2^y}, C_{2^z}, C_3^{xyz}, C_3^{x\bar{y}\bar{z}}, C_3^{\bar{x}\bar{y}z}, C_3^{\bar{x}y\bar{z}}, C_3^{xyz^2}, C_3^{x\bar{y}\bar{z}^2}, C_3^{\bar{x}\bar{y}z^2}, C_3^{\bar{x}y\bar{z}^2}\right\}$$

generated from the symmetry elements comprising four C_3's and three C_2's. Now T is a subgroup of O and the latter a subgroup of O_h. The genesis of the O_h group can be understood from the following scheme.

$$\begin{array}{l} \text{T (group)} \\ \quad \big\downarrow \text{generator } C_4^{\,z} \\ [\text{T(subgroup)+coset } C_4^{\,z}\text{T}] \equiv \text{O group} \\ \qquad\qquad\qquad + \big\downarrow \text{generator } i \\ \qquad\qquad\qquad [\text{O subgroup+coset } i\text{ O}] \equiv \text{O}_\text{h} \end{array}$$

That is

$$\mathrm{O}_\hbar \equiv \mathrm{T} + \text{coset } \mathrm{C}_4^z\mathrm{T} + \text{coset } i\ \mathrm{T} + i(\text{coset } \mathrm{C}_4^z\mathrm{T}) \tag{2.1}$$

Hence the totality of the group elements emerging from the four right hand terms are the symmetry operations under $\mathrm{O}_\hbar$. Since T is given above, we first evaluate coset $\mathrm{C}_4^2\mathrm{T}$.

$$\begin{aligned} &\text{Coset}\quad \mathrm{C}_4^z\mathrm{T} \equiv \mathrm{C}_4^zE,\ \mathrm{C}_4^z\mathrm{C}_2^x,\ \mathrm{C}_4^z\mathrm{C}_2^y,\ \mathrm{C}_4^z\mathrm{C}_2^z,\ \mathrm{C}_4^z\mathrm{C}_3^{xyz},\ \mathrm{C}_4^z\mathrm{C}_3^{\bar{x}y\bar{z}} \\ &\mathrm{C}_4^z\mathrm{C}_3^{\bar{x}\bar{y}z},\ \mathrm{C}_4^z\mathrm{C}_3^{\bar{x}y\bar{z}},\ \mathrm{C}_4^z\mathrm{C}_3^{xyz2},\ \mathrm{C}_4^z\mathrm{C}_3^{x\bar{y}\bar{z}2},\ \mathrm{C}_4^z\mathrm{C}_3^{\bar{x}\bar{y}z2},\ \mathrm{C}_4^z\mathrm{C}_3^{\bar{x}y\bar{z}2} \end{aligned}$$

The technique of working out some of these products is shown below

$$\text{(i) } \mathrm{C}_4^z\mathrm{C}_2^x = \mathrm{C}_2^{xy}\text{since} \begin{pmatrix} x \\ y \\ z \end{pmatrix} \xrightarrow{\mathrm{C}_2^x} \begin{pmatrix} x \\ -y \\ -z \end{pmatrix} \xrightarrow{\mathrm{C}_4^z} \begin{pmatrix} y \\ x \\ -z \end{pmatrix} \equiv \mathrm{C}_2^{xy} \begin{pmatrix} x \\ y \\ z \end{pmatrix}$$

$$\text{(ii) } \mathrm{C}_4^z\mathrm{C}_3^{x\bar{y}\bar{z}} = \mathrm{C}_4^x\text{since} \begin{pmatrix} x \\ y \\ z \end{pmatrix} \xrightarrow{\mathrm{C}_3^{x\bar{y}\bar{z}}} \begin{pmatrix} -y \\ z \\ -x \end{pmatrix} \xrightarrow{\mathrm{C}_4^z} \begin{pmatrix} x \\ z \\ -y \end{pmatrix} \equiv \mathrm{C}_4^x \begin{pmatrix} x \\ y \\ z \end{pmatrix}$$

Thus the whole coset $\mathrm{C}_4^z\mathrm{T}$ may be found out by using a small cube as a model and noting changes in the xyz coordinate axes. When coset $\mathrm{C}_4^z\mathrm{T}$ is completely worked out, further exercises with the cube model enable us to find coset $i\mathrm{T}$ and $i(\text{coset } \mathrm{C}_4^z\mathrm{T})$. Some specimen products (two from each) are indicated below:

$$\text{(i) } i\mathrm{C}_3^{xyz} = \overline{\mathrm{S}}_6^{xyz} \text{ since} \begin{pmatrix} x \\ y \\ z \end{pmatrix} \xrightarrow{\mathrm{C}_3^{xyz}} \begin{pmatrix} y \\ z \\ x \end{pmatrix} \xrightarrow{i} \begin{pmatrix} -y \\ -z \\ -x \end{pmatrix} \equiv \overline{\mathrm{S}}_6^{xyz} \begin{pmatrix} x \\ y \\ z \end{pmatrix}$$

$$\text{(ii) } i\mathrm{C}_3^{xyz^2} = \mathrm{S}_6^{xyz} \text{ since} \begin{pmatrix} x \\ y \\ z \end{pmatrix} \xrightarrow{\mathrm{C}_3^{xyz^2}} \begin{pmatrix} z \\ x \\ y \end{pmatrix} \xrightarrow{i} \begin{pmatrix} -z \\ -x \\ -y \end{pmatrix} \equiv \mathrm{S}_6^{xyz} \begin{pmatrix} x \\ y \\ z \end{pmatrix}$$

$$\text{(iii) } i\mathrm{C}_4^z = \overline{\mathrm{S}}_4^z \text{ for } \begin{pmatrix} x \\ y \\ z \end{pmatrix} \xrightarrow{\mathrm{C}_4^z} \begin{pmatrix} y \\ -x \\ z \end{pmatrix} \xrightarrow{i} \begin{pmatrix} -y \\ x \\ -z \end{pmatrix} \equiv \overline{\mathrm{S}}_4^z \begin{pmatrix} x \\ y \\ z \end{pmatrix}$$

$$\text{(iv) } i\mathrm{C}_2^{yz} = \sigma^{yz} \text{ for } \begin{pmatrix} x \\ y \\ z \end{pmatrix} \xrightarrow{\mathrm{C}_2^{yz}} \begin{pmatrix} -x \\ z \\ y \end{pmatrix} \xrightarrow{i} \begin{pmatrix} x \\ -z \\ -y \end{pmatrix} \equiv \sigma^{yz} \begin{pmatrix} x \\ y \\ z \end{pmatrix}$$

In this way, O_h group elements are obtained and these are embodied in the table 2.1. In compiling the table counterclockwise rotation is adopted.Rotations with bar signs, $\overline{\mathrm{C}}_4$, $\overline{\mathrm{S}}_4$, carry a sense of clockwise operation.

Table 2.1: O_h group

T	E	C_{2^x}	C_{2^y}	C_{2^z}	$C_{3^{xyz}}$	$C_{3^{\bar x\bar y z}}$	$C_{3^{x\bar y\bar z}}$	$C_{3^{\bar x y\bar z}}$	$C_{3^{xyz}}^2 \Vert \bar C_{3^{xyz}}$	$C_{3^{x\bar y\bar z}}^2 \Vert \bar C_{3^{x\bar y\bar z}}$	$C_{3^{\bar x\bar y z}}^2 \Vert \bar C_{3^{\bar x\bar y z}}$	$C_{3^{\bar x y\bar z}}^2 \Vert \bar C_{3^{\bar x y\bar z}}$	
Coset C_{4^z} T		C_{4^z}	$C_{2^{xy}}$	$C_{2^{\bar x y}}$	$\bar C_{4^z}$	$C_{2^{yz}}$	C_{4^x}	$C_{2^{\bar y z}}$	$\bar C_{4^x}$	$\bar C_{4^y}$	$C_{2^{\bar z x}}$	C_{4^y}	$C_{2^{zx}}$
Coset iT		i	σ^x	σ^y	σ^z	$\bar S_{6^{xyz}}$	$\bar S_{6^{xyz}}$	$\bar S_{6^{\bar x\bar y z}}$	$\bar S_{6^{\bar x y\bar z}}$	$S_{6^{xyz}}$	$S_{6^{x\bar y\bar z}}$	$S_{6^{\bar x\bar y z}}$	$S_{6^{\bar x y\bar z}}$
i(Coset $C_4{}^z$T)		$\bar S_{4^z}$	σ^{xy}	$\sigma^{\bar x y}$	S_{4^z}	σ^{yz}	$\bar S_{4^x}$	$\sigma^{\bar y z}$	S_{4^x}	S_{4^y}	$\sigma^{\bar z x}$	$\bar S_{4^y}$	σ^{zx}

Tetrahedral group

The addition of a generator $\overline{\mathrm{S}}_4^z$ to T yields the elements of T_d, i.e., $\mathrm{T}_d \equiv$T+Coset $\overline{\mathrm{S}}_4^z$T. Table 2.2 embodies these elements.

Table 2.2–$\mathrm{T_d}$ group

T	E	C_{2^x}	C_{2^y}	C_{2^z}	$C_{3^{xyz}}$	$C_{3^{x\bar y\bar z}}$	$C_{3^{\bar x y\bar z}}$	$C_{3^{\bar x y\bar z}}$	$\overline{C}_{3^{xyz}}$	$\overline{C}_{3^{x\bar y\bar z}}$	$\overline{C}_{3^{\bar x y\bar z}}$	$\overline{C}_{3^{\bar x y\bar z}}$
Coset $\overline{S}_4{}^z T$	$\bar S_4^z$	$\sigma^{x\bar y}$	$\sigma^{x\bar y}$	S_4^z	σ^{yz}	$\overline{S}_4^z$	$\sigma^{y\bar z}$	S_4^x	S_4^y	$\sigma^{z\bar x}$	$\overline{S}_4^y$	σ^{zx}

Apart from giving a deeper insight and highlighting the specific locations of the symmetry elements and results of the corresponding operations, this detailed break-up becomes necessary for drawing up the group multiplication table. For practical applications, classwise bunched-up group elements, as shown before, are adequate. This remark applies to the simpler axial groups also.

2.3.5 Special Groups of Linear Molecules

Linear molecules like HCl, CO, CO_2, N_2O can squarely be classed under two groups $C_{\propto v}$ and $D_{\propto \hbar}$ depending on whether the molecule lacks any inversion centre (as in CO, N_2O) or possesses the latter (e.g., CO_2). Molecules falling under $C_{\propto v}$ have the internuclear line as the $C_\propto$ axis about which any rotation ranging from an infinitesimally small value to 2π will be a symmetry operation. Such molecules also have an infinite number of σ_v's all containing the $C_\propto$ axis. The $D_{\propto \hbar}$, type of molecules have the internuclear line as the $C_\propto$ axis and are also additionally characterised by their possession of infinite number of C_2' axes all normal to $C_\propto$ axis. Both $C_{\propto v}$ and $D_{\propto \hbar}$ are infinite groups.

2.3.6 Molecules of Very High Symmetry

Almost all known molecules, excepting the linear molecules $(C_{\alpha,\nu}, D_{\alpha,h})$ do not have symmetry higher than that of the cubic groups. Rarely does one come across molecules of even higher symmetry groups $(I, I_\hbar)$ i.e., the icosahedral group. The dodecahedral ion $(B_{12}H_{12})^{2-}$ synthesized in the sixties of the last century was the only species known to belong to I_h point group. The ion has 12 vertices of a regular icosahedron occupied each by a boron atom and twenty equilateral triangular faces. Very recently, the so - called buckyballs (fullerenes) with molecules containing an agglomerate of 60 carbon atoms arranged in icosahedral symmetry (I_h) have been synthesized. If the 12 vertices of a regular icosahedron are chopped off suitably, this will result in 12 regular pentagons interconnected by intervening regular hexagons. This truncated structure has still the I_h symmetry. In buckyball molecules the vertices of the twelve pentagons are each occupied by a carbon atom, Fullerenes in the solid state may be considered as an allotropic modification of carbon besides diamond and graphate The molecules, being of I_h symmetry have the element $6C_5$, $10C_3$, $6S_{10}$, $10S_6$ $15C_2'$, $15\sigma_v$ and i permitting altogether 120 symmetry operations.

Free atoms have the symmetry of a sphere. Their symmetry and group theoretical treatment are considered in chapter 11.

2.3.7 Point Groups - Molecules and Crystals, Schonflies and Hermann Mauguin Symbols

The objective method (Sec. 2.3.2) of building up the point groups for molecules is also applicable to crystals. Owing, however, to the restriction placed upon crystals for its three dimensional repeated arrangemsnt and space filling requirements, the crystallographic point groups are less numerous than the molecular point groups. All the nonaxial point groups, the axial point groups upto symmetry axis C_6 (excepting D_{4d}, D_{6d}, S_8 and those having C_5 symmetry axis) and all the cubic groups are valid crystallographic point groups.

The symmetry symbolisms used in crystal literature, called Hermann Mauguin symbols, differ from the Schonflies symbols used in the previous pages for the molecules. A few examples of the different notations are provided below:

		Schonflies symbol	Hermann Mauguin symbol
Symmetry elements	Proper axis n–fold	C_n	n
	proper axis accom– pained by C_2' axis	D_n	n2
	symmetry plane :		
	vertical	σ_v	m
	horizontal	$\sigma_\hbar$	|m
Point group		$C_2, C_{3v}\ C_{6\hbar}$	$2, 3m, 6/m$
		$D_4, D_{3d}\ D_{2\hbar}$	$422, 3m2, 2, 22$
		S_4	
		$T, T_d, O, O_\hbar$	$32, \overline{4}3m, 432, m3m$

There are also alternative notations for T_d and $O_\hbar$ group crystals.

(Additional matter to be added at the end of the Sec 2.3.7)

It should be noted that while S_4 represents a roto- reflection axis of order four in schonflies symbol, a bar sign, such as $\overline{4}$, is indicative a roto- inversion axis in international symbol. Also, the point groups of molecules are more numerous then those of crystals for which the number is thirtytwo. The following is the full list of crystallographic point groups given in both schonflies and international symbols

1	(a) Crystal System		Tri–clinic	Mono–clinic	Ortho–rhombic
	(b) Crystallo–graphic	Schonflies	C_1 C_i	C_s C_2 C_{2h}	C_{2v} D_2 D_{2h}
2	Point groups	*International*	$1\ \bar{1}$	m 2 2/m	2mm 222 mmm

1		Trigonal	Tetragonal
2	a	C_3 S_6 C_{3v} D_3 D_{3h}	C_4 S_4 D_{2d} C_{4h} C_{4v} D_4 D_{4h}
	b	$3\ \bar{3}$ 3m 32 $\bar{6}$2m	$4\ \bar{4}$ $\bar{4}$2m 4/m 4mm 42 4/mmm

1		Hexagonal	Cubic
2	*a*	C_6 C_{3h} D_{3d} C_{6h} C_{6v} D_6 D_{6h}	T T_h T_d O O_h
	b	6 $\bar{6}$ $\bar{3}$m 6/m 6mm 62 6/mmm	23 m3 $\bar{4}$3m 43 m3m

Molecules of cubic symmetry, other than those of T_d and O_h, do not exist. All the remaining point groups listed here together with many more (not listed here) are also molecular point groups.

2.3.8 Direct Product and Generation of Groups

The ideas of generators and incidentally of subgroup and its cosets have been extensively employed in the construction of point groups (Sec. 2.3.2). It may be remarked that the idea of direct product groups (Sec. 2.2.8) provides a parallel avenue leading to bigger groups from two smaller ones with no element in common and with commutability of the elements of one group with those of the other. Higher axial groups $C_{n\hbar}$, $D_{n\hbar}$ and the cubic group $O_\hbar$ can be built up in this way as the direct product of nonaxial groups and an axial or a cubic group, e.g.,

$$C_s \bigotimes C_n = C_{n\hbar}$$

$$C_s \bigotimes D_n = D_{n\hbar}$$

$$C_s \bigotimes O = O_\hbar$$

Of course, the nonaxial group C_i could also have been used in the direct product relations for only the odd values of n.

2.3.9 Point Groups and Flexibility of Molecules

All the descriptions classifying molecules under different point groups are based on the assumption of rigid nuclear frameworks. But molecules and ions are flexible to varying extents undergoing free rotations around single bonds where the rotation barriers are not high. This actual state of molecules thus militates against the point group concept so laboriously constructed on the basis of rigid frame work. To keep an account of the group properties resulting from their nonrigidity, Longuet-Higgins has developed the ideas of symmetry operations in the new context.

We conclude this section with an explanation for the use of the term 'point group' for the molecular symmetry groups and crystal classes. In all the symmetry operations under the various groups, there is at least one point (if not more) which remains unmoved from its original location. For H_2O molecule it is the oxygen atom, for C_6H_6 it is the centre of the ring, for eclipsed C_2H_6 it is the midpoint of the C-C linkage which remain stationary under the respective relevant symmetry operations. Such point groups valid also for crystal classes are to be distinguished from the space groups which involve translational motions as additional symmetry operations possible only in crystals.

2.4 Recognition of Point Groups of Molecules

If a molecule be given, how to recognise its point symmetry group? This may be ascertained by following the sequence below.

(i) Conclude whether the molecule belongs to the non-axial, cubic or the special group. This is easy to decide remembering the simplicity of the nonaxial groups, the plurality of proper highest fold axes in cubic groups (Sec. 2.3.1) and the linear structure of the molecules that only can belong to special groups.

(ii) If the molecule does not belong to any of the above categories, look out for the occurrence of the minimum basic symmetry elements, e.g., C_n, (C_n+C_2') or S_{2n}, required to act as generators for the group C_n, D_n or S_{2n} respectively. This will at once characterise the molecule as having the minimum basic symmetry of either C_n or D_n or S_{2n}.

(iii) After deciphering its minimum symmetry, look out for any additional generator, searching first for $\sigma_{\hbar}$. Its presence will categorise the molecule as one of the group $C_{n\hbar}/D_{n\hbar}$ compatible with (ii) above. If, however, $\sigma_{\hbar}$ be lacking, but one σ_v or σ_d generator be detected, the result becomes evident, viz., C_{nv}/D_{nd} as may be inferred by taking (ii) into account. If σ_v be also absent, the conclusion will rest on $C_n/D_n/S_{2n}$ as dictated by (ii).

(iv) If finally the choice is narrowed down to C_n or S_{2n}, attention should be paid to ascertain whether the C_n axis also serves as an improper S_{2n} axis or not. If the improper axis be lacking, only then it is labelled C_n.

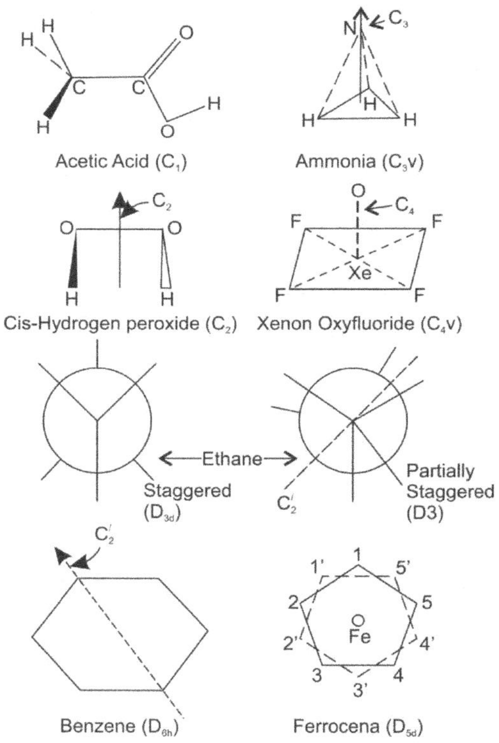

Fig. 2.7: Molecules of different point groups.

A host of molecules and ions with known structures and configurations are now examined to decide their point groups. The assortment comprises (i) H_2O (ii) CH_3COOH (iii) HCOOH (iv) NH_3. (v) H_2O_2 (iv) $XeOF_4$ (vii) SF_5C1 (viii) C_2H_6 (eclipsed, staggered and intermediate orientation) (ix) Naphthalene (x) Benzene (xi) Ruthanocene (xii) Ferrocene (xiii) $[PtCl_4]^{-2}$ (xiv) Cyclohexane (chair form) (xv) spiro [trans 2, 3-7, 8 tetramethyl] nonane.

(i) H_2O: The structure is planar and it does not belong to the nonaxial, cubic or the special group. While a C_2 axis in the molecular plane is evident, no C_2' is there. Looking for $\sigma_\hbar$, one does not find it. Following up the instruction sequence, one searches next for σ_v. A σ_v normal to the molecular plane and bisecting the HOH angle is too apparent to be missed. Without searching for any further element, one can label the group as C_{2v}.

(ii) CH_3COOH: The molecule as a whole (Fig. 2.7) is bereft of all symmetry elements excepting the trivial axis C_1. Hence it belongs to the non-axial group C_1.

(iii) HCOOH: This molecule, being planar, has just one reflection plane coincident with the molecular plane. Its group is C_s.

(iv) NH_3: The molecule is pyramidal (Fig. 2.7) with N-atom at the apex. It has a C_3 passing through N, no C_2' or $\sigma_\hbar$. Search for a σ_v shows one such passing through N, one H atom and the C_3 axis. Of course, there are others. But one can readily assign the molecule to C_{3v} point group without waiting to find the other elements.

(v) H_2O_2: The equilibrium form of cis H_2O_2 is not planar. If the two oxygen atoms be supposed to lie in the plane of the paper, the two OH bonds separately project above and below the plane making an angle θ each with the paper plane (Fig.2.7). There is one C_2 axis, bisecting the O-O bond, in the paper plane and no C_2' or $\sigma_\hbar$. Next the search for σ_v ends in no success. Noting, that this C_2 axis does not serve simultaneously as an improper S_4 (see instruction sequence), the point group is C_2.

(vi) $XeOF_4$: The molecule (Fig.2.7) is a square pyramid with the four fluorine atoms constituting the square and Xe and O- atoms lying on a C_4 axis. C_2' is absent, and so is $\sigma_\hbar$. There is a σ_v (in fact a number of them) containing the C_4 axis and bisecting the square base. Hence its point group is C_{4v}.

(vii) SF_5C1: The molecule is a square bipyramid with S atom at the centre. The apical SCl and SF bonds are somewhat different. The point group is C_{4v} as arguments of (vi) above apply to it as well.

(viii) C_2H_6: Two tetrahedra one placed above the other. A ball and stick model of the molecule helps appraisal of the symmetry beautifully.

Eclipsed C_2H_6: C_3 and C_2' being present, the molecule has the minimum symmetry of D_3. The presence of $\sigma_\hbar$ normal to C_3 axis is too evident to be missed. Hence it belongs to the group $D_{3\hbar}$.

Staggered C_2H_6: A ball and stick model or reference to projection formula (Fig.2.7) shows the evident presence of C_3, C_2' affording it a minimum symmetry of D_3. Furthermore, search for $\sigma_\hbar$ being abortive, one then looks for σ_v (or σ_d). There are some and each passes through one H atom at the top, its antipodal H atom at the bottom and the pair of C atoms. The relevant group is D_{3d}.

C_2H_6, intermediate orientation: Projection diagram (Fig.2.7) suggests the existence of C_3 and C_2' (only one C_2' is shown), though the latter is somewhat difficult to perceive. Absence of $\sigma_\hbar$ or σ_d suggests the point group D_3 for this molecule.

(ix) Naphthalene: The molecule is planar. The highest fold C_2 axis is perpendicular to the plane. There are C_2' axes also in the molecular plane which is also a $\sigma_\hbar$. Hence the molecular point group is $D_{2\hbar}$.

(x) Benzene: This planar molecule (Fig.2.7) has one C_6 and a number of C_2' (only one is indicated). The molecular plane serves as a $\sigma_\hbar$. The point group is $D_{6\hbar}$

(xi) Ruthanocene: $Ru(C_5H_5)_2$, regular pentagonal prism. The presence of C_5 through Ru and normal to the two pentagonal rings is evident. C_2' perpendicular to C_5 and passing through Ru and bisecting also the line joining one C atom and a C atom vertically below, is also present. A horizontal reflection plane $\sigma_\hbar$, normal; to C_5 axis, bisects the molecule. Therefore, the point group is $D_{5\hbar}$.

(xii) Ferrocene: $(C_6H_5)_2Fe$, a pentagonal antiprism in which the two pentagonal rings are in parallel layers, but in staggered orientation (Fig.2.7). A C_5 axis passes through Fe atom. A C_2' axis (of course there are more) is also detectable. Thus the molecule has the minimum basic symmetry D_5. The search for $\sigma_\hbar$ being fruitless, one finds σ_d. Each σ_d contains in it an upper ring C-atom, its antipodal (lower ring) C-atom, the central Fe atom as well as the C_5 axis. The point group is, hence, D_{5d}.

(xiii) $[PtCl_4]^{2-}$: This has a square planar structure provided with both C_4 and C_2's. Besides the presence of $\sigma_\hbar$, which is the ionic plane, earmarks it for the $D_{4\hbar}$ category.

(xiv) C_6H_{12}: The chair form of cyclohexane.

The six carbon atoms are distributed in two parallel planes. Each plane comprises three C-atoms forming the vertices of an imaginable equilateral triangle. Two such equilateral triangles in the two parallel planes are in relatively staggered orientation reminding one of the case of staggered ethane (see viii above). The molecular point group is thus D_{3d}.

(xv) Spiro [trans 2,3,7,8-tetramethyl] nonane: This molecule has two pentagonal rings placed crosswise at right angle (Fig.2.8).

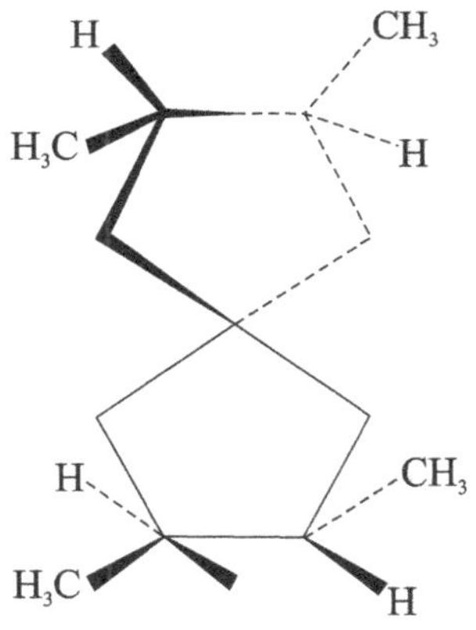

Fig. 2.8: Spiro [trans 2, 3, 7, 8-tetramethyl] nonane.

If the lower ring is in the paper plane, the upper one is perpendicular to the plane of the paper with half of it projecting above the plane.

The molecule has a C_2 axis, but no C_2', $\sigma_\hbar$ or σ_v. But it should be noted that the C_2 axis also functions as an improper axis S_4. The point group of the molecule is thus S_4.

A trial examination with Dreiding model helps ready recognition of the symmetry groups of the chairform of cyclohexane and of the spiro molecule just discussed.

Before concluding this section a few remarks on molecules possessing minimum basic symmetry of D_n seem to be in order. Molecules of D_n

and D_{nd} groups are somewhat difficult to recognise. It is generally found that such molecules can usually be split into two similar halves satisfying one or the other of the following two arrangements:

(1) The two halves (neglecting the central atom or atoms, if any) are arranged in two parallel layers in fully eclipsed, staggered or intermediate configurations pointing to $D_{n\hbar}$, D_{nd} or D_n groups respectively. Example: Ruthanocene ($D_{5\hbar}$), Ferrocene (D_{5d}), eclipsed ethane ($D_{3\hbar}$), staggered ethane (D_{3d}), ethane with intermediate configuration (D_3).

(2) The planes of the two halves are arranged in crosswise (transverse) orientations, one above the other with an inclination of $\frac{\pi}{2}$ for the D_{2d} group or an arbitrary inclination of θ for D_2. Examples: Allene (D_{2d}), Spirononane (D_{2d}), Biphenyl (D_2), Ethylene distorted from the planar configuration (D_2).

Chapter 3

Vector Spaces, Matrices and Transformations

The topics to be discussed briefly in this chapter include the linear spaces, vector spaces, metrics, some special matrices, theorems and transformations in vector spaces. These concepts form the fundament of the group representation theory (Chapters 4, 5) that constitutes the link leading from the symmetries of molecules to their practical applications. A general acquaintance with the handling of matrices will, however, be assumed.

3.1 Linear Spaces and Basis Vectors

In physical sciences and mathematics we may come across a bunch of entities (infinite in number) that satisfy the following two criteria.

(i) The sum of any two entities (called vectors) is a vector present within the bunch.

(ii) The multiplication by a scalar C of any vector is also a member of the bunch.

This collection of entities satisfying the closure law with respect to addition and multiplication by a scalar quantity constitutes a linear space L. A few concerete examples will be helpful at this stage.

(a) All the position vectors (i.e., the lines drawn from the origin to the respective locations) in the three dimensional physical space constitute a linear space. Since any vector can be expressed in terms of their components, the sum of any two vectors $\vec{p}$ and $\vec{q}$ (fig.3.1) must be some other vector $\vec{r}$ where

$$\begin{aligned} \vec{r} = \vec{p} + \vec{q} &= (p_1 + q_1)\vec{i} + (p_2 + q_2)\vec{j} + (p_3 + q_3)\vec{k} \\ &= r_1\vec{i} + r_2\vec{j} + r_3\vec{k} \end{aligned} \tag{3.1}$$

Similarly, multiplying by a scalar C, we have

$$C\vec{q} = Cq_1\vec{i} + Cq_2\vec{j} + Cq_3\vec{k} = q_1'\vec{i} + q_2'\vec{j} + q_3'\vec{k} = \vec{q}\,' \tag{3.2}$$

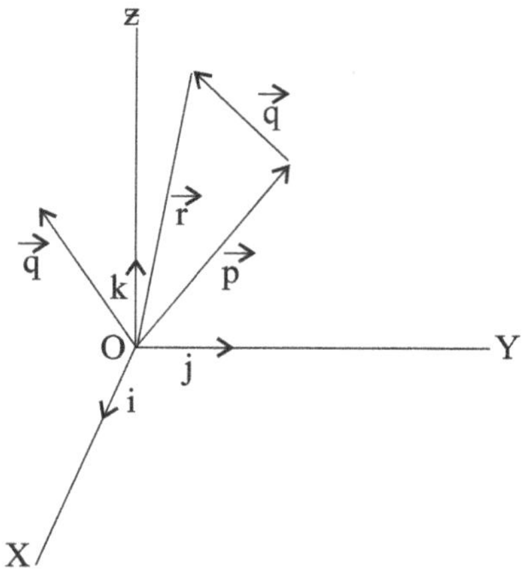

Fig. 3.1: Linear space of position vectors P, q, r, ...

The foregoing relations are true for any set of vectors $\vec{p}$, $\vec{q}$ in the three dimensional space. The following incidental remarks in connection with the present example of linear space are worthy of attention.

1. The entities or the vectors here are purely vectorial quantities which, however, will not be a general feature in many other linear spaces. Notwithstanding this, the term 'vector' will apply to such entities.
2. The linear space L, i.e., its contents are after all, expressed in terms of the unit vectors i, j, k and the relevant components. So instead of writing down the infinite number of individual vectors to describe L, we may, instead, succinctly describe the latter as the space spanned by the unit base vectors $\vec{i}$, $\vec{j}$, $\vec{k}$. Thus ($\vec{i}$, $\vec{j}$, $\vec{k}$) constitute the basis spanning the space L.
3. The basis vectors are not unique. These need not be orthogonal nor need necessarily comprise unit base vectors. We may choose a basis of non-coplanar (and linearly independent) base vectors $\vec{a}$, $\vec{b}$, $\vec{c}$ where the latter are not either unit base vectors or a cartesian set of axes. What is even more, $\vec{a}$, $\vec{b}$, $\vec{c}$, may or may not be mutually orthogonal. Under any circumstances the linearly independent basis set ($\vec{a}$, $\vec{b}$, $\vec{c}$) may span the linear space L enabling its mem-

bers $\vec{p}$, $\vec{q}$ to be expressible in terms of $\vec{a}$, $\vec{b}$, $\vec{c}$ and their relevant components.

4. Often for convenience, one would choose the unit vectors ($\vec{e}_1$, $\vec{e}_2$, $\vec{e}_3$), that are mutually orthogonal (this aspect really comes under vector spaces and metrics dealt with later), to span the linear space L of the position vectors.
5. Like the vector fields of position vectors, one can have linear spaces of other vector fields such as force fields, momentum fields etc.

(b) We can easily carry over this idea of three dimensional linear spaces to multidimensional linear spaces, a specific example of which is the configuration space of a molecule having, say, n moving electrons. The instantaneous configuration of the molecule is described by either n points (with 3n coordinates) in the ordinary physical space or by just a single point in the 3n-dimensional configuration space which, of course, cannot be drawn but easily imagined. The configurational vector is given by imagining an hyperline joining the origin and the point itself. At other times the molecule will have different configurations. This set of configurational vectors satisfy the closure properties of the 3n-dimensional linear space spanned by the unit base vectors ($\vec{e}_1$, $\vec{e}_2, \ldots \vec{e}_k \ldots \vec{e}_n$, $\vec{e}_{n+1} \ldots \vec{e}_{3n}$)

(c) The set of infinite number of solutions of an ordinary second order homogeneous differential equation constitutes a linear function space L. For example, the simple harmonic oscillator equation $\frac{d^2x}{dt^2} = -w^2x$ has the general solution, x=a sin wt+b cos wt. Hence, the linear space is a two-dimensional function space spanned by the basis functions, a sin wt and b cos wt, or any other pair of two linearly independent solutions.

Functional solutions of quantum mechanical linear homogeneous differential eigenvalue equations provide plenty of such instances of function spaces. The H-atom problem has a series of non-spin wave functions ψ_1, ψ_2, ψ_3... satisfying the equation $H\psi_1=E_1\psi_1$, $H\psi_2=E_2\psi_1$ $H\psi_3=E_3\psi_3$.... for the first, second, third...... quantum states. Many of these energy eigenstates E_2, E_3, E_4...... are degenerate. For instance the one s and three p functions in the second quantum state, viz., ψ_{2s}, ψ_{2p1}, ψ_{2p0}, ψ_{2p-1} are all characterised by the same energy value E_2. It can be shown that the closure property is obeyed by the infinite manifold (i.e., infinite dimensional) function space spanned by the basis functions $(\psi_1,\ \psi_{2s},\ \psi_{2p1},\ \psi_{2p0}\cdots\psi_{3s}\cdots\psi_{3p1}\cdots\psi_{3d1}\cdots)$.

A more useful idea of the function space is that all the k degenerate atomic wave functions corresponding to a particular quantum number pair (n, l) and energy E_k form the basis functions for the k-dimensional space. The addition rule is satisfied as shown below.

Let $\phi_{1k}\phi_{2k}\cdots\phi_{kk}$ be the k number of degenerate wave functions corresponding to a particular energy E_k. Let any linear combination of these wave functions be $\psi = (C_1\phi_{1k} + C_2\phi_{2k} + \cdots C_k\phi_{kk})$

$$\begin{aligned} H\psi &= H(C_1\phi_{1k} + C_2\phi_{2k} + \cdots + C_k\phi_{kk}) \\ &= C_1H\phi_{1k} + C_2H_{\phi}2_k + \cdots C_kH\phi_{kk} \\ &= E_k(C_1\phi_{1k} + C_2\phi_{2k} + \cdots + C_k\phi_{kk}) = E_k\psi \end{aligned} \tag{3.3}$$

Relation (3.3) shows ψ also forms a solution of Schrodinger equation and is therefore a member of the function space. It is in this context that the three degenerate *np* atomic orbitals or the five degenerate *nd* atomic orbitals form the basis functions for spanning three dimensional and five dimensional function spaces respectively. It may be remarked incidentally that the molecular orbitals obtained by LCAO method are just certain specific members of the function space spanned by a given basis set of atomic orbitals.

(d) Eigenvectors can also form a basis set to span an eigenvector space. An acquaintance with the principle of projection operator vis a vis eigenvectors is necessary for which the reader may consult a standard text of quantum mechanics.

(e) The set of three Pauli matrices,

$$\sigma_x = \begin{pmatrix} 0 & 1 \\ 1 & 0 \end{pmatrix} \sigma_y = \begin{pmatrix} 0 & -i \\ i & 0 \end{pmatrix} \sigma_z = \begin{pmatrix} 1 & 0 \\ 0 & -1 \end{pmatrix},$$

span a three dimensional linear space of all 2×2 matrices with vanishing trace. Any 2×2 matrix with zero trace, $\mathbf{h} = \begin{pmatrix} h_{11} & h_{12} \\ h_{21} & h_{22} \end{pmatrix}$ can be expressed as a linear combination $C_1\sigma_x + C_2\sigma_y + C_3\sigma_z$ where $C_1,\ C_2,\ C_3$ are coefficients that can be related to $h_{11},\ h_{12}$ etc. of the matrix elements of $\mathbf{h}$.

3.1.1 Matrix Forms of Vectors In Linear Spaces

Having cited the different illustrations of linear spaces we are to digress a little to consider the practice of expressing vectors of linear spaces in the form of matrices. Consider any vector $\overrightarrow{r_k}$ in an n-di mensional linear space spanned by the unit base vectors $\overrightarrow{e_1}$, $\overrightarrow{e_2}$, $\overrightarrow{e_3} \cdots \overrightarrow{e_n}$ so that

$$\overrightarrow{r}_k = C_{1k}\overrightarrow{e}_1 + C_{2k}\overrightarrow{e}_2 + \cdots + C_{nk}\overrightarrow{e}_n \tag{3.4}$$

If we now agree to a convention of writing the base vectors in the form of a row matrix, $\mathbf{e} = (e_1\ e_2 \cdots e_n)$ and the respective components (i.e., the coefficients) as a column matrix, viz.,

$$\mathbf{C}_k = \begin{pmatrix} C_{1k} \\ C_{2k} \\ \vdots \\ C_{nk} \end{pmatrix}, \text{ then Eq. (3.4) can be written as}$$

$$\overrightarrow{r}_k = \sum_j \overrightarrow{e}_j c_{jk} = \mathbf{e}\mathbf{C}_k \tag{3.5}$$

In terms of a second basis set $\overrightarrow{e}'_1$, $\overrightarrow{e}'_2$, $\overrightarrow{e}'_3 \ldots\ldots \overrightarrow{e}'_n$ spanning the same space, one can express

$$\overrightarrow{r}_k = \mathbf{e}'\mathbf{C}'_k \tag{3.6}$$

where $\mathbf{e}'$ and $\mathbf{C}'_k$ represent the row and column matrices respectively.

Anyone of the unit vectors of $\mathbf{e}'$ basis serves as a general vector in the space spanned by $\mathbf{e}$. Thus, following the relation (3.5), one can write

$$\overrightarrow{e}'_i = \sum_k \overrightarrow{e}_k a_{ki} = \mathbf{e}\mathbf{A}_i \tag{3.7}$$

where $\mathbf{A}_i$ is a column matrix similar to $\mathbf{C}_k$

All the individual base vectors of $\mathbf{e}'$, when expressed in forms similar to Eq. (3.7), will cast the n numbers of such base vectors in the form of n matrix relations which can all be combined into a single matrix relation

$$\mathbf{e}' = \mathbf{e}\mathbf{A} \tag{3.8}$$

$$\text{where the matrix } \mathbf{A} = (\mathbf{A}_1\mathbf{A}_2 \cdots \mathbf{A}_i \cdots \mathbf{A}_n)$$
$$= \begin{pmatrix} a_{11} & a_{12} & a_{13} & \cdots a_{1i} & \cdots & a_{1n} \\ a_{21} & a_{22} & a_{23} & \cdots a_{2i} & & a_{2n} \\ \vdots & \vdots & \vdots & \vdots & & \vdots \\ a_{n1} & a_{n2} & a_{n3} & \cdots a_{ni} & \cdots & a_{nn} \end{pmatrix}$$

a square matrix of the coefficients of the transformation equations analogous to (3.7). The reverse transformation from $\mathbf{e}'$ basis to $\mathbf{e}$ basis set is simply.

$$\mathbf{e} = \mathbf{e}'\mathbf{A}^{-1} \tag{3.9}$$

The pair of relations (3.8) and (3.9) show the interrelation of the basis sets. Let us now inquire how the components $\mathbf{C}_k$ and $\mathbf{C}'_k$ of the general vector $\overrightarrow{\mathrm{r}}_k$ relative to the basis sets $\mathbf{e}$ and $\mathbf{e}$' are related. Evidently we can write for $\overrightarrow{\mathrm{r}}_k$ as

$$\sum_i \overrightarrow{\mathrm{e}}'_i \mathrm{c}'_{ik} = \overrightarrow{\mathrm{r}}_k = \sum_j \overrightarrow{\mathrm{e}}_j \mathrm{c}_{jk}$$

Now using (3.7), we have from above,

$$\sum_i \sum_j \overrightarrow{\mathrm{e}}_j \mathrm{a}_{ji} \mathrm{c}'_{ik} = \sum_j \overrightarrow{\mathrm{e}}_j \mathrm{c}_{jk} \tag{3.10}$$

Hence for a given j,

$$\mathrm{c}_{jk} = \sum_i \mathrm{a}_{ji} \mathrm{c}'_{ik} \tag{3.11}$$

If we thus recoup all the elements of the matrix $\mathbf{C}_k$ for different j's, all the linear relations like (3.11) can be compacted into a single matrix relation

$$\mathbf{C}_k = \mathbf{A}\mathbf{C}'_k \tag{3.12}$$

We carefully note from Eqs. (3.8), (3.9), (3.12) that while the transformation matrix $\mathbf{A}$ changes the $\mathbf{e}$ basis set to $\mathbf{e}'$ basis set through, post multiplication, the same matrix transforms, through premultiplication, the components relative to $\mathbf{e}'$ to those relative to $\mathbf{e}$. This characteristic feature is to be borne in mind. Additionally one should particularly note the specific sequence of the subscripts while expressing a vector in the form of a linear combination of the basis vectors [eqs. (3.5), (3.7), (3.10)]. In many instances of practical applications and problems, we shall often find the $\mathbf{A}$ matrix orthogonal or unitary. This will enable us immediately to write down the inverse transformation matrix $\mathbf{A}^{-1}$ as simply the transposed $\mathbf{A}$ ($\widetilde{\mathbf{A}}$) or the adjoint of $\mathbf{A}$ (i.e., $\mathbf{A}^{\neq}$) respectively.

3.2 Linear Subspaces and Linear Product Spaces

An n-dimensional linear space L, spanned by base vectors $\overrightarrow{\mathrm{e}}_1$, $\overrightarrow{\mathrm{e}}_2, \ldots \mathrm{e}_n$ can be regarded as composed of one dimensional subspaces L_1 spanned by $\vec{e}_1$, L_2 carried by $\vec{e}_2$ and so on. In other words L is the direct sum of the subspaces, viz.

$$L = L_1 \bigoplus L_2 \bigoplus \cdots \bigoplus L_n \tag{3.13}$$

This mode of subdivision effectively means that we are considering linear subspaces of vectors parallel to $\vec{e}_1$ to $\vec{e}_2$ etc., respectively. It is the viewpoint itself (here suitable projections) that decides the splitting of L into the corresponding subspaces. The viewpoint may also be different demanding a different set of subspaces. For example if in the ordinary 3-dimensional linear space, spanned by unit (and orthogonal) vectors $\vec{e}_1$, $\vec{e}_2$, $\vec{e}_3$ we are interested in vectors lying in the XY plane irrespective of their mutual inclinations and a second set parallel to the z-axis, then L can be expressed as a direct sum:

$L=L_1 \bigoplus L_2$, where the two dimensional L_1 is spanned by $\vec{e}_1$, $\vec{e}_2$ and $\vec{e}_3$ sustains the one dimensional L_2. These considerations will also apply to function spaces.

As will be realised afterwards in group theoretical methods, the subdivision of a linear space (including the function space) L will be dictated by the perspective (i.e., the viewpoint) of a given set of symmetry operations characteristic of a symmetry group. A five dimensional d-orbital function space will be divisible into subspaces differently depending on the symmetry group in view.

Let us consider the transformation from one basis to another for the two subspaces. Let n dimensional L_1 be spanned by the set $\vec{e}_1$, $\vec{e}_2 ... \vec{e}_n$ and (m-n) dimensional L_2 be carried by the set $\vec{e}_{n+1}$, $\vec{e}_{n+2}, ... \vec{e}_m$. This evidently means that L is m dimensional. A change to the primed set enables us to write (sec 3.1.1.)

$$\vec{e}'_i = \sum_{j=1}^{n} \vec{e}_j a_{ji} \text{ and } \vec{e}'_k = \sum_{l=n+1} \vec{e}_l b_{lk} \tag{3.14}$$

Considering the entire linear space L, any general primed base vector

$$\vec{e}'_p = \sum_{j=1}^{m} \vec{e}_j c_{jp} \tag{3.15}$$

where the matrix element C_{jp} is either one of the a_{ji}'s or one of the b_{lr}'s. These two alternatives can be accommodated by writing the transformation matrix **C** in the form of a partitioned matrix, viz., $C= \left(\begin{array}{c|c} A & O \\ \hline O & B \end{array}\right)$

which is another way of writing $\mathbf{C}=\mathbf{A}\bigoplus\mathbf{B}$, a direct sum of two transformation matrices. In such block diagonalised matrix **C**, the top right and the bottom left corner submatrices are null matrices.

We can extend this idea to the formation of direct product space of two linear spaces. Let L_1 and L_2 be two linear spaces which may be independent of each other or may be two subspaces of a function space. The first is spanned by the basis functions f_1, f_2,....f_n and the second by the basis set g_1, g_2,...g_m. A change of basis to the primed set enables us to write

$$f_i' = \sum_{j=1}^{n} f_j a_{ji} \text{ and } g_k' = \sum_{l=1}^{m} g_l b_{lk} \tag{3.16}$$

The basis functions in the product space are given by $f_i g_k$'s=h_{ik}'s where l and k run from 1 to n and 1 to m respectively. Hence a shift to the primed basis set for the product space means

$$\begin{aligned} h_{ik}' = f_i' g_k' &= \sum_{j=1}^{n} f_i a_{ji} \sum_{l=1} g_l b_{lk} \\ &= \sum_{j=1}^{n}\sum_{l=1}^{m} f_j g_l a_{ji} b_{lk} = \sum_{j=1}^{n}\sum_{l=1}^{m} h_{jl} a_{ji} b_{lk} \end{aligned} \tag{3.17}$$

The elements of the transformation matrix **C** in the product space are just all possible products of $a_{ji}b_{lk}$'s with j and l taking up all the permissible values. In detail it works out to a supermatrix, viz.,

$$\mathbf{C} = \begin{pmatrix} a_{11}\mathbf{B} & a_{12}\mathbf{B}.... & a_{1n}\mathbf{B} \\ a_{21}\mathbf{B} & a_{22}\mathbf{B}.... & a_{2n}\mathbf{B} \\ \vdots & \vdots & \\ a_{n1}\mathbf{B} & a_{n2}\mathbf{B}.... & a_{nn}\mathbf{B} \end{pmatrix} \tag{3.18}$$

This type of matrix has an alternative notation called direct product matrix, viz.,

$$\mathbf{C} = \mathbf{A} \bigotimes \mathbf{B} \tag{3.19}$$

and the product space is, hence, called a direct product space. The ideas of subgroups and direct product groups (sec 2.2.2 and 2.2.8) are closely allied to those of the linear subspaces and the direct product spaces. (See also Sec. 5.7)

3.2.1 Vector Space And Metrical Matrix

A linear space L is called a vector space if, in the addition to the usual attribute of the closure property (sec. 3.1) of the vectors, an inner product is defined and preserved under a transformation for any two vectors of the space. Our ordinary concept of the inner product of two vectors $\vec{\mathbf{p}}$, $\vec{\mathbf{q}}$ is a scalar represented by

$$\vec{\mathbf{p}}.\vec{\mathbf{q}} = p_x q_x + p_y q_y + p_z q_z = \sum_{i=x,y,z} p_i q_i \tag{3.20}$$

and the norm of the vector $\vec{p}$ is

$$\vec{p}.\vec{p} = p_x^2 + p_y^2 + p_z^2 \tag{3.21}$$

To cover the real and complex vectors, a more general definition of the innerproduct is needed. For a vector space L spanned by the base vectors $\vec{e}_1$, $\vec{e}_2$, ...$\vec{e}_n$, the innerproduct of $\vec{p}$ and $\vec{q}$ is

$$\begin{aligned} <p \mid q>= \vec{p}^*.\vec{q} &= (p_1^*\vec{e}_1^* + p_2^*\vec{e}_2^* + ... + p_n^*\vec{e}_n^*)(q_1\vec{e}_1 + \\ &\qquad q_2\vec{e}_2 + ...q_n\vec{e}_n) \\ &= \sum_{ij} p_i^* q_j(\vec{e}_i.\vec{e}_j) = \sum_{ij} p_i^* q_j M_{ij} \end{aligned} \tag{3.22}$$

A slight reorganisation of (3.22) enables one to express the inner product of vectors in matrix product form.

$$< p \mid q >= \vec{p}^*.\vec{q} = \sum_{ij} p_i^* q_j M_{ij} = \mathbf{P}^{\neq}\mathbf{MQ} \tag{3.23}$$

where **P** and **Q** are (n×1) column matrices of the components and **M** is an (n×n) square matrix with elements M_{ij}'s i.e., $\vec{e}_i^*.\vec{p}_j$'s. $\mathbf{P}^{\neq}$, the hermitian conjugate matrix (i.e., the adjoint of **P** matrix) has been used in (3.23) to render the matrices conformable for matrix multiplication. The matrix **M** is called metrical matrix and is .indicative of the nature of the basis set.

$$\mathbf{M} = (M_{ij}) = \begin{pmatrix} \vec{e}_1^*.\vec{e}_1 & \vec{e}_1^*.\vec{e}_2 & \cdots & \vec{e}_1^*.\vec{e}_n \\ \vec{e}_2^*.\vec{e}_1 & \vec{e}_2^*.\vec{e}_2 & \cdots & \vec{e}_2^*.\vec{e}_n \\ \vdots & \vdots & & \vdots \\ \vec{e}_n^*.\vec{e}_1 & \vec{e}_n^*.\vec{e}_2 & \cdots & \vec{e}_n^*.\vec{e}_n \end{pmatrix} \tag{3.24}$$

$$\mathbf{M} = \begin{pmatrix} \overrightarrow{e}_1^{\,*} \\ \overrightarrow{e}_2^{\,*} \\ \vdots \\ \overrightarrow{e}_n^{\,*} \end{pmatrix} (\overrightarrow{e}_1 \; \overrightarrow{e}_2 \; ...... \overrightarrow{e}_n) = \mathbf{e}^{\neq}\mathbf{e} \tag{3.25}$$

When the unit base vectors constitute an orthogonal basis set, the matrix elements $\overrightarrow{e}_i^*.\overrightarrow{e}_j = \delta_{ij}$ (Kronecker delta). Hence from (3.24) and (3.25), we can write

$$\mathbf{M} = 1 = \mathbf{e}^{\neq}\mathbf{e} \tag{3.26}$$

which indicates that the metrical matrix is a unit matrix only when the basis set is orthonormal.

The inner product of functions in the function space spanned by the basis functions, f_1, f_2, $f_3...f_n$, can similarly be expressed in general notational form as well as in matrix product form. If ϕ and ψ be any two functions of the space, the inner product.

$$< \phi \mid \psi >= \int_\tau \phi^* \psi d\tau = \mathbf{P}^{\neq}\mathbf{M}\mathbf{Q} \tag{3.27}$$

The inner product is thus given by an integral. In the expression (3.27), $\mathbf{P}^{\neq}$ is the adjoint of the column matrix $\mathbf{P}$ of the coefficients occurring in the linear combination of the basis set. The matrix elements of the metric $\mathbf{M}$ of (3.27) are all definite integrals of the type $\int_\tau f_i^* f_i d\tau$. If the basis functions are an orthonormal set, then

$$\mathbf{M} = \mathbf{1}(\text{unit matrix}) = \mathbf{f}^{\neq}\mathbf{f} \tag{3.28}$$

3.3 Matrices and Diagonalisation

We start here listing down a few types of matrices with their symbolic (element-based) characterisation and a mention of a few of their properties. Consider a nonsingular square matrix $\mathbf{A}$ with elements a_{ij}'s. From this matrix, the following matrices may be built up.

(i) $\widetilde{\mathbf{A}}$, called transposed $\mathbf{A}$, has elements $(\widetilde{\mathbf{A}})_{ij}=a_{ji}$ of the $\mathbf{A}$ matrix. This matrix is obtained by transposition, i.e., changing the rows of $\mathbf{A}$ into corresponding columns.

(ii) $\mathbf{A}^{\neq}$, termed adjoint of $\mathbf{A}$, has the ij-th element $(\mathbf{A}^{\neq})_{ij}=(\widetilde{\mathbf{A}})^*_{ij} = \mathrm{a}^*_{ji}$. $\mathbf{A}^{\neq}$ is formed from $\mathbf{A}$ by the twin actions of transposition and complex conjugation of the elements of $\mathbf{A}$.

(iii) $\mathbf{A}^{-1}$, called the inverse of $\mathbf{A}$, has the property $\mathbf{A}\mathbf{A}^{-1}=\mathbf{A}^{-1}\mathbf{A}=1$. There are a number of wellknown processes of obtaining the inverse of a nonsingular square matrix.

Further the matrix $\mathbf{A}$ is

(i) hermitian or self adjoint if $\mathbf{A}=\mathbf{A}^{\neq}$. In other words each element $\mathrm{a}_{ij}=(\mathbf{A}^{\neq})_{ij}=(\widetilde{\mathbf{A}})^*_{ij}=\mathrm{a}^*_{ji}$

(ii) orthogonal if $\widetilde{\mathbf{A}}=\mathbf{A}^{-1}$, i.e., $\mathbf{A}\widetilde{\mathbf{A}}=\mathbf{A}\mathbf{A}^{-1}=\mathbf{1}$

(iii) unitary, if $\mathbf{A}^{\neq}=\mathbf{A}^{-1}$, i.e., $\mathbf{A}\mathbf{A}^{\neq}=\mathbf{A}\mathbf{A}^{-1}=\mathbf{1}$

We shall now describe some of the properties of orthogonal, unitary and hermitian matrices.

I. The rows (or columns) of an orthogonal matrix $\mathbf{A}$ behave like the components of orthogonal vectors. The validity of the above proposition can be justified in the following way.

Since $\mathbf{A}$ is orthogonal $\mathbf{A}\widetilde{\mathbf{A}}=1$, the ik-th element of the unit matrix is δ_{ik}, (Kronecker delta)

$\delta_{ik} = \sum_j \mathrm{a}_{ij}(\widetilde{\mathbf{A}})_{jk} = \sum_j \mathrm{a}_{ij}\mathrm{a}_{kj}$ = the sum of the products of the corresponding elements in the i-th and k-th rows running over all the columns of $\mathbf{A}$. For i$\neq$k, this sum equals zero, which indicates that the two rows behave as if these are orthogonal. If i=k then $\delta_{ik} = 1 = \sum_j \mathrm{a}_{ij}\mathrm{a}_{ij} = \sum_j \mathrm{a}^2_{ij}$, which shows that each row is normalised like a similar vector.

The conclusions reached in respect of rows are also valid in respect of columns. To demonstrate this one merely starts from $\widetilde{\mathbf{A}}\mathbf{A}=\mathbf{1}$

II. The above process may be adopted to show that the rows and columns of a unitary matrix behave like orthonormal vectors.

Some simple specimens of orthogonal and unitary matrices satisfying the above properties are cited in the next page.

$$\underbrace{\begin{pmatrix} \cos\theta & -\sin\theta & 0 \\ \sin\theta & \cos\theta & 0 \\ 0 & 0 & 1 \end{pmatrix} : \begin{pmatrix} \frac{1}{\sqrt{3}} & \frac{1}{\sqrt{2}} & \frac{1}{\sqrt{6}} \\ \frac{1}{\sqrt{3}} & -\frac{1}{\sqrt{2}} & \frac{1}{\sqrt{6}} \\ \frac{1}{\sqrt{3}} & 0 & -\sqrt{\frac{2}{3}} \end{pmatrix}}_{\text{OrthogonalMatrices}} : \underbrace{\begin{pmatrix} 0 & -i \\ i & 0 \end{pmatrix}}_{\text{UnitaryMatrix}}$$

III. Eigenvector Equations and Diagonalisation of Matrices: Consider a set of n linear homogenous equations:

$$\begin{array}{ccccccccc} \mathrm{C}_1\mathrm{a}_{11} & + & \mathrm{C}_2\mathrm{a}_{12} & + & \cdots & + & \mathrm{C}_n\mathrm{a}_{1n} & = & \lambda\mathrm{C}_1 \\ \mathrm{C}_1\mathrm{a}_{21} & + & \mathrm{C}_2\mathrm{a}_{22} & + & \cdots & + & \mathrm{C}_n\mathrm{a}_{2n} & = & \lambda\mathrm{C}_2 \\ \vdots & & \vdots & & & & \vdots & & \vdots \\ \mathrm{C}_1\mathrm{a}_{n1} & + & \mathrm{C}_2\mathrm{a}_{n2} & + & \cdots & + & \mathrm{C}_n\mathrm{a}_{nn} & = & \lambda\mathrm{C}_n \end{array} \tag{3.29}$$

where C_1, C_2... C_n are all unknown and λ is also an unknown parameter. The set of Eqs. (3.29) can be recast in a single matrix equation

$$\mathbf{AC} = \lambda\mathbf{C} \tag{3.30}$$

where the matrix $\mathbf{A} = \begin{pmatrix} a_{11} & a_{12} & \dots & a_{1n} \\ a_{21} & a_{22} & \dots & a_{2n} \\ \vdots & \vdots & & \vdots \\ a_{n1} & a_{n2} & \dots & a_{nn} \end{pmatrix}$ square matrix and $\mathbf{C} = \begin{pmatrix} \mathrm{C}_1 \\ \mathrm{C}_2 \\ \vdots \\ \mathrm{C}_n \end{pmatrix}$ a column matrix of the unknown coefficients. The matrix equations of type (3.30), often encountered in quantum mechanics, are called eigenvalue eigenvector equations where λ is any specific eigenvalue and the **C** matrix, the corresponding eigenvector. In quantum mechanics the homogeneous equations (3.29) or equivalently the equation (3.30) are often accompanied by one additional relation, viz.,

$$\sum_{ik} \mathrm{C}_i^*\mathrm{C}_k = \delta_{ik} \text{ (Kronecker delta)} \tag{3.31}$$

The Eqs. (3.30) or (3.29) have non-trivial solution provided

$$\det \mid \mathbf{A}-\lambda 1 \mid \; i.e. \begin{vmatrix} (a_{11}-\lambda) & a_{12} & \cdots & a_{1n} \\ a_{12} & (a_{22}-\lambda) & \cdots & a_{2n} \\ \vdots & & & \\ a_{n1} & a_{n2} & \cdots & (a_{nn}-\lambda) \end{vmatrix} \text{ vanishes}$$

This determinantal equation together with relation (3.31) enables evaluations of n set of $\mathbf{C}_1\ \mathbf{C}_2 ... \mathbf{C}_n$ eigenvector matrices corresponding to the nondegenerate eigenvalues of λ, viz., $\lambda_1,\ \lambda_2 ... \lambda_k ... \lambda_n$. Here the eigenvector matrix has the explicit form $\mathbf{C}_k = \begin{pmatrix} C_{1k} \\ C_{2k} \\ \vdots \\ C_{nk} \end{pmatrix}$ In terms of the solutions obtained, the single matrix equation (3.30) may be split into n matrix equations:

$$\left.\begin{aligned} \mathbf{AC}_1 &= \lambda_1 \mathbf{C}_1 \\ \mathbf{AC}_2 &= \lambda_2 \mathbf{C}_2 \\ &\vdots \\ \mathbf{AC}_n &= \lambda_n \mathbf{C}_n \end{aligned}\right\} \tag{3.32}$$

We now define two matrices:

(i) a square matrix $\mathbf{C} = (\mathbf{C}_1 \mathbf{C}_2 ... \mathbf{C}_k ... \mathbf{C}_n) = \begin{pmatrix} C_{11}... & C_{1k}... & C_{1n} \\ C_{21}... & C_{2k}... & C_{2n} \\ \vdots & \vdots & \\ C_{n1}... & C_{nk}... & C_{nn} \end{pmatrix}$

and

(ii) a diagonal matrix $\wedge = \begin{pmatrix} \lambda_1 & & & \\ & \lambda_2 & & \\ & & \ddots & \\ & & & \lambda_n \end{pmatrix}$ of n × n order where all the off diagonal elements are zero and the diagonal elements comprise the specific eigenvalues of Eqs. (3.32). In terms of these newly defined matrices, Eqs. (3.32) can all be written in the form

$$\mathbf{AC} = \wedge \mathbf{C} \qquad \mathbf{C}^{-1}\mathbf{AC} = \wedge \tag{3.33}$$

The matrix $\mathbf{A}$ is thus diagonalised to its eigenvalue matrix by the eigenvector matrix $\mathbf{C}$. This type of reduction of a square matrix to the diagonal form represented by (3.33) is called a similarity transformation or similarity reduction. Very often it will be found that the **C** matrix is orthogonal or unitary. The similarity transformation (3.33) in such a case bears the name orthogonal or unitary reduction respectively.

It should be noted that there exist other modes, not involving the eigenvector matrix, for reduction of a square matrix **A** to a diagonal form but not necessarily to eigenvalue diagonal matrix. The reader may try the diagonalisation of the matrix $\mathbf{A} = \begin{pmatrix} 2 & 2 & 0 \\ 2 & 4 & 2 \\ 0 & 2 & 2 \end{pmatrix}$ assuming the validity of Eq. (3.31) and ascertain for himself whether the similarity transformation is really an orthogonal one or not.

The ideas of similarity transformation, the definition of which occurs incidentally in (3.33), are of extreme importance in the theory of groups. The usefulness of this transformation and of the allied ones, viz., orthogonal and unitary reductions together with relevant theorems will recur in many places of group representation theory (Chapters 4,5) and in matrix formalism of expressing conjugate elements of a class (sec. 2.2.6).

IV. Hermitian Matrices-A Few Theorems:

Theorem 3.1. *The eigenvalues of a hermitian matrix are real.*

Let **A** be an n×n hermitian matrix and λ any eigenvalue. The determinantal equation, det $(\mathbf{A} - \lambda \mathbf{1}) = 0$ (Sec 3.3.) gives rise to a characteristic nth order equation in λ when the determinant is split. The same characteristic equation would also be obtained if one starts from the transposed matrix $\widetilde{\mathbf{A}}$. Hence the eigenvalue $\lambda = \widetilde{\lambda}$. Again $\widetilde{\mathbf{A}}$ and $\mathbf{A}^{\neq}$ have the same form of characteristic equation except for conjugation of the elements. Therefore, their eigenvalues will also be similar except for conjugation. In other words, $\lambda^* = \widetilde{\lambda} = \lambda$ which demands λ to be real.

Theorem 3.2. *The eigenvector matrix that diagonalises a hermitian matrix is unitary in nature.*

Let **U** be the matrix which diagonalises the hermitian matrix **A**. Hence, referring to eq. (3.33), $\mathbf{U}^{-1}\mathbf{A}\mathbf{U} = /\!\!\Lambda$ which on taking adjoint, becomes $\left(\mathbf{U}^{-1}\mathbf{A}\mathbf{U}\right)^{\neq} = /\!\!\Lambda^{\neq} = /\!\!\Lambda$ (the eigenvalues λ_k's are real).

$$\therefore \mathbf{U}^{\neq}\mathbf{A}^{\neq}\mathbf{U}^{-1\neq} = /\!\!\Lambda = \mathbf{U}^{-1}\mathbf{A}\mathbf{U}$$

Now since $\mathbf{A}$ is hermitian, one may write $\mathbf{U}^{\neq}\mathbf{A}\mathbf{U}^{-1\neq} = /\!\!\wedge = \mathbf{U}^{-1}\mathbf{A}\mathbf{U}$, which necessitates $\mathbf{U}^{\neq} = \mathbf{U}^{-1}$ That is, the transformation matrix $\mathbf{U}$ is unitary.

Theorem 3.3. *A nondegenerate hermitian matrix has orthogonal eigenvectors.*

Let $\mathbf{C}_i$, $\mathbf{C}_j$ be two nondegenerate eigenvectors of $\mathbf{A}$ with eigenvalues λ_i and λ_j respectively.

$$\mathbf{A}\mathbf{C}_i = \lambda_i \mathbf{C_i} \text{ and } \mathbf{A}\mathbf{C_j} = \lambda_j \mathbf{C_j}$$
$$\text{or } \mathbf{C}^{\neq}_{\mathbf{j}}\mathbf{A}^{\neq} = \lambda_{\mathbf{j}}\mathbf{C}^{\neq}_j (\lambda_j = \text{real})$$

Premultiplying the first equation by $\mathrm{C}^{\neq}_j$ and post multiplying the second by $\mathbf{C}_i$, we have $\mathbf{C}^{\neq}_j\mathbf{A}\mathbf{C}_i = \lambda_i\mathbf{C}^{\neq}_j\mathbf{C}_i$ and $\mathbf{C}^{\neq}_j\mathbf{A}\mathbf{C}_i = \lambda_j\mathbf{C}^{\neq}_j\mathbf{C}_i$ ($\mathbf{A}$ is hermitian). On equating, $(\lambda_i - \lambda_j)\mathbf{C}^{\neq}_j\mathbf{C}\,_i = \mathbf{O}$ (null matrix). Since $\lambda_i \neq \lambda_j$, it follows the eigenvectors $\mathbf{C}_j$, $\mathbf{C}_i$ are orthogonal.

3.4 Transformations In Vectorspaces and Matrices

Prolific use of matrices is made to represent the geometrical operations on vectors in vector spaces which, amongst others, may be physical spaces or function spaces (sec. 3.1). These geometrical transformations include rotations, reflection, inversion or combination of operations on a vector. Alternatively (and equivalently) the set of transformations may be effected on the basis vectors or functions spanning the space. In the first phase, transformations in physical spaces will be discussed followed by transformations in function spaces (Sec. 3.5).

3.4.1 Rotations In Physical Spaces

Let $\overrightarrow{p}$ be a vector in the space spanned by orthogonal base vectors $\overrightarrow{e}_1$, $\overrightarrow{e}_2$, $\overrightarrow{e}_3$ and let C symbolise rotation operation. A rotation C of the basis $\mathbf{e}$ in the anti clock wise sense through an angle θ about $\overrightarrow{e}_3$ would produce $\mathbf{e}'$ matrix (sec. 3.1.1) $\mathbf{e}' = \mathbf{e}\mathbf{C}$, where the $\mathbf{C}$ matrix is the mathematical equivalence of the rotation operation in the basis set $\mathbf{e}$. The elements of the $\mathbf{C}$ matrix can be obtained in the following way (See Fig. 3.2).

$$\mathbf{e}' \;=\; (e'_1\ e'_2\ e'_3) = \mathrm{C}(e_1\ e_2\ e_3)$$

$$\begin{aligned} \vec{e}'_1 &= \vec{Ce}_1 = \vec{e}_1 \cos\theta + \vec{e}_2 \sin\theta + \vec{e}_3.0 \\ \vec{e}'_2 &= C\vec{e}_2 = -\vec{e}_1 \sin\theta + \vec{e}_2 \cos\theta + \vec{e}_3.0 \\ \vec{e}'_3 &= C\vec{e}_3 = \vec{e}_1.0 + \vec{e}_2.0 + \vec{e}_3 \end{aligned} \tag{3.34}$$

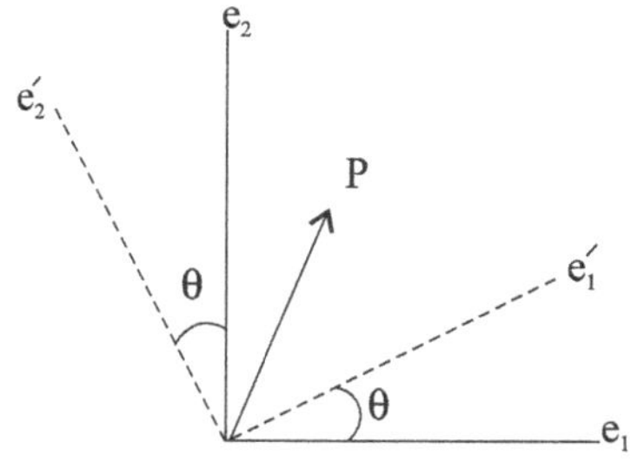

Fig. 3.2: Rotation operation on the basis set.

$$\mathbf{e}' = (e'_1\ e'_2\ e'_3) = (e_1\ e_2\ e_3) \begin{pmatrix} \cos\theta & -\sin\theta & 0 \\ \sin\theta & \cos\theta & 0 \\ 0 & 0 & 1 \end{pmatrix} = \mathbf{eC}$$

Resorting to active mode of transformation (sec 1.7), if the vector $\vec{p}$ be rotated through θ in the clockwise sense (θ = negative), the scalar components of $\vec{p}$ and $\vec{p}'$ (rotated $\vec{p}$) can be easily correlated (Fig. 3.3).

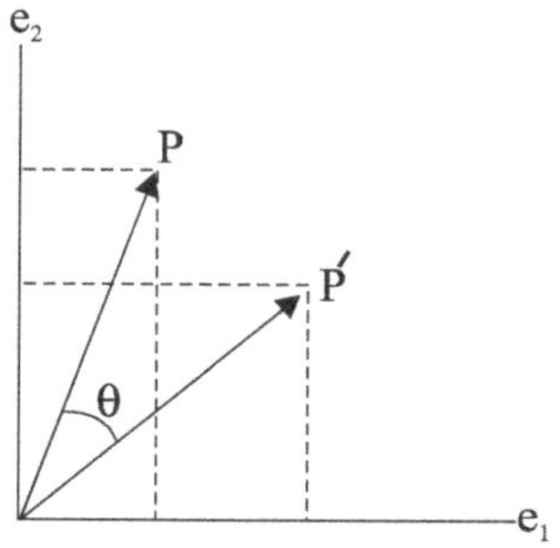

Fig. 3.3: Rotation of the vector $\vec{p}$.

$$\vec{p}' = C\vec{p} = C\,(x_1\vec{e}_1 + x_2\vec{e}_2 + x_3\vec{e}_3)$$

$$x'_1\vec{e}_1 + x'_2\vec{e}_2 + x'_3\vec{e}_3 = (x_1 C\vec{e}_1 + x_2 C\vec{e}_2 + x_3 C\vec{e}_3)$$

Remembering θ to be negative, $C\vec{e}_1$, $C\vec{e}_2$ and $C\vec{e}_3$ may be evaluated. On equating the coefficients of $\vec{e}_1$, $\vec{e}_2$ and $\vec{e}_3$ and sticking to the convention of writing the scalar components in column matrices, we have

$$\begin{pmatrix} x_2' \\ x_3' \end{pmatrix} = \begin{pmatrix} \cos\theta & \sin\theta & 0 \\ -\sin\theta & \cos\theta & 0 \\ 0 & 0 & 1 \end{pmatrix} \begin{pmatrix} x_1 \\ x_2 \end{pmatrix} = \mathbf{C'X} \quad (3.35)$$

A little examination will reveal that the transformation matrix C is orthogonal and that $\widetilde{\mathbf{C}} = \mathbf{C}'$. Therefor $\mathbf{C}' = \mathbf{C}^{-1}$.

We shall use active or passive modes of transformations in future as will suit our convenience.

3.4.2 Rotations about an Arbitrary Axis

While considering molecular symmetry groups (Sec 3.2, 3.2.2.) it has been noted that many of the C_2 operations imply rotations about axes not coincident with the x, y or z axes. Rotations about such selective axes are special cases of eulerian rotations (ϕ, θ, η) However, one can skirt the use of eulerian angles and yet find the transformation matrices in many cases where the rotating axes lie in the planes of ($\vec{e}_1$, $\vec{e}_2$) or ($\vec{e}_2$, $\vec{e}_3$) or ($\vec{e}_3$, $\vec{e}_1$). It is noted from Eq. (3.34) that the matrix for transformation of the basis set (e_1 e_2 e_3) is obtained by collecting independently the effects of C operation on $\vec{e}_1$, $\vec{e}_2$ and $\vec{e}_3$. Such independent collection means that the common angle θ associated with $\vec{e}_1$ and $\vec{e}_2$ during rotation may be substituted by different angles of separate rotations if the circumstances so demand. To illustrate let us consider (Fig. 3.4) an anticlockwise rotation π about an arbitrary axis AB inclined at an angle θ with the $\vec{e}_1$ vector and lying in the plane of the base vectors ($\vec{e}_1$, $\vec{e}_2$). For $\vec{e}_1$ of which the final rotated position is $\vec{e}\,'_1$, the rotation through π about AB is formally equivalent to an effective anticlock sweep through 2θ in the $\vec{e}_1$, $\vec{e}_2$ plane about the $\vec{e}_3$ vector. The $\vec{e}_3$ vector, however, undergoes a rotation of π in the $\vec{e}_2\vec{e}_3$ plane about $\vec{e}_1$ vector would then be different from θ. Thus a rotation of π about AB splits into three independent linear relations.

$$\begin{aligned} \vec{e}\,'_1 &= C\vec{e}_1 = \vec{e}_1\cos 2\theta + \vec{e}_2\sin 2\theta + \vec{e}_3.0 \\ \vec{e}\,'_2 &= C\vec{e}_2 = -\vec{e}_1\ \sin(\pi + 2\theta) + \vec{e}_2\cos(\pi + 2\theta) + \vec{e}_3.0 \\ \vec{e}\,'_3 &= C_\pi\vec{e}_3 = \vec{e}_1.0 - \vec{e}_2\sin\pi + \vec{e}_3\cos\pi \end{aligned} \quad (3.36)$$

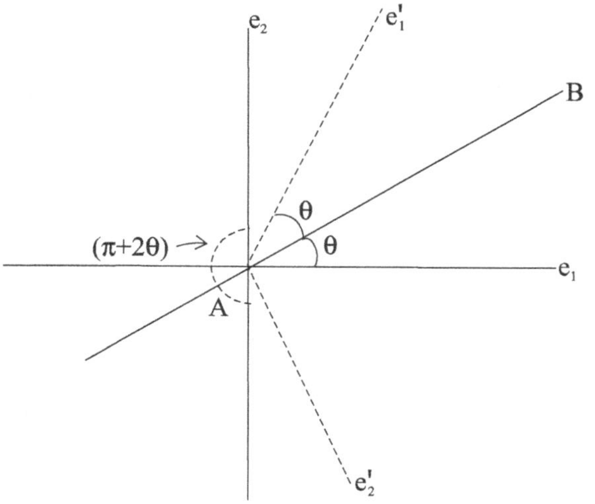

Fig. 3.4: Rotation about an arbitrary axis AB in the plane of e_1 and e_2 vectors .

These, when recast in matrix form, lead to

$$(e_1' \; e_2' \; e_3') = (e_1 \; e_2 \; e_3) \begin{pmatrix} \cos 2\theta & -\sin(\pi + 2\theta) & 0 \\ \sin 2\theta & \cos(\pi + 2\theta) & -\sin \pi \\ 0 & 0 & \cos \pi \end{pmatrix}$$

$$\text{Hence, } \mathbf{C} = \begin{pmatrix} \cos 2\theta & \sin 2\theta & 0 \\ \sin 2\theta & -\cos 2\theta & 0 \\ 0 & 0 & -1 \end{pmatrix} \quad (3.37)$$

A different rotation matrix **C** would be obtained with a different basis set in as much as the angle between the rotating axis AB and the base vector would then be different from θ.

3.4.3 Reflections, Inversion and Improper Rotations

All these geometrical operations lead to results that are obtainable as the integrated effect of separate piecemeal rotations of $\overrightarrow{e}_1$, $\overrightarrow{e}_2$ and $\overrightarrow{e}_3$ vectors about suitable axes as depicted in the previous section 3.4.2.

A general matrix σ_v for mirroring or reflection across any plane containing the $\overrightarrow{e}_3$ vector (i.e., the z-axis) and making an angle θ with the

$\vec{e}_1$ vector (See Fig. 3.4) is found to be

$$\sigma_{\mathbf{v}} = \begin{pmatrix} \cos 2\theta & \sin 2\theta & 0 \\ \sin 2\theta & -\cos 2\theta & 0 \\ 0 & 0 & 1 \end{pmatrix}$$

Similarly a rotation-based approach toward reflection in the $(\vec{e}_1, \vec{e}_2)$ plane normal to $\vec{e}_3$ vector yields

$$\sigma_{\hbar} = \begin{pmatrix} 1 & 0 & 0 \\ 0 & 1 & 0 \\ 0 & 0 & -1 \end{pmatrix} \tag{3.38}$$

The inversion matrix

$$\mathbf{i} = \begin{pmatrix} -1 & 0 & 0 \\ 0 & -1 & 0 \\ 0 & 0 & -1 \end{pmatrix} \tag{3.39}$$

An improper rotation consisting of a rotation of θ degree about $\vec{e}_3$ vector followed by reflection in the plane perpendicular to $\vec{e}_3$ vector is thus a matrix product $\sigma_{\hbar}\mathbf{C}_{\theta}$. Hence the transformation matrix is, using eqs.(3.34, 3.38),

$$\begin{aligned} \mathbf{S}_n &= \begin{pmatrix} 1 & 0 & 0 \\ 0 & 1 & 0 \\ 0 & 0 & -1 \end{pmatrix} \begin{pmatrix} \cos\theta & -\sin\theta & 0 \\ \sin\theta & \cos\theta & 0 \\ 0 & 0 & 1 \end{pmatrix} \\ &= \begin{pmatrix} \cos\theta & -\sin\theta & 0 \\ \sin\theta & \cos\theta & 0 \\ 0 & 0 & -1 \end{pmatrix} \end{aligned} \tag{3.40}$$

3.5 Matrices And Transformations In Function Spaces:

The discussion on transformations has uptil now been confined to physical spaces. How would rotation, reflection, inversion etc., of functions in function spaces be interpreted and mathematically expressed? A meaning can be adduced to such operations by following a convention set by Winger.

Consider a function F(X) where X collectively stands for the coordinates (x_1, x_2, x_3). This function attributes a definite numerical value to P with coordinates (x_1, x_2, x_3). By a rotation C of the function F(X), we mean a new function F′, i.e., CF. The form of the new function should be such that it assigns the same numerical value to P′ with coordinates (x'_1, x'_2, x'_3, *i.e.*, X') as the original function does at P, the relation between P′ and P, i.e., between X′ and X being a rotation C of X to X′ in physical space. Or, explicitly, F′(X′)=CF(X′)=F(X), where **X′=CX**, i.e.

$$\begin{pmatrix} x'_1 \\ x'_2 \\ x'_3 \end{pmatrix} = \mathbf{C} \begin{pmatrix} x_1 \\ x_2 \\ x_3 \end{pmatrix} \tag{3.41}$$

In other words, to find the new rotated function one has to think of (i) the rotation C of X to CX, i.e., to X′ in ordinary space, (ii) to implant these transformed coordinates in the functional from to get CF (CX) i.e., CF (X′), and (iii) to finally demand

$$\begin{aligned} \mathrm{F}'(\mathrm{X}') &= \mathrm{CF}(\mathrm{X}') = \mathrm{F}(\mathrm{X}) \\ &= F'(CX') = F(X) \end{aligned} \tag{3.42}$$

for all possible X in the original function.

What has been described above in terms of rotation C, will all apply to other forms of operations, viz. reflections and inversions in function spaces. Also, as in the earlier examples of vector spaces, we can equivalently rotate the basis set of functions instead of rotating the general functions.

It is now necessary that we illustrate the concepts introduced above with typically important basis functions. Let us consider the rotation C about $\overrightarrow{e}_3$ of the basis functions comprising the three degenerate p-orbitals spanning a function space. Here rotation of three individual functions are involved. These are

$$\left. \begin{aligned} p_1 &= f(r)\tfrac{x_1}{r} \\ p_2 &= f(r)\tfrac{x_2}{r} \\ p_3 &= f(r)\tfrac{x_3}{r} \end{aligned} \right\} \tag{3.43}$$

Individually,

$$\left. \begin{aligned} \mathrm{C}p_1(x'_1, x'_2, x'_3) &= p_1(x_1, x_2, x_3) = f(r)\tfrac{x_1}{r} \\ \mathrm{C}p_2(x'_1, x'_2, x'_3) &= p_2(x_1, x_2, x_3) = f(r)\tfrac{x_2}{r} \\ \mathrm{C}p_3(x'_1, x'_2, x'_3) &= p_3(x_1, x_2, x_3) = f(r)\tfrac{x_3}{r} \end{aligned} \right\} \tag{3.44}$$

Now we substitute x_1, x_2 and x_3 in eq. (3.44) by their equivalents x'_1, x'_2 and x'_3 obtainable from rotation effect in physical. Anti-clock rotation of functions means anti-clock rotation of X

$$\therefore\ \mathbf{X}' = \begin{pmatrix} x'_1 \\ x'_2 \\ x'_3 \end{pmatrix} = \begin{pmatrix} \cos\theta & -\sin\theta & 0 \\ \sin\theta & \cos\theta & 0 \\ 0 & 0 & 1 \end{pmatrix} \begin{pmatrix} x_1 \\ x_2 \\ x_3 \end{pmatrix} \tag{3.45}$$

which on inversion yields

$$\left.\begin{array}{l} x_1 = x'_1\cos\theta + x'_2\sin\theta \\ x_2 = -x'_1\sin\theta + x'_2\cos\theta \\ \text{and} \quad x_3 = x'_3 \end{array}\right\} \tag{3.46}$$

Substitution of the above results in (3.44) yields

$$\left.\begin{array}{l} Cp_1(x'_1, x'_2, x'_3) = \frac{f(r')}{r'}(x'_1\cos\theta + x'_2\sin\theta) \\ Cp_2(x'_1, x'_2, x'_3) = \frac{f(r')}{r'}(-x'_1\sin\theta + x'_2\cos\theta) \\ Cp_3(x'_1, x'_2, x'_3) = \frac{f(r')}{r'}x'_3 \end{array}\right\} \tag{3.47}$$

Since on both sides of eq.(3.47), the same set of primed coordinates of a point occurs, the relations will in general be true for the coordinates of any other point. Hence the primes may be removed for replacement by x_1, x_2 and x_3 to yield

$$\left.\begin{array}{l} Cp_1 = \frac{f(r)}{r}x_1\ \cos\theta + \frac{f(r)}{r}x_2\sin\theta \\ \qquad = p_1\ \cos\theta + p_2\ \sin\theta \\ \text{and similarly} \\ Cp_2 = -p_1\ \sin\theta + p_2\ \cos\theta \\ Cp_3 = p_3 \end{array}\right\} \tag{3.48}$$

which represent the rotated basis set.

As a second example, let us consider the complex d-orbitals basis functions

$$\left.\begin{array}{l} d_{+2} = \sqrt{\frac{5}{32\pi}}(x+iy)^2;\ d_{+1} = -\sqrt{\frac{15}{8\pi}}(x+iy)z; \\ d_0 = \sqrt{\frac{5}{4\pi}\cdot\frac{1}{2}}(3z^2 - r^2);\ d_{-1} = -\sqrt{\frac{15}{8\pi}}(x-iy)z; \\ d_{-2} = \sqrt{\frac{5}{32\pi}}(x-iy)^2 \end{array}\right\} \tag{3.49}$$

and subject this basis set to a rotation of $\pi/2$ (i.e., $C\pi/2$) about the z-axis. It should be noted that in writing the relations (3.49), the phase convention of Condon and Shortley has been used.

Drawing upon the interrelations (3.46) of the previous example it follows (remembering $\theta = \pi/2$)

$$\left.\begin{aligned} x &= x'\cos\theta + y'\sin\theta = y' \\ y &= -x'\sin\theta + y'\cos\theta = -x' \\ z &= z' \end{aligned}\right\} \tag{3.50}$$

Hence, using (3.42) and (3.50)

$$C\pi/2d_{+2}(x',\ y',\ z') = \sqrt{\frac{5}{32\pi}}(x+iy)^2 = \sqrt{\frac{5}{32\pi}}(y'-ix')^2$$

For any general set of coordinates (removal of primes), it yields

$$C\pi/2d_{+2}(x',\ y',\ z) = \sqrt{\frac{5}{32\pi}}(-i)^2\left(x - \frac{y}{i}\right)^2 = -d_{+2}$$

A similar process applied to d_{+1} yields

$$\begin{aligned} C\pi/2d_{+1}(x',\ y',\ z') &= -\sqrt{\frac{15}{8\pi}}(x+iy)z \\ &= -\sqrt{\frac{15}{8\pi}}(y'-ix')z' \\ C\pi/2d_{+1}(x,\ y,\ z) &= -id_{+1},\ \text{on removal of primes} \end{aligned}$$

Rotational effects on the remaining d-orbital basis functions can similarly be ascertained.

The foregoing description and application of equating the newly formed function F′(X′)=f(X) where $X'' = \begin{pmatrix} x' \\ y' \\ z' \end{pmatrix}$ and $X = \begin{pmatrix} x \\ y \\ z \end{pmatrix}$ with $\mathbf{x}'=R\mathbf{x}$, may be substituted by an equivalent alternative process. Since $F'(\mathbf{x}') = F'(R\mathbf{x}) = f(\mathbf{x})$, we may write

$$F'(R^{-1}\mathbf{x}') = F'(\mathbf{x}) = f(R^{-1}\mathbf{x}) = f(\mathbf{x}'')$$

The last three terms enable us to say that the value of the new function F' at the old coordinate set (x, y, z) be equal to that of the old function f at the new coordinate set (x" y" z")

Let us illustrate this once again with the $C_{\frac{\pi}{2}}$ operation on $d_{+2}=\sqrt{\frac{15}{32\pi}}$ $(x+iy)^2$ and upon $d_{-1} = \frac{15}{8\pi}(x\text{-}iy)z$ now,

$$\begin{pmatrix} x'' \\ y'' \\ z'' \end{pmatrix} = \begin{pmatrix} \cos\theta & \sin\theta & 0 \\ -\sin\theta & \cos\theta & 0 \\ 0 & 0 & 1 \end{pmatrix}\begin{pmatrix} x \\ y \\ z \end{pmatrix} \quad \begin{array}{l} \text{with } \theta = \frac{\pi}{2} \\ R^{-1}\text{anticlock}) = R \text{ (clockwise)} \end{array}$$

$$x'' = -y,\ y'' = x \text{ and } z'' = z \tag{3.50 a}$$

we now write

$$\begin{aligned} C_{\pi/2}d_{+2} &= d'_{+2}(x,y,z) = d_{+2}(x''y''z'') = \sqrt{\frac{15}{32\pi}}(x''+iy'')^2 \\ &= \sqrt{\frac{15}{32\pi}}(-y+2x)^2 = (i)^2\sqrt{\frac{15}{32\pi}}(x+iy)^2 = -d_{+2} \end{aligned}$$

on using (3.50 a),
Similarly,

$$\begin{aligned} C_{\pi/2}d_{-1} &= d'_{-1}(x,y,z) = d_{-1}(x'',y'',z'') = \sqrt{\frac{15}{8\pi}}(x''-iy'')z'' \\ &= \sqrt{\frac{15}{8\pi}}(-y-ix)x = -i\sqrt{\frac{15}{8\pi}}(x+\frac{y}{i})z = -id_{-1} \end{aligned}$$

3.6 Transformations In Other Spaces

We have seen how the operations of rotations, reflections a inversion can be represented by matrices in physical space or function spaces. We can similarly think of translational space, rotational space and vibrational space for matrix representations rotations, reflections and others. We use as the basis set a triplet of translational vectors and find how these transform under the specific operations. Similar use may be made of unit rotational vector R_x, R_y, and R_z, the directions of the vectors being perpendicular to the rotational planes YZ, ZX and XY respectively. The positive directions of these vectors are decided in the sense of advancement of a righthanded cork screw. The transformations in vibrational are liberally illustrated in Chapter 8.

We shall cite here a very simple example of reflection in the one dimensional rotation space spanded by the R_z vector. Let us set up laboratory fixed coordinate axes XYZ for a body undergoing rotation in the XY plane in a counterclock direction. The R_z vector is thus collinear with and points along the ve Z direction. A reflection of the rotation across YZ plane can be followed from the Fig. 3.5. It is seen that

$R'_z = \sigma^{yz}R_z = -1R_z$. Thus the matrix $\sigma^{yz} = (-1)$
But a reflection in the horizontal plane will yield

$$R'_z = \sigma_h R_z = 1.\ R_z \text{ leading to } \sigma_h = (1)$$

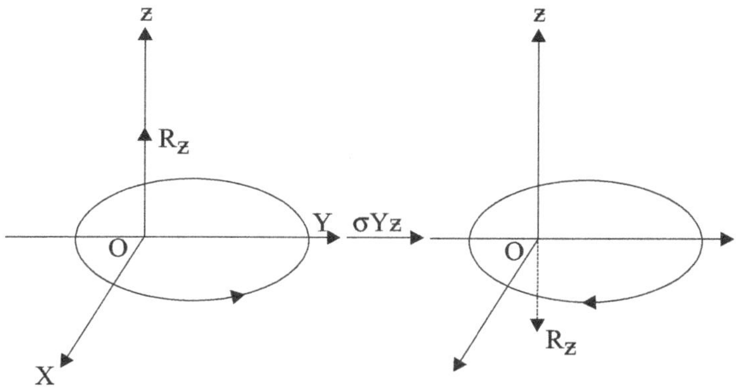

Fig. 3.5: Reflection of rotational motion across the YZ plane.

[Reflection effect is to be considered only on the rotational motion and not on the laboratory fixed axes OX, OY and OZ]. The R_z vector mimics $e_3$vector in some operations but may be quite different in many others (as under σ^{yz}, σ_h etc).

Hilbert Space

In dealing with quantamechanical problems relating to symmetry groups of atoms molecules and ions, we often resort to using the wave functions, i.e., the quantum eigenkets and their possible linear combinations as the basis functions. All these together form a linear space termed the Hilbert space. In theory, the Hilbert space may be of infinite dimension, but in practice we only utilize a portion of the space to effect group theoretical transformations of the original form of the chosen part of the space into a new transformed form, desirably diagonal or the block diagonal matrix form (see chaps 4, 5 and 6). The original form of the chosen part of the space comprises a definite number of eigenkets and the transformed equivalent form is sustained by some or other specific linear combinations of the eigenkets. The ideas will start becoming clearer as we wade into details of the forthcoming chapter.

3.7 Rotations about Arbitrary Axes. Euler Angles

We shall presently examine rotations in three dimensional space about an arbitrary axis, defined by a unit vector $\overrightarrow{n}$ passing through the origin of the laboratory fixed coordinate system XYZ. Two methods will be discussed in this connection.

(a) Let OP be a vector with components (x, y, z). If the vector be rotated (anticlockwise) through an angle θ about $\vec{n}$ the final vector will be OP$'$ with components (x', y', z') in the same axial system XYZ.

We can construct the transformation matrix T_θ (vide infra) such that $\begin{pmatrix} x' \\ y' \\ z' \end{pmatrix} = \mathbf{T}_{-\theta} \begin{pmatrix} x \\ y \\ z \end{pmatrix}$ provided we know both θ and the direction cosines of $\vec{n}$

One method to obtain $\mathbf{T}_\theta$ is to use the Euler angles of rotation in the passive sense that is the vector OP is kept fixed and the axes are rotated successively through three suitable angles α, β & γ (anticlockwise) to arrive at a final rotated axis system where the coordinates of OP would be (x', y', z'), i.e., the same as $\mathbf{T}_\theta$ effect for rotation about **n**.

To describe this process, let us consider a flexible cartesian axial system (X_f, Y_f, Z_f) initially coincident with the fixed (X, Y, Z) system. The flexible system is now rotated through an angle α about Z_f axis giving tise to the first rotated axial system (ξ', η', ζ') so that the components of OP in this axial system can be written as

$$\begin{pmatrix} \xi_1 \\ \eta_1 \\ \zeta_1 \end{pmatrix} = \underline{\mathbf{D}} \begin{pmatrix} x \\ y \\ z \end{pmatrix} = \begin{pmatrix} \cos\alpha & \sin\alpha & 0 \\ -\sin\alpha & \cos\alpha & 0 \\ 0 & 0 & 1 \end{pmatrix} \begin{pmatrix} x \\ y \\ z \end{pmatrix} \tag{3.51}$$

We next rotate the (ξ', η', ζ') system through β (still anticlockwise) about ξ' axis to arrive at (ξ'', η'', ζ'') axial system (note $\xi' \equiv \xi''$) where upon the coordinates of OP are (ξ_2, η_2, ζ_2) enabling us to write

$$\begin{pmatrix} \xi_2 \\ \eta_2 \\ \zeta_2 \end{pmatrix} = \underline{\mathbf{C}} \begin{pmatrix} \xi_1 \\ \eta_1 \\ \zeta_1 \end{pmatrix} = \begin{pmatrix} 1 & 0 & 0 \\ 0 & \cos\beta & \sin\beta \\ 0 & -\sin\beta & \cos\beta \end{pmatrix} \begin{pmatrix} \cos\phi & \sin\phi & 0 \\ -\sin\phi & \cos\phi & 0 \\ 0 & 0 & 1 \end{pmatrix} \begin{pmatrix} x \\ y \\ z \end{pmatrix} \tag{3.52}$$

In the final action a rotation through γ (anticlockwise) about ζ'' axis results in a rotated axial system (ξ''', η''', ζ''') where $\zeta''' \equiv \zeta''$ and the final coordinates of OP in this system and XYZ system are the same.

Thus we can write

$$\begin{pmatrix} x' \\ y' \\ z' \end{pmatrix} = \underline{\mathrm{B}} \begin{pmatrix} \xi_2 \\ \eta_2 \\ \zeta_2 \end{pmatrix} = \begin{pmatrix} \cos\gamma & \sin\gamma & 0 \\ -\sin\gamma & \cos\gamma & 0 \\ 0 & 0 & 1 \end{pmatrix} \begin{pmatrix} 1 & 0 & 0 \\ 0 & \cos\beta & \sin\beta \\ 0 & -\sin\beta & \cos\beta \end{pmatrix} \begin{pmatrix} \cos\alpha & \sin\alpha & 0 \\ -\sin\alpha & \cos\alpha & 0 \\ 0 & 0 & 1 \end{pmatrix} \begin{pmatrix} x \\ y \\ z \end{pmatrix} \tag{3.53}$$

relation (3.53) provides us with the $\mathbf{T}_\theta$ matrix

(b) The second method

Since the rotations of vectors or function about the fixed XYZ axis are easier to imagine and formulate, the whole episode and the net effect depicted under (a), can be equivalently substituted by rotating the vector OP successively through γ, β and α about Z, X and again Z axis respectively (all in anticlockwise manner), the rotation matrices being R(γ), R(β) and R(α). The components of the final OP' vector and of the original OP in the (XYZ) system can be immediately related

$$\begin{pmatrix} x' \\ y' \\ z' \end{pmatrix} = \underline{\mathrm{R}}(\alpha)\underline{\mathrm{R}}(\beta)\underline{\mathrm{R}}(\gamma) \begin{pmatrix} x \\ y \\ z \end{pmatrix} \tag{3.54}$$

with the $\underline{\mathrm{R}}$'s having their familiar forms. (Transposed forms of the former set in 3.53)

It is to be noted that $\underline{\mathrm{R}}(\gamma)$ is to operate first followed by $\underline{\mathrm{R}}(\beta)$ and $\underline{\mathrm{R}}(\alpha)$ operations, the sequence being reverse of that in (a).

The chosen angles α, β and γ are called Euler angles.

It is evident that the use of Euler angles to substitute the direct rotation of θ about **a** depends on the suitable discreet choice of their magnitudes. It turns out that in most problems involving point group symmetries of molecules and atoms, the respective Euler angles (α, β, γ) can be arrived at after a little careful reflection.

In case we want to formally evaluate α, β and γ, we should have the transformation matrix $\mathbf{T}_\theta$ for direct rotation of OP about **a**. Then T_θ=R(α)R(β) R(γ) of Eq. (3.54).
The structure of $\mathbf{T}_\theta$ is obtained from the solutions x', y', z' from the

following three equations.

$$\begin{aligned} lx' + my' + nz' &= lx + my + nz \\ x'^2 + y'^2 + z'^2 &= x^2 + y^2 + z^2 \\ xx' + yy' + zz' &= (x^2 + y^2 + z^2)\cos\theta + (lx + my + nz)^2(1 - \cos\theta) \end{aligned} \tag{3.55}$$

Here (x, y, z) and $(x'y'z')$ are the coordinates of the points -P and P' before and after rotation of OP about $\vec{a}$ through an angle θ. The unit vector $\vec{a}$ has the direction cosines, l, m, n in the axial system XYZ. There will be only one set of x', y', z' values which will be physically admisiible being consistent with the known values of l,m,n and the angle of rotation θ. The arrival at the set of equations (3.56), though straight forward in principle, is nevertheless a bit involved and is not discussed here.

As an example of Euler rotations, let us find the C_3 transformation matrix for a rotation of $\frac{2\pi}{3}$ about the body diagonal of a cube passing through the points viz., the origin (000) and (111). The sides of the cube, each of unit length, define the X-, Y- and Z-axes respectively.

Here the Eulerangles are chosen to be $\alpha = \frac{\pi}{2}$, $\beta = \frac{\pi}{2}$, $\gamma = 0$ after a little reflection on the effect of a direct rotation of $\frac{2\pi}{3}$ about the C_3 -axis. Putting these values of rotations in (3.54), R(α), R(β) and R(γ) are respectively found to be

$$\begin{pmatrix} \cos\alpha & -\sin\alpha & 0 \\ \sin\alpha & \cos\alpha & 0 \\ 0 & 0 & 1 \end{pmatrix} \begin{pmatrix} 1 & 0 & 0 \\ 0 & \text{Cos}\beta & -\sin\beta \\ 0 & \sin\beta & \cos\beta \end{pmatrix} \text{ and } \begin{pmatrix} \cos\gamma & -\sin\gamma & 0 \\ \sin\gamma & \cos\gamma & 0 \\ 0 & 0 & 1 \end{pmatrix}$$

The insertion of the specific values of α, β and γ will give the final rotation matrix,

$$\mathbf{C_3} = \begin{pmatrix} 0 & 0 & 1 \\ 1 & 0 & 0 \\ 0 & 1 & 0 \end{pmatrix} \tag{3.56}$$

A more involved calculation concerns an overall rotations of p_x, p_y and p_z orbitals. This overall rotation, when analyzed in terns of Euler rotations, gives $\alpha = \frac{\pi}{2}$, $\beta = 0$ and $\gamma = \frac{\pi}{4}$. Taking the forms of p_x, p_y and p_z, following Condon Shortly's phase convention it turns out $p'_x = Cp_x$ $= \frac{1}{\sqrt{2}}(p_y - p_z)$. Readers may verify this an additionally find out p'_y and p'_z to arrive at the final transformation matrix.

Chapter 4

Representation Of Groups, Equivalent Representations And Reducible Representations

We know that the molecular point groups comprise group elements which are symmetry operations. These symmetry operations have been viewed as mere geometrical operations of rotations reflections and inversions. How are we to translate such mental viewing and perception of symmetry operations into their mathematical analogues? A suitable way of mathematical representation of the group elements (i.e., symmetry operations) is suggested by the transformation processes in vector spaces of the last chapter.

4.1 Representation Of Geometrical Operations By Matrices:

It was shown (Sec 3.4-3.4.3) that a full geometrical operation on vectors in a vector space or its basis set can be expressed by a matrix product relation such as

$$\hat{\mathrm{C}}\mathbf{e} = \mathbf{eC},\ \ \hat{\mathrm{C}}\mathbf{X} = \mathbf{CX}\ \text{(Eqs. 3.33, 3.34) and others.} \tag{4.1}$$

The components of the matrix product are thus a transformation matrix (e.g., $\mathbf{C}$, σ, $\mathbf{i}$, $\mathbf{C}$' etc.) and a matrix, $\mathbf{e}$, of the basis set or a column matrix, $\mathbf{X}$ of the components. Sometimes, as has been seen, the base vectors may be functions also. In this context we may, therefore, express a geometrical operation by a transformation matrix in a given basis set. In doing so, it should be remembered that there still exists the possibility of different dimensional representations of the same geometrical operation depending on the dimensionality of the vector space.

4.2 Representations Of Group symmetry Operations And Of Groups:

The group elements of a point group, being symmetry operations (and hence geometrical operations), may then be reasonably considered as operators and thus associated with transformation matrices (Sec. 4.1) in a given basis set. Let there be a point group, {P, Q, R, S, ...} $\in$ G. Each such individual symmetry operation P, Q, ... may have in the basis set **f** their transformation matrices $\mathbf{D}^{f}$ (P), $\mathbf{D}^{f}/(Q)$,...... with elements D_{ij} (P), D_{ij} (Q), ... respectively. The superscript f in $\mathbf{D}^{f}$ (P) is suggestive of the basis set used. This superscript is dropped in cases where there is no need for explicit reference to the basis set. Now while we understand the logic that **D**(P), **D**(Q), **D**(R).....are the representative matrices for the individual group elements, we are not sure whether the set of matrices as a whole is a representation of the entire group G. The latter will imply that these proxies, viz., the matrices should also satisfy the product rules that define a group. It is necessary, therefore, that the following features be satisfied by the matrices:

(i) If the group G is such that PR=T, RS=Q etc. then **D**(P) **D**(R) **D**(PR)=**D**(T), **D**(R)**D**(S)=**D**(Q) and so on for other pairs.

(ii) Each such matrix should have its inverse. This means, provided PS=E, $\mathbf{D}(P)=\mathbf{D}(S)^{-1}$, etc.

(iii) The necessary provision that **D**(P)**D**(R)=**D**(T) etc., indicates that there should exist some correlations between the matrix elements of **D**(T) and those of **D**(P) **D**(R). To exhibit what it is like, let us suppose that these matrices are of n×n order in the basis set $\mathbf{f}=(f_1 f_2 ... f_k ... f_n)$. Choosing any basis vector f_k it is evident that Tf_k is a vector in that space. Hence,

$$\begin{aligned} Tf_k &= \sum_j f_j D_{jk}(T); \; PRf_k = P(Rf_k) = P\sum_l f_l D_{lk}(R) \\ &= \sum_i fiD_{ik}(T) \qquad = \sum_i \sum_l f_i D_{il}(P) D_{lk}(R) \end{aligned} \tag{4.2}$$

Since $Tf_k = PRf_k$ it follows from 4.2

$$D_{ik}(T) = \sum_l D_{il}(P) D_{lk}(R) \tag{4.3}$$

We thus see that the transformation matrices built up to represent the individual symmetry (i.e., geometrical) operations of a point group should also collectively obey the above features.

The discussion so far in the present section has been only in symbolic terms. The net upshot is that the elements of a point group are just operators in the form of transformation matrices in any given basis set and that these matrices are to obey the product rules of the group. The obedience of the group law by the matrices can be checked (subjectively) by taking any specific point group and developing the corresponding matrices for any vector space. To do it let us consider the group D_3 and the physical vector space spanned by $\overrightarrow{e_1}$, $\overrightarrow{e_2}$ and $\overrightarrow{e_3}$. A suitable species of D_3 group is ethane molecule, which is neither completely eclipsed nor fully staggered, but is only slightly staggered. A more workable example will be planar BCl_3 molecule if we arbitrarily think it to be divested of $\sigma_\hbar$ and σ_v's. We start with a trace of the BCl_3 molecule (Fig. 4.1) in which the Cl atoms, numbered 1, 2 and 3, lie at the vertices of an

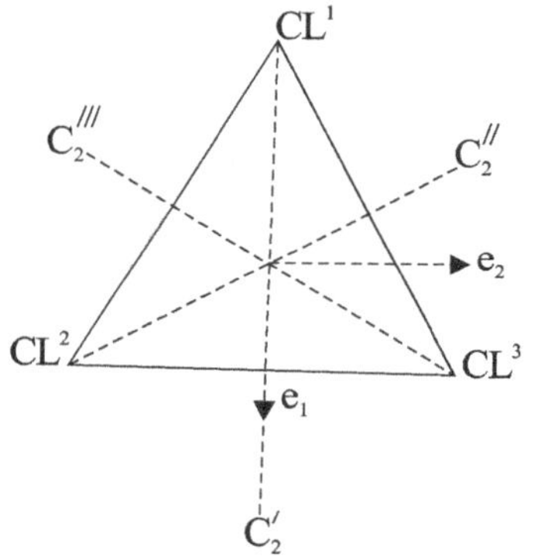

Fig. 4.1: Symmetry elements and base vectors in planar BCl_3 arbitrarily supposed to bc divested of $\sigma_\hbar$ and σ_v's

equilateral triangle with boron atom at the centroid which is taken to be the origin of the coordinate system. The group elements are six in number E, C_3, C_3^2, C_2', C_2'', C_2'''. Some of the symmetry elements and the base vectors are indicated in Fig. 4.1. Now the matrices can easily be constructed. To find $\mathbf{D}(C_2')$, $\mathbf{D}(C_2'')$ and $\mathbf{D}(C'''_2)$, one may follow the technique indicated in Sec. 3.4.2 keeping an eye on the effective counter clock rotation of $\overrightarrow{e_1}$ and $\overrightarrow{e_2}$ vectors in the $(e_1\ e_2)$ plane

$$\left.\begin{array}{lll}
\mathbf{D}(\mathrm{E}) & = & \begin{pmatrix} 1 & 0 & 0 \\ 0 & 1 & 0 \\ 0 & 0 & 1 \end{pmatrix}, \ \mathbf{D}(\mathrm{C}_3) = \begin{pmatrix} \cos 120^\circ & -\sin 120^\circ & 0 \\ \sin 120^\circ & \cos 120^\circ & 0 \\ 0 & 0 & 1 \end{pmatrix} \\
 & = & \begin{pmatrix} -\frac{1}{2} & -\frac{\sqrt{3}}{2} & 0 \\ \frac{\sqrt{3}}{2} & -\frac{1}{2} & 0 \\ 0 & 0 & 1 \end{pmatrix} \\
\mathbf{D}(\mathrm{C}_3^2) & = & \begin{pmatrix} -\frac{1}{2} & \frac{\sqrt{3}}{2} & 0 \\ -\frac{\sqrt{3}}{2} & -\frac{1}{2} & 0 \\ 0 & 0 & 1 \end{pmatrix}, \ \mathbf{D}(\mathrm{C}_2') = \begin{pmatrix} 1 & 0 & 0 \\ 0 & -1 & 0 \\ 0 & 0 & -1 \end{pmatrix} \\
\mathbf{D}(\mathrm{C}'''_2) & = & \begin{pmatrix} \cos 120^\circ & -\sin 300^\circ & 0 \\ \sin 120^\circ & \cos 300^\circ & 0 \\ 0 & 0 & \cos\pi \end{pmatrix} = \begin{pmatrix} -\frac{1}{2} & \frac{\sqrt{3}}{2} & 0 \\ \frac{\sqrt{3}}{2} & \frac{1}{2} & 0 \\ 0 & 0 & -1 \end{pmatrix}, \\
\mathbf{D}(\mathrm{C}''_2) & = & \begin{pmatrix} \cos 240^\circ & -\sin 60^\circ & 0 \\ \sin 240^\circ & \cos 60^\circ & 0 \\ 0 & 0 & \cos\pi \end{pmatrix} = \begin{pmatrix} -\frac{1}{2} & -\frac{\sqrt{3}}{2} & 0 \\ -\frac{\sqrt{3}}{2} & \frac{1}{2} & 0 \\ 0 & 0 & -1 \end{pmatrix}
\end{array}\right\}$$

In finding the $\mathbf{C}_2$ matrices of (4.4) one might also have directly used the $\mathbf{C}$ matrix of (3.36) in Sec. 3.4.2, taking care of θ in each case. This set of matrices forms a representation of the group D_3 in the basis set $\mathbf{e}$. The product rules are satisfied as shown below.

From the group multiplication table of D_3, it is known $C_2''C_3=C_2'''$. To check in terms of the matrices, we have

$$\begin{aligned}
\mathbf{D}(\mathrm{C}_2'')\mathbf{D}(\mathrm{C}_3) &= \begin{pmatrix} -\frac{1}{2} & -\frac{\sqrt{3}}{2} & 0 \\ -\frac{\sqrt{3}}{2} & \frac{1}{2} & 0 \\ 0 & 0 & -1 \end{pmatrix} \begin{pmatrix} -\frac{1}{2} & -\frac{\sqrt{3}}{2} & 0 \\ \frac{\sqrt{3}}{2} & -\frac{1}{2} & 0 \\ 0 & 0 & 1 \end{pmatrix} \\
&= \begin{pmatrix} -\frac{1}{2} & \frac{\sqrt{3}}{2} & 0 \\ \frac{\sqrt{3}}{2} & \frac{1}{2} & 0 \\ 0 & 0 & -1 \end{pmatrix} = \mathbf{D}\,(\mathrm{C}_2''')
\end{aligned}$$

Trial examinations of other pair products of the matrices lead to one or the other of the remaining matrices in conformity with the expectation of the product of the corresponding group elements. Secondly each matrix has it inverse present within the set. Thus$\mathbf{D}$ (C_2'') is its own inverse and $\mathbf{D}(\mathrm{C}_3)$ that of $\mathbf{D}(\mathrm{C}_3^2)$.

To show the validity of Eq. (4.3), a feature requirement as stated under (iii) of this section, we can check $D_{21}(C_2''')$ element in terms of elements of $\mathbf{D}(\mathrm{C}_2'')$ and $\mathbf{D}(\mathrm{C}_3)$ since $\mathbf{D}(\mathrm{C}_2''')=\mathbf{D}(\mathrm{C}_2'')\mathbf{D}(\mathrm{C}_3)$. According to Eq. (4.3), it is demanded that

$$D_{21}(C_2''') = \sum_{i=1} D_{2i}(C_2'')D_{i1}(C_3)$$

$$= \left(-\frac{\sqrt{3}}{2}\right)\left(-\frac{1}{2}\right) + \left(\frac{1}{2}\right)\left(\frac{\sqrt{3}}{2}\right) + 0 \times 0$$
$$= \frac{\sqrt{3}}{2}$$

which is really the corresponding element found in the representative matrix $\mathbf{D}(C_2''')$.

This retention of the group properties by the representative matrices in a given basis set is a general truth for any vector space of finite dimension and for all finite groups. Having thus devised a way of forming a representation, Γ, of any group, one should be aware of several disconcerting points to be elucidated and sorted out later. These are

(i) Non-uniqueness or multiplicity of representations in a given vector space.

(ii) Variable dimensionality of the representative matrices depending on the dimensionality of the vector space.

(iii) Correlations, if any, among representations in different vector spaces. These points will be attended to later in appropriate places (Sec 4.3, 4.4, 4.5 and 5.6).

4.3 Multiplicity Of Representations, Similarity Transformations And Equivalent Representations.

While in the preceding section we have shown how a symmetry group can be represented, we harbor some uneasy feeling over the possibility of the non-uniqueness of the representations even in the same vector space. To what extent can we get over this aspect of non-uniqueness and make the representation or, at least, some property of the representation more definite and precise for the same group? We shall try to answer this question in this section.

Let us confine ourselves to the symmetry group D_3 and the physical vector space. A representation of this group in the basis set $\vec{e_1}$, $\vec{e_2}$ and $\vec{e_3}$ has been built up (Sec. 4.2). Two other basis sets. spanning the same physical space will be taken up after developing the implications of similarity transformations relevant for such processes involving changes in basis sets.

Similarity transformation :– The construction of a representation with respect to a new basis set becomes easier if we know the transformation to the new basis set from the old one with respect to which the representation is already known.

Consider the basis sets $\mathbf{f}=(f_1 f_2 \ldots f_k \ldots f_n)$ and $\mathbf{g}=(g_1 g_2 \ldots g_k \ldots g_n)$ spanning the same space. Suppose $\mathbf{g}=\mathbf{fA}$, where $\mathbf{A}$ is the transformation leading to the new basis $\mathbf{g}$.

$$\mathbf{f} = \mathbf{gA}^{-1} = \mathbf{gB} \text{ where } \mathbf{B} = \mathbf{A}^{-1}$$

If $\mathbf{R}$ be any group element, then

$$\mathrm{Rf_i} = \sum_j \mathrm{f_j D^f_{ji}(R)} \text{ and } \mathrm{Rg_k} = \sum_l \mathrm{g_l D^g_{lk}(R)} \tag{4.5}$$

where $\mathrm{D^f_{ji}(R)}$ and $\mathrm{D^g_{lk}(R)}$ are the corresponding matrix elements of the representations of $\mathbf{R}$ in the $\mathbf{f}$ and the $\mathbf{g}$ basis sets respectively.
Again $\mathrm{g_k} = \sum_i \mathrm{f_i a_{ik}}$, $\mathrm{a_{ik}}$ is the (ik)th. element of $\mathbf{A}$

$$\begin{aligned} \mathrm{Rg_k} = \sum_i \mathrm{Rf_i a_{ik}} &= \sum_i \sum_j \mathrm{f_j D^f_{ji}(R) a_{ik}} \\ &= \sum_i \sum_j \sum_l \mathrm{g_l b_{lj} D^f_{ji}(R) a_{ik}} \end{aligned} \tag{4.6}$$

In the last step, the $\mathrm{f_j}$ vector is expressed in the $\mathbf{g}$ basis and b's are the elements of the matrix $\mathbf{B}$.

On comparing (4.5) and (4.6) and equating the coefficients,

$$\mathrm{D^g_{lk}(R)} = \sum_i \sum_j \mathrm{b_{lj} D^f_{ji}(R) a_{ik}} = \left(\mathbf{BD}^{\mathrm{f}}(\mathrm{R})\mathbf{A}\right)_{\mathrm{lk}} \tag{4.7}$$

From this elementwise relation, we can write the full matrix relation

$$\mathbf{D}^{\mathrm{g}}(\mathrm{R}) = \mathbf{BD}^{\mathrm{f}}(\mathrm{R})\mathbf{A} = \mathbf{A}^{-1}\mathbf{D}^{\mathrm{f}}(\mathrm{R})\mathbf{A} \tag{4.8}$$

Hence, the representations in the new and the old basis sets are connected through a similarity transformation. The important point to know is the transformation matrix $\mathbf{A}$ and its inverse. $\mathbf{A}^{-1}$ If. however, $\mathbf{A}$ is unitary, then (4.8) can be written in the form

$$\mathbf{D}^{\mathrm{g}}(\mathrm{R}) = \mathbf{A}^{\neq}\mathbf{D}^{\mathrm{f}}(\mathrm{R})\mathbf{A}$$

Two theorems will now be proved in connection with the similarity transformation.

Theorem 4.1. The product rules obeyed by a representation of a group in a basis set remain unaffected by a similarity transformation.

Let {P, Q, R, S, T, ...} $\in$ G and let **PR**=T, so that **D**(P)**D**(R)=**D**(T). If **A** be the transformation matrix involved in the change of basis sets,

$$\begin{aligned} \mathbf{A}^{-1}\mathbf{D}(\mathrm{T})\mathbf{A} &= \mathbf{A}^{-1}\mathbf{D}(\mathrm{P})\mathbf{D}(\mathrm{R})\mathbf{A} \\ &= \{\mathbf{A}^{-1}\mathbf{D}(\mathrm{P})\mathbf{A}\}\{\mathbf{A}^{-1}\mathbf{D}(\mathrm{R})\mathbf{A}\} \\ \mathbf{D}^{\mathrm{n}}(\mathrm{T}) &= \mathbf{D}^{\mathrm{n}}(\mathrm{P})\mathbf{D}^{\mathrm{n}}(\mathrm{R}) \end{aligned} \tag{4.9}$$

where $\mathbf{D}^{\mathrm{n}}$ is the matrix representative in the new basis set **n**. Hence the new representation also obeys the product rules.

Theorem 4.2. The trace (character) of a matrix remains invariant under a similarity transformation.

Consider a matrix **C** and its similarity transform by the matrix **T**, i.e. $\mathbf{T}^{-1}\mathbf{C}\mathbf{T}$

$$\mathrm{Tr}\ \mathbf{C} = \sum_{\mathrm{j}} c_{\mathrm{jj}} \tag{4.10}$$

$$\begin{aligned} \mathrm{Tr}\ \mathbf{T}^{-1}\mathbf{C}\mathbf{T} &= \sum_{\mathrm{i}}\sum_{\mathrm{j}}\sum_{\mathrm{k}}(\mathbf{T}^{-1})_{\mathrm{ij}}c_{\mathrm{jk}}(\mathbf{T})_{\mathrm{ki}} \\ &= \sum_{\mathrm{j}}\sum_{\mathrm{k}}\sum_{\mathrm{i}}(\mathbf{T})_{\mathrm{ki}}(\mathbf{T}^{-1})_{\mathrm{ij}}c_{\mathrm{jk}} \\ &= \sum_{\mathrm{j}}\sum_{\mathrm{k}}\left(\mathbf{T}\mathbf{T}^{-1}\right)_{\mathrm{kj}}\ \delta_{\mathrm{kj}}c_{\mathrm{jk}}, \qquad \mathbf{T}\mathbf{T}^{-1} = 1 \\ &= \sum_{\mathrm{j}} c_{\mathrm{jj}} \end{aligned} \tag{4.11}$$

From Eqs. (4.10) and (4.11) it follows Tr **C**=Tr$\mathbf{T}^{-1}$ **CT**. This very important relation demands that even when a change of basis is effected, the character of the representation of any particular element remains unaltered. That is

Tr $\mathbf{D}^{\mathrm{g}}$(R)=Tr.$\mathbf{D}^{\mathrm{f}}$(R), or to use a more conventional symbol for character,

$$\chi^{\mathrm{g}}(\mathrm{R}) = \chi^{\mathrm{f}}(\mathrm{R}) \tag{4.12}$$

The representation of the D_3 group in the basis set **e**, obtained in Sec. 4.2, may now be utilized in building up those with respect to two other basis sets in the same physical space. Of course, the construction of the representations, though can be done independently, becomes easier and mechanical if the representation $\mathbf{D}^e$ be used.

Example 4.1.

Consider the basis $\vec{l_1}$, $\vec{l_2}$, $\vec{l_3}$ which are of unit magnitude and directed along C_2', C_2'' and C_3 symmetry axes of BCl_3 molecule, arbitrarily supposed to be devoid of σ_v's and σ_h elements (Fig. 4.2)

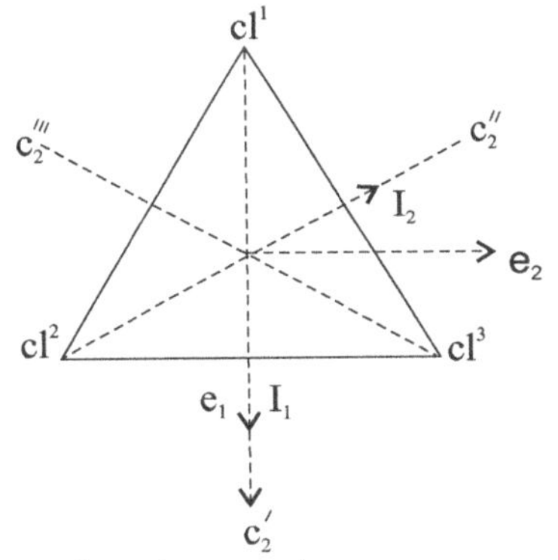

Fig. 4.2: Choice of $\vec{l}_1$, $\vec{l}_2$ and $\vec{l}_3$ base vectors in BCl_3 molecule.

Using a little of trigonometry, the transformation matrix between **l** and **e** is found out. It is seen that $\vec{l_1} = \vec{e_1}$; $\vec{l_2} = -\vec{e_1}\sin 30 + \vec{e_2}\cos 30°$ and $\vec{l_3} = \vec{e_3}$.

$$(l_1\ l_2\ l_3) = (e_1\ e_2\ e_3)\begin{pmatrix} 1 & -\frac{1}{2} & 0 \\ 0 & \frac{\sqrt{3}}{2} & 0 \\ 0 & 0 & 1 \end{pmatrix}$$

Thus the transformation matrix $\mathbf{A} = \begin{pmatrix} 1 & -\frac{1}{2} & 0 \\ 0 & \frac{\sqrt{3}}{2} & 0 \\ 0 & 0 & 1 \end{pmatrix}$ and

$$\mathbf{A}^{-1} = \begin{pmatrix} 1 & \frac{1}{\sqrt{3}} & 0 \\ 0 & \frac{2}{\sqrt{3}} & 0 \\ 0 & 0 & 1 \end{pmatrix}$$

The different matrices in the basis **1** are now obtained by using the similarity relation (4.8) and the representation with respect to **e** given in relation (4.4).

$$\mathbf{D}^{\mathrm{l}}(\mathrm{E}) = \begin{pmatrix} 1 & 0 & 0 \\ 0 & 1 & 0 \\ 0 & 0 & 1 \end{pmatrix}$$

$$\begin{aligned}\mathbf{D}^{\mathrm{l}}(\mathrm{C}_3) &= \begin{pmatrix} 1 & \frac{1}{\sqrt{3}} & 0 \\ 0 & \frac{2}{\sqrt{3}} & 0 \\ 0 & 0 & 1 \end{pmatrix} \begin{pmatrix} -\frac{1}{2} & -\frac{\sqrt{3}}{2} & 0 \\ \frac{\sqrt{3}}{2} & -\frac{1}{2} & 0 \\ 0 & 0 & 1 \end{pmatrix} \begin{pmatrix} 1 & -\frac{1}{2} & 0 \\ 0 & \frac{\sqrt{3}}{2} & 0 \\ 0 & 0 & 1 \end{pmatrix} \\ &= \begin{pmatrix} 0 & -1 & 0 \\ 1 & -1 & 0 \\ 0 & 0 & 1 \end{pmatrix}\end{aligned} \tag{4.13}$$

$$\mathbf{D}^{\mathrm{l}}\left(\mathrm{C}_3^2\right) = \begin{pmatrix} -1 & 1 & 0 \\ -1 & 0 & 0 \\ 0 & 0 & 1 \end{pmatrix}, \mathbf{D}^{\mathrm{l}}(\mathrm{C}_2') = \begin{pmatrix} 1 & -1 & 0 \\ 0 & -1 & 0 \\ 0 & 0 & -1 \end{pmatrix}$$

$$\mathbf{D}^{\mathrm{l}}\left(\mathrm{C}_2''\right) = \begin{pmatrix} -1 & 0 & 0 \\ -1 & 1 & 0 \\ 0 & 0 & -1 \end{pmatrix}, \mathbf{D}^{\mathrm{l}}(\mathrm{C}_2'''') = \begin{pmatrix} 0 & 1 & 0 \\ 1 & 0 & 0 \\ 0 & 0 & -1 \end{pmatrix}$$

Example 4.2

We now consider the D_3 subgroup of the point group of the BCl_3 molecule with another non-orthogonal basis set where the unit base vectors $\vec{d_1}$, $\vec{d_2}$, $\vec{d_3}$ all project from the origin (Boron atom) such that the C_3 symmetry axis, C_2' symmetry axis and $\vec{d_1}$ are in one plane. So also the trio, viz. C_3, C_2'' axes and $\vec{d_2}$ lie in one plane. The third set, comprising C_3, C_2'' axes and $\vec{d_3}$, is also coplanar. Fig. 4.3 displays the $\vec{d_1}$, $\vec{d_2}$, $\vec{d_3}$ vectors along with $\vec{e_1}$, $\vec{e_2}$ and $\vec{l_1}$ and $\vec{l_2}$. Let $\vec{d_1}$, $\vec{d_2}$ and $\vec{d_3}$ make an angle α each with the C_2', C_2'' and C_2''' symmetry axes respectively.

$$\left.\begin{aligned}
\vec{d_1} &= -\vec{e_1}\,\cos\alpha - \vec{e_3}\,\sin\alpha \\
\vec{d_2} &= \vec{n_2}\cos\alpha - \vec{e_3}\sin\alpha \\
&= (\vec{e_1}\cos 60^\circ - \vec{e_2}\sin 60^\circ)\cos\alpha - \vec{e_3}\sin\alpha \\
&= \tfrac{1}{2}\cos\alpha\vec{e_1} - \tfrac{\sqrt{3}}{2}\cos\alpha\vec{e_2} - \vec{e_3}\sin\alpha \\
\vec{d_3} &= \vec{n_3}\cos\alpha - \vec{e_3}\sin\alpha \\
&= (\vec{e_1}\cos 60^\circ + \vec{e_2}\sin 60^\circ)\cos\alpha - \vec{e_3}\sin\alpha \\
&= \tfrac{1}{2}\cos\alpha\vec{e_1} + \tfrac{\sqrt{3}}{2}\cos\alpha\vec{e_2} - \vec{e_3}\sin\alpha
\end{aligned}\right\} \tag{4.14a}$$

$\vec{n_2}$ and $\vec{n_3}$ in the foregoing relations represent unit vectors in the requisite directions.

$$(\mathrm{d}_1\ \mathrm{d}_2\ \mathrm{d}_3) = (\mathrm{e}_1\ \mathrm{e}_2\ \mathrm{e}_3)\begin{pmatrix} -\cos\alpha & \frac{1}{2}\cos\alpha & \frac{1}{2}\cos\alpha \\ 0 & -\frac{\sqrt{3}}{2}\cos\alpha & \frac{\sqrt{3}}{2}\cos\alpha \\ -\sin\alpha & -\sin\alpha & -\sin\alpha \end{pmatrix}$$

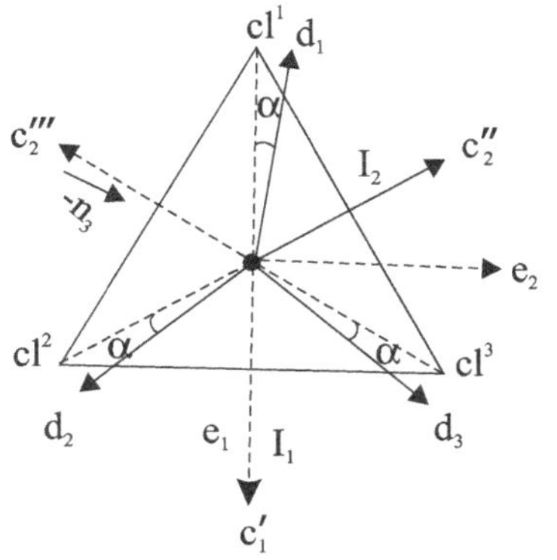

Fig. 4.3: New set $\vec{d}_1$, $\vec{d}_2$. $\vec{d}_3$ of non-orthogonal axes in the molecule BCl_3. d_1, d_2 and d_3 project below the triangular plane from the boron atom.

α can have any value depending on the choice.

If the vectors d_1, d_2, d_3 are to be mutually orthogonal then $\cot\alpha = \sqrt{2}$ (as can be shown from eqs. 4.14a) which demands $\cos\alpha = \sqrt{\frac{2}{3}}$ and $\sin\alpha = \sqrt{\frac{1}{3}}$. On using these values the transformation matrix, **A**, and its inverse, $\mathbf{A}^{-1}$. become

$$\mathbf{A} = \begin{pmatrix} -\sqrt{\frac{2}{3}} & \frac{1}{\sqrt{6}} & \frac{1}{\sqrt{6}} \\ 0 & -\frac{1}{\sqrt{2}} & \frac{1}{\sqrt{2}} \\ -\frac{1}{\sqrt{3}} & -\frac{1}{\sqrt{3}} & -\frac{1}{\sqrt{3}} \end{pmatrix} \text{ and } \mathbf{A}^{-1} \begin{pmatrix} -\sqrt{\frac{2}{3}} & 0 & -\frac{1}{\sqrt{3}} \\ \frac{1}{\sqrt{6}} & -\frac{1}{\sqrt{2}} & -\frac{1}{\sqrt{3}} \\ \frac{1}{\sqrt{6}} & \frac{1}{\sqrt{2}} & -\frac{1}{\sqrt{3}} \end{pmatrix} \tag{4.14}$$

The matrices (4.14), utilized in the form of similarity transformation upon the $\mathbf{D}^e$ matrices (Eq. 4.4.), i.e., $\mathbf{A}^{-1}\mathbf{D}^e\mathbf{A}$ (ci. Eq. 4.8), yield all the necessary representative matrices of the D_3 group in the basis set (d_1, d_2, d_3). For the orthogonal d-basis set as considered above, the matrices are

$$\mathbf{D}^d(\mathrm{E}) = \begin{pmatrix} 1 & 0 & 0 \\ 0 & 1 & 0 \\ 0 & 0 & 1 \end{pmatrix} \mathbf{D}^d(\mathrm{C}_3) = \begin{pmatrix} 0 & 0 & 1 \\ 1 & 0 & 0 \\ 0 & 1 & 0 \end{pmatrix}$$

$$\mathbf{D}^d(\mathrm{C}_3^2) = \begin{pmatrix} 0 & 1 & 0 \\ 0 & 0 & 1 \\ 1 & 0 & 0 \end{pmatrix}$$

$$\mathbf{D}^d(\mathrm{C'}_2) = \frac{1}{3}\begin{pmatrix} 1 & -2 & -2 \\ -2 & -2 & 1 \\ -2 & 1 & -2 \end{pmatrix} \mathbf{D}^d(\mathrm{C''}_2) = \frac{1}{3}\begin{pmatrix} -2 & -2 & 1 \\ -2 & 1 & -2 \\ 1 & -2 & -2 \end{pmatrix} \tag{4.15}$$

$$\text{and } \mathbf{D}^{\mathrm{d}}(\mathrm{C}_2''') = \frac{1}{3}\begin{pmatrix} -2 & 1 & -2 \\ 1 & -2 & -2 \\ -2 & -2 & 1 \end{pmatrix}$$

Before arriving at any conclusion we may tabulate certain relevant data obtained by comparing the three representations (4.4. 4,13, 4.15), viz., Γ^{e}, Γ^{l} and Γ^{d} of the point group D_3.

Table 4.1

Point Group D_3 – Representation

Group Elements	Character of the matrices in the basis set			Interrelationship of the elements and same character value (Sec. 2.26, 2.3)
	e	l	d	
E	3	3	3	Separate class
C_3	0	0	0	Conjugate elements in the same class
C_3^2	0	0	0	
C_2'	−1	−1	−1	Conjugate elements of a single class
C_2''	−1	−1	−1	
C_2'''	−1	−1	−1	

Since any n×n (here 3×3) representation of a group in different basis sets of the same vector space represents the transformation behavior of the basis sets or equivalently of a general vector in that space, these representations, inspite of their elementwise distinctions in the matrices, are regarded as equivalent representations. Moreover,

(i) As in the present instance, we find that such equivalent representations are connected through a similarity transformation (4.8).

(ii) The invariance of χ (R), i.e., the character in different representations as demanded by the theorem (4.2) and supported by the data in Table 4.1, is also characteristic of equivalent representations.

(iii) An additional feature of the equivalent representations, evident from Table 4.1, is that the group elements in the same class have the same character value (see. also sec. 5.3).

What has been stated with regard to D_3 group is also valid for any other point group. It is thus seen that one can bunch up a whole lot of $n \times n$ order matrix representations of a point group in different basis sets of the same space as basically equivalent in view of the uniqueness of the character property, e.g., $\chi^e(R) = \chi^l(R) = \chi^d(R)$. It is also noted that a certain measure of unification of the representations, but not a total measure, is thus achieved. We are still fettered by reference to basis sets in the same vector space. Can we travel beyond this vector space and still carry forward the idea of equivalent representations? We shall find an answer after we consider the representations of D_3 in function spaces.

4.4 Representations In Function Spaces. Extension Of The Idea Of Equivalent Representations.

Consider first the three-dimensional p orbital function space carried by p_1, p_2 and p_3 orbiials of Sec 3.5. A representation of the point group D_3 in this basis set is easily written down by noting-that its transformation properties are the same as the base vectors e_1, e_2 and e_3 of the physical space. Hence the transformation matrices of D_3 group in the basis set $(p_1\ p_2\ p_3)$ are identical with those of relation (4.4) in the basis set **e**.

A second basis set in this function space comprises p_{+1}, p_{-1} and p_0 which are complex p wave functions of the type.

$$\begin{aligned}
p_{+1} &= Nf(r)\sin\theta e^{i\phi} = \frac{Nf(r)}{r}(x_1 + ix_2) = \frac{1}{\sqrt{2}}(p_1 + ip_2) \\
p_{-1} &= Nf(r)\sin\theta e^{-i\phi} = \frac{1}{\sqrt{2}}(p_1 - ip_2) \\
p_0 &= Nf(r)\cos\theta = p_3
\end{aligned} \tag{4.16}$$

where N is the normalizing factor and $p_1\ p_2\ p_3$ are the conventional p orbitals treated in Sec. 3.5.

There exist two methods of building up the representation in the basis set $(p_{+1}\ p_{-1}\ p_0)$. The first is to use the similarity transformation technique (4.8) of the previous section and the known representation in the basis set $(p_1\ p_2\ p_3)$. In the present example, the transformation matrix A and its inverse A^{-1} to be utilized in the similarity relation follow from (4.16).

$$\mathbf{A} = \begin{pmatrix} \frac{1}{\sqrt{2}} & \frac{1}{\sqrt{2}} & 0 \\ \frac{i}{\sqrt{2}} & -\frac{i}{\sqrt{2}} & 0 \\ 0 & 0 & 1 \end{pmatrix}, \quad \mathbf{A}^{-1} = \begin{pmatrix} \frac{1}{\sqrt{2}} & \frac{i}{\sqrt{2}} & 0 \\ \frac{1}{\sqrt{2}} & -\frac{i}{\sqrt{2}} & 0 \\ 0 & 0 & 1 \end{pmatrix} \tag{4.17}$$

It is now easy to write down the representative matrices. For example, using $\mathbf{D}^{p}(C_3)$ and relation (4.17),

$$\mathbf{D}^{p}c(C_3) = \begin{pmatrix} \frac{1}{\sqrt{2}} & \frac{i}{\sqrt{2}} & 0 \\ \frac{1}{\sqrt{2}} & -\frac{i}{\sqrt{2}} & 0 \\ 0 & 0 & 1 \end{pmatrix} \begin{pmatrix} -\frac{1}{2} & -\frac{\sqrt{3}}{2} & 0 \\ \frac{\sqrt{3}}{2} & -\frac{1}{2} & 0 \\ 0 & 0 & 1 \end{pmatrix} \begin{pmatrix} \frac{1}{\sqrt{2}} & \frac{1}{\sqrt{2}} & 0 \\ \frac{i}{\sqrt{2}} & -\frac{i}{\sqrt{2}} & 0 \\ 0 & 0 & 1 \end{pmatrix}$$

the superscript p_c stands for the complex p basis functions.

However, this similarity transformation process having been profusely illustrated previously, the alternative independent method is conveniently chosen for the basis functions $(p_{+1}\ p_{-1}\ p_0)$. Using the convention of symmetry operations in function space (cf. Sec 3.5) and Eq. (4.16) one may write

$$\begin{aligned} (p'_{+1}p'_{-1}p'_0) &= C_3(p_{+1}p_{-1}p_0)\{r',\ \theta',\ \phi'\} = (p_{+1}p_{-1}p_0) \\ &= f(r,\ \theta,\ \phi) = f(r',\ \theta',\ \phi' - \alpha) \end{aligned}$$

where $\phi' = \phi + \alpha$, α being $\frac{2\pi}{3}$ under C_3 rotation.
Individually,

$$\begin{aligned} p'_{+1} = C_3 p_{+1}(r',\ \theta',\ \phi') &= p_{+1} = Nf(r)\sin\theta e^{i\phi} \\ &= Nf(r')\sin\theta' e^{i(\phi'-\alpha)} \end{aligned}$$

Dropping the primes, which are common to the variables on both sides, $p'_{+1}=p_{+1}e^{-i\alpha}$. Similarly $p'_{-1}=p_{-1}e^{i\alpha}$ and $p'_0 = p_0$

$$\mathbf{D}^{pc}(C_3) = \begin{pmatrix} e^{-i\alpha} & 0 & 0 \\ 0 & e^{i\alpha} & 0 \\ 0 & 0 & 1 \end{pmatrix} = \begin{pmatrix} e^{-i2\pi/3} & 0 & 0 \\ 0 & e^{i2\pi/3} & 0 \\ 0 & 0 & 1 \end{pmatrix}$$

with $\chi^{p}c(C_3) = 1 + e^{-i2\pi/3} + e^{i2\pi/3} = 0$

To find $\mathbf{D}^{p}c(C'_2)$ it is noted that under a rotation of π about C'_2 axis (i.e., $\vec{e_1}$ vector Fig. 4.1), the following relations amongst the variables will hold good.

$$r' = r,\ \theta' = \pi - \theta \text{ and } \phi' = 2\pi - \phi \tag{4.18}$$

Using eq. (4.18),

$$p'_{+1} = C'_2\{p_{+1}(r',\ \theta',\ \phi')\} = p_{+1}(r,\ \theta,\ \phi) = Nf(r)\sin\theta e^{i\phi}$$
$$= Nf(r)\sin(\pi-\theta')e^{i(2\pi-\varphi)'}$$

$$p'_{+1} = p_{-1}e^{2i\pi} \text{ (on removing the primes from the variable.)}$$
$$p'_{-1} = p_{+1}e^{-i2\pi}$$
$$\text{and } p'_0 = -p_0 \text{ since, } [\cos(\pi-\theta') = -\cos\theta']$$

It is thus clear that $\mathbf{D}^{p}c(C'_2) = \begin{pmatrix} 0 & 1 & 0 \\ 1 & 0 & 0 \\ 0 & 0 & -1 \end{pmatrix}$ having the character $\chi^{p}c(C'_2) = -1$ With $\mathbf{D}^{p}c(C_3)$, $\mathbf{D}^{p}c(C'_2)$ and the apparent $\mathbf{D}^{p}c(E)$ of the symmetry group D_3 in the basis set $(p_{+1}\ p_{-1}\ p_0)$ thus already known, we have here three representatives of three elements of the three distinct classes of the group. For comparison, this is enough and a Table 4.2 similar to the previous one is given below.

Table 4.2
Point Group D_3 - Partial Representation

Group elements of there classes	Basis functions in function space		Basis vectors in Physical space
	$p-$ Orbitals	Complex $p-$ wave functions	**e**
E	3	3	3
C_3	0	0	0
C'_2	-1	-1	-1

It may be observed that the characters of the representations in three dimensional different vector spaces, as chosen here, are the same for each typical class element. All three representations (Table 4.1 and 4.2) of the same dimensionality are called equivalent representations in as much as the characters, $\chi(R)$'s, are the same for all R's. We are thus no longer confined to a single vector space and to a similarity relation between the representations of the same vector space in defining equivalent representations, but have moved out of the bounds of a single vector space in doing so. Same dimensionality and identity of characters of the matrices for the different group elements are only to be considered for equivalence of representations in different vector spaces.

4.5 Representations Of Variable Dimensions. Reducible And Irreducible Representations.

Of the three questions raised at the conclusion of Sec. 4.2, the first one has been sorted out fully and the third one only partially in the last two sections. A fuller answer to the latter will have to be gleaned from the scattered patches in the present section and in the next chapter. An elucidation of the variable dimensional representations of a group as raised in the second query will be taken up here.

When one tries to derive a representation, one essentially starts with a basis set of a vector space. In the examples of Sec. 4.3 and 4.4, different three-dimensional vector spaces were chosen with the ultimate turn out of the three-dimensional representations of the D_3 group. One might have chosen instead a bigger basis set, e.g., the five-dimensional (d-orbital space, to get a five-dimensional representation or in the other extreme, his fancy might have fallen on just a single basis vector, say $\overrightarrow{e_3}$, spanning a one-dimensional physical space.

In the face of the possibilities of multitude of choices, is there any general unifying pattern or principle that serves to knit together the equivalent representations more closely? The answer lies in the development of the concept of irreducible representations which will be dealt with qualitatively in the present chapter and more quantitatively in the next.

Let us go back to the actual representations, Γ's, of the group D_3 given in relations (4.4), (4.13), (4.15) and in Sec. 4.4 with the basis sets **e**, **l**, **d** and $p_\mathbf{C}$. All the matrices of any representation in the basis set **e** (or **l** or $p_\mathbf{C}$) are block diagonalised, i.e., these are of the type $\left(\begin{array}{c:c} \mathbf{L}_2 & \mathbf{O} \\ \hline \mathbf{O} & \mathbf{L}_1 \end{array}\right)$ The total matrix thus occurs neatly partitioned into two matrices a (2×2) $\mathbf{L}_2$ submatrix and a one-dimensional $\mathbf{L}_1$ submatrix. The other elements of the total matrix besides $\mathbf{L}_1$ and $\mathbf{L}_2$ are all zeros. It is also evident that the set of $(2 \times 2)\mathbf{L}_2$ matrices in any given representation, say Γ^e of (4.4), would also form a two-dimensional representation of D_3 in the basis set of e_1, e_2 vectors. Specifically, $\mathbf{L}_2$ matrices of this latter representation are

$$\left.\begin{array}{l} \mathbf{D}_2(\mathrm{E}) = \begin{pmatrix} 1 & 0 \\ 0 & 1 \end{pmatrix}, \ \mathbf{D}_2(\mathrm{C}_3) = \begin{pmatrix} -\frac{1}{2} & -\frac{\sqrt{3}}{2} \\ -\frac{\sqrt{3}}{2} & -\frac{1}{2} \end{pmatrix}, \\ \mathbf{D}_2(\mathrm{C}_3^2) = \begin{pmatrix} -\frac{1}{2} & \frac{\sqrt{3}}{2} \\ -\frac{\sqrt{3}}{2} & -\frac{1}{2} \end{pmatrix}, \ \mathbf{D}_2(\mathrm{C}_2') = \begin{pmatrix} 1 & 0 \\ 0 & -1 \end{pmatrix}, \\ \mathbf{D}_2(\mathrm{C}_2'') = \begin{pmatrix} -\frac{1}{2} & -\frac{\sqrt{3}}{2} \\ -\frac{\sqrt{3}}{2} & \frac{1}{2} \end{pmatrix}, \ \mathbf{D}_2(\mathrm{C}_2''') = \begin{pmatrix} -\frac{1}{2} & \frac{\sqrt{3}}{2} \\ \frac{\sqrt{3}}{2} & \frac{1}{2} \end{pmatrix}, \end{array}\right\} \tag{4.19}$$

with $\chi(\mathrm{E}) = 2$, $\chi(\mathrm{C}_3) = \chi(\mathrm{C}_3^2) = -1$ and $\chi(\mathrm{C}_2') = \chi(\mathrm{C}_2'') = \chi(\mathrm{C}_2''') = 0$. In a like manner the set of $\mathbf{L}_1$ submatrices, forming another representation of the group D_3, are

$$\left.\begin{array}{l} \mathbf{D}_1(\mathrm{E}) = (1), \ \mathbf{D}_1(\mathrm{C}_3) = (1), \ \mathbf{D}_1(\mathrm{C}_3^2) = (1) \\ \mathbf{D}_1(\mathrm{C}_2') = (-1), \ \mathbf{D}_1(\mathrm{C}_2'') = (-1), \ \mathbf{D}_1(\mathrm{C}_2''') = (-1) \end{array}\right\} \tag{4.20}$$

With $\chi(\mathrm{E}) = \chi(\mathrm{C}_3) = \chi(\mathrm{C}_3^2) = 1$ and $\chi(\mathrm{C}_2') = \chi(\mathrm{C}_2'') = \chi(\mathrm{C}_2''') = -1$

The following conclusions can be drawn from the above description.

(1) The representation, Γ^{e}, with all its matrices appearing in similar block diagonal forms, is said to be in reduced form amenable to splitting into a one-dimensional and a two-dimensional irreducible representations (henceforward abbreviated as IR's and symbolised by Γ_{i}'s). This means that each original matrix representative, e.g., $\mathbf{D}(\mathrm{C}_3)$ of the representation Γ^{e} is the direct sum of the matrices $\mathbf{D}_1(\mathrm{C}_3)$ and $\mathbf{D}_2(\mathrm{C}_3)$, i.e.,

$$\mathbf{D}(\mathrm{C}_3) = \mathbf{D}_1(\mathrm{C}_3) + \mathbf{D}_2(\mathrm{C}_3) \tag{4.21}$$

$$\text{Hence the net representation } \Gamma^{\mathrm{e}} = \Gamma_1 \oplus \Gamma_2 \tag{4.22}$$

where Γ_1 is the IR standing for the matrices of (4.20) and Γ_2, the IR represented by the matrices of (4.19). Thus the vectors $\overrightarrow{\mathrm{e}_1}$, $\overrightarrow{\mathrm{e}_2}$ span the two-dimensional invariant subspace and $\overrightarrow{\mathrm{e}_3}$ alone spans the one-dimensional subspace of the larger representation space.

(2) A similar splitting into two IR's, Γ_1 and Γ_2 with similar set of χ_{i} values, as stated before, is also characteristic of the representations in the basis sets **1** or $\mathrm{p_c}$. In the case of **1**, the IR. Γ_1 is carried by the basis $\overrightarrow{\mathrm{l}_3}$ and Γ_2 by the basis $(\mathrm{l}_1\ \mathrm{l}_2)$. For the complex $\mathrm{p_c}$ basis, Γ_1 is carried by the base function p_0 and Γ_2 by the basis $(\mathrm{p}_{+1}\ \mathrm{p}_{-1})$. The total representation and the IR's in these cases follow a direct summation relation of the type (4.22).

(3) The three (2×2) IR's in the basis sets $(e_1\ e_2)$, $(l_1\ l_2)$ and $(p_{+1}\ p_{-1})$ are different matrices but have the identical set of characters for all elements of the group. Therefore, these IR's are all equivalent. The three one-dimensional IR's, Γi's, in the three basis sets are also equivalent for reasons of identity of character values.

We may now discuss some further pertinent points.

(a) An attempt to build up a one-dimensional representation of D_3 by choosing a single base vector $\vec{e_1}$ (or $\vec{l_1}$ or p_{+1}) instead of $\vec{e_3}$ (or $\vec{l_3}$ or p_0) will not succeed. The group symmetry operations of D_3 (excepting E) will invariably end in a linear mixing up of $\vec{e_1}$ and $\vec{e_2}$. Similar will be the case for the vector pair in the $(l_1\ l_2)$ set or the function pair in p_c set. Therefore, any and every base vector may not, by itself alone, form a basis of an IR.

(b) If we try a basis of five d-orbitals, viz., $d(x^2\text{-}y^2)$, d_{xy}, d_{xz}, d_{yz}, d_z^2, the (5×5) matrices constituting the five-dimensional representation of D_3 will each appear in similar block diagonal form. The representation, Γ, is, therefore, already in reduced form. There will appear three IR's satisfying

$$\Gamma = \Gamma_1 \oplus \Gamma_2 \oplus \Gamma_3 \tag{4.23}$$

in which Γ_1 is one-dimensional and the remaining two are two-dimensional each. d_z2 is the basis for the first and the pair of functions $(d_{x^2-y^2}, d_{xy})$ and $(d_{xz},\ d_{yz})$ constitute the basis for the remaining two. For this Γ_1,

$$\begin{aligned}\text{the } \chi_1(E) &= \chi_1(C_3) = \chi_1(C_3^2) = \chi_1(C_2') \\ &= \chi_1(C_2'') = \chi_1(C_2''') = 1\end{aligned}$$

This shows that the set of χ_i values are not totally identical elementwise with the χ_1 values of Γ_1 (with the base vectors $\vec{e_3}$ or $\vec{l_3}$ or p_0. Hence these two IR's, though both one-dimensional, are non-equivalent and different representations. The Γ_2 and Γ_3 of the d-orbital space have the same set of χ_i values elementwise as those of the Γ_2's of the former examples with the basis sets $(e_1,\ e_2)$, $(l_1,\ l_2)$ or $(p_{+1},\ p_{-1})$. These are, therefore, all equivalent representations.

(c) Since increasingly multidimensional representations are possible by choosing bigger and bigger basis sets, is there for any group no limit to the number of IR's resulting from the bigger and bigger representations? The answer to this for any finite group will be provided via a theorem in the next chapter. It may, however, be mentioned in advance that there exist only a finite number of non-equivalent and distinct IR's for any finite group. So, however big a basis set be chosen, the initially constructed representation when transformed to block diagonalised or reduced form (Sec. 4.6) will yield non-equivalent IR's the number of which cannot certainly exceed that stipulated by the theorem for that group. The most that one can expect in the handling of such big basis sets is the final appearance of any particular or a few IR's in plural number, but the number of distinct IR's emanating from it wiil either be equal to or less than the maximum number permissible by the theorem. If the basis set chosen is far too small, the initial representation will yield a few of the distinct IR's but not all. For example the basis set (e_l e_2 e_3) yields Γ splitting into Γ_1 and Γ_2 of the group D_3 but not the non-equivalent Γ_1 obtainable with the basis function d_z^2. In this sense a small enough basis is able to unravel the properties of a group only partially.

(d) We have been so long talking of representations, Γ, which are already in reduced or block diagonal form. Looking at the representative matrices of D_3 with the basis set **d** (4.15), it is observed that unlike the other cases its matrices are not in the block diagonal forms. Prior to obtaining the possible IR's from the initial representation Γ^d, it is necessary to reduce the representation to the block diagonal form. Methods exist (Sec. 4.6 and Chapter 5) for effecting this change over of Γ^d to the reduced form. Such a representation (Γ^d) is called a reducible representation.

4.6 Reduction of RepresentationsQualitative Outline:

Consider a group in which {P, Q, R, S, ...} $\in$G. Let **P**, **Q**, **R** etc. be the matrices of a representation Γ of the group in a certain basis set g in which the matrices of Γ are not block diagonalised.

Suppose by some means we hit upon some orthogonal basis set **f** in the same vector space and in this basis set the matrices representing the group are in block diagonal forms. Let **B** be the transformation matrix linking **f** and **g**, i.e., **f**=**gB**. Hence all the matrices of Γ in the **g** basis should be related to the matrices of the representation in the **f** basis through a common similarity relation (Eq. 4.8). Therefore, we may write

$\mathbf{P}$'f$=\mathbf{B}^{-1}\mathbf{PB}$, $\mathbf{Q}^{\text{'f}}=\mathbf{B}^{-1}\ \mathbf{QB}$, $\mathbf{R}^{\text{'f}}=\mathbf{B}^{-1}\ \mathbf{RB}$ and so on, where $\mathbf{P}^{\text{'f}}$, $\mathbf{Q}^{\text{'f}}$ etc. are the representative matrices in the **f** basis and all are in block diagonal forms. For illustrations

$$\left.\begin{aligned}\mathbf{B}^{-1}\mathbf{PB} = \mathbf{P}'^{\text{f}} &= \begin{pmatrix} \boxed{P_1'} & & & \\ & \boxed{P_2'} & & \\ & & \boxed{P_3'} & \\ & & & \boxed{P_n'} \end{pmatrix} \\ \mathbf{Q}'^{\text{f}} &= \begin{pmatrix} \boxed{Q_1'} & & & \\ & \boxed{Q_2'} & & \\ & & \boxed{Q_3'} & \\ & & & \boxed{Q_n'} \end{pmatrix}\end{aligned}\right\} \tag{4.24}$$

The set $\{\mathbf{P}_1',\ \mathbf{Q}_1',\ \mathbf{R}_1'\ ...\}$ forms a representation Γ_1' and, similarly, the set $\{\mathbf{P}_2',\ \mathbf{Q}_2',\ \mathbf{R}_2'\ ...\}$ constitutes the representation Γ_2'. Likewise the other representations, which are all IR's, may be picked. (If any particular set of blocks is not already in irreducible form, a second similarity transformation should be applied to this representation to split the matrices into irreducible forms). Following the direct sum rule, we write conventionally,

$$\Gamma^{\text{red}} = \Gamma_1' \oplus \Gamma_2' \oplus \Gamma_3' \oplus \tag{4.25}$$

although the representations on the right hand side spring from Γ' and not directly from Γ^{red}. All the same it should be remembered that due to the invariance of the trace under similarity transformation (Theo. 4.2), character of any matrix of a reducible representation is equal to the sum of the characters of the different IR's into which the reducible matrix can be split. It is from this viewpoint that the summation sign acquires significance apart from the fact that the total representation space is a direct sum of the invariant subspaces. We thus hit upon a very important relation emerging from (4.25), viz.,

$\chi^{\text{red}}(\text{R}) = \chi'_1(\text{R}) + \chi'_2(\text{R}) + \chi'_3(\text{R}) + \ ...$ for all R's of the point group.

Two relevant points which naturally arise in the foregoing qualitative approach merit mention.

(1) How is one to select the orthonormal basis set that would lead to block diagonalisation of the reducible representation?
(2) How is it that if, under a similarity transformation, **P** is block diagonalised, this similarity transformation of **Q**, **R** etc. with the same matrix **B** leads to blockings that are exactly similar dimensionally and in number (cf. Eq. 4.24)?

The answer to the first is obtained in a quantitative treatment of reduction of a representation employing the methods of projection operators (Chap. 5 Sec. 5.6).

As to the second, the reason for similar block diagonalisation can be intitutively understood. For if in the group, **PR**=**T**, then after block diagonalisations of both,

$$\mathbf{P'R'} = (\mathbf{B}^{-1}\mathbf{PB})(\mathbf{B}^{-1}\mathbf{RB}) = \mathbf{B}^{-1}\mathbf{PRB} = \mathbf{B}^{-1}\mathbf{TB} = \mathbf{T'}$$

This condition, which points to the obedience of the product rules of the group, demands the block diagonal form of **P**' and **R**' as well as of **T**' be similar dimensionally in order to make the multiplication feasible between conformable matrices.

We conclude this section by stating the formal definitions of reducible and irreducible representations of a group.

A reducible representation of a symmetry group in a certain basis set is the set of matrices imitating the behavior of group elements in their product laws as expressed in the multiplication table of the group. Such a representation can always be shown as a direct sum of representations of smaller dimensions by a suitable change of basis set. That is the representation can be block-diagonalised through a proper similarity transformation.

An irreducible representation in a certain basis is a set of matrices reflecting the product laws of the group elements and the matrices are nonamenable to further being expressed as a direct sum of smaller matrices by any change of basis set.

4.7 Representations of Groups C_{4v} and C_{3h}

In order to have ready on hand some concrete sets of representations other than those of D_3 to introduce variety in the incidental further references and also as further examples of reduced forms of representations, let us compile those of the symmetry groups C_{4v}, and $C_{3\hbar}$, in the five-dimensional d-orbital function space. The normalised angular wave functions of the five d-orbitals are obtainable as suitable linear combinations of the normalised spherical harmonics.

These in cartesian coordinates are

$$\left.\begin{aligned} d_z{}^2 = \frac{\sqrt{5}}{4r^2\sqrt{\pi}}(3z^2-1);\ d_{xz} = \frac{\sqrt{15}}{2r^2\sqrt{\pi}}(xz);\ d_{yz} = \frac{\sqrt{15}}{2r^2\sqrt{\pi}}(yz) \\ d_{xy} = \frac{\sqrt{15}}{4r^2\sqrt{\pi}}(xy);\ d_{x^2-y^2} = \frac{\sqrt{15}}{4r^2\sqrt{\pi}}(x^2-y^2) \end{aligned}\right\} \tag{4.26}$$

The symmetry operations of the group C_{4v} are E, C_4, C_2 C_4^3, $\sigma_v^{(xz)}$, $\sigma_v^{(yz)}$, σ_v', σ_v'' (Fig. 4.4).

To assess the effect of symmetry operations on the basis functions we need to know first how **X** and $\hat{R}$**X** (i.e., **X**'), where $\mathbf{X}=\begin{pmatrix} x \\ y \\ z \end{pmatrix}$, are related and then employing the inverse matrix, **X** is expressed in terms of **X**' and finally the original functional forms are expressed in terms of

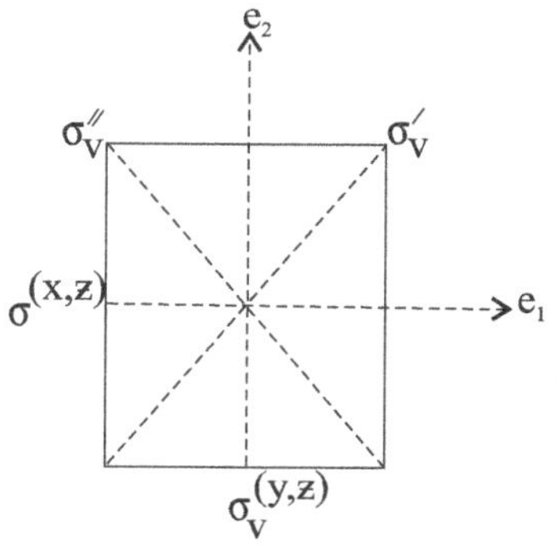

Fig. 4.4: Choice of axes and location of symmetry elements of C_{4v} symmetry group.

X' and the primes then dropped (Sec, 3.5). Thus for

$$\left.\begin{array}{llll}
(1)\ C_4^{-1}\mathbf{X}, & (2)\ C_2^{-1}\mathbf{X}', & (3)\ (C_4^3)^{-1}\mathbf{X}' & (4)\ [\sigma_v^{(xz)}]^{-1}\mathbf{X}', \\
\left.\begin{array}{r} x = y' \\ y = -x' \\ z = z' \end{array}\right\} & \left.\begin{array}{r} x = -x' \\ y = -y' \\ z = z' \end{array}\right\} & \left.\begin{array}{r} x = -y' \\ y = x' \\ z = z' \end{array}\right\} & \left.\begin{array}{r} x = x' \\ y = -y' \\ z = z' \end{array}\right\} \\
(5)\ [\sigma_v^{(yz)}]^{-1}\mathbf{X}' & (6)\ (\sigma_v')^{-1}\mathbf{X}' & (7)\ (\sigma_v'')^{-1}\mathbf{X}' & \\
\left.\begin{array}{r} x = -x' \\ y = y' \\ z = z' \end{array}\right\} & \left.\begin{array}{r} x = y' \\ y = x' \\ z = z' \end{array}\right\} & \left.\begin{array}{r} x = -y' \\ y = -x' \\ z = z' \end{array}\right\} &
\end{array}\right) \tag{4.27}$$

Using (4.27), one finds after dropping of primes,

(i) $C_4d_{z^2} = d_{z^2}$, $C_4d_{xz} = d_{yz}$, $C_4d_{yz} = -d_{xz}$, $C_4d_{xy} = -d_{xy}$,
$C_4d(x^2 - y^2) = -d(x^2 - y^2)$

(ii) $C_2d_{z^2} = d_{z^2}$, $C_2d_{xz} = -d_{xz}$, $C_2d_{yz} = -d_{yz}$, $C_2d_{xy} = -d_{xy}$,
$C_2d(x^2 - y^2) = -d(x^2 - y^2)$

(iii) $C_{4^3}d_{z^2} = d_{z^2}$, $C_4^3d_{xz} = -d_{yz}$, $C_4^3d_{yz} = d_{xz}$, $C_4^3d_{xy} = -d_{xy}$
$C_{4^3}d_{(x^2-y^2)} = -d(x^2 - y^2)$

(iv) $\sigma_v^{(xz)}d_{z^2} = d_{z^2}$, $\sigma_v^{(xz)}d_{xz} = d_{xz}$, $\sigma_v^{(xz)}d_{yz} = -d_{yz}$,
$\sigma_v^{(xz)}d_{xy} = -d_{xy}$, $\sigma_v^{(xz)}d_{x^2-y^2} = d_{(x^2-y^2)}$

(v) $\sigma_v^{(yz)}d_{z^2} = d_{z^2}$, $\sigma^{(yz)}d_{xz} = -d_{xz}$, $\sigma_v^{(yz)}d_{yz} = d_{yz'}$
$\sigma_v^{(yz)}d_{xy} = -d_{xy}$, $\sigma_v^{(yz)}d_{(x^2-y^2)} = d_{(x^2-y^2)}$

(vi) $\sigma_v'd_{z^2} = d_{z^2}$, $\sigma_v'd_{xz} = d_{yz}$, $\sigma_v'd_{yz} = d_{xz}$, $\sigma_v'd_{xy} = d_{xy}$,
$\sigma_v'd_{x^2-y^2} = -d_{(x^2-y^2)}$.

(vii) (g) $\sigma_v''d_{z^2} = d_{z^2}$, $\sigma_v''d_{xz} = -d_{yz}$, $\sigma_v''d_{yz} = -d_{xz}$,
$\sigma_v''d_{xy} = d_{xy}$, $\sigma_v''d_{x^2-y^2} = -d_{(x^2-y^2)}$

Remembering that the basis functions form a row matrix, the representative matrices are

$$\mathbf{C}_4 = \begin{pmatrix} 1 & 0 & 0 & 0 & 0 \\ 0 & 0 & -1 & 0 & 0 \\ 0 & 1 & 0 & 0 & 0 \\ 0 & 0 & 0 & -1 & 0 \\ 0 & 0 & 0 & 0 & -1 \end{pmatrix}, \quad \mathbf{C}_2 = \begin{pmatrix} 1 & 0 & 0 & 0 & 0 \\ 0 & -1 & 0 & 0 & 0 \\ 0 & 0 & -1 & 0 & 0 \\ 0 & 0 & 0 & 1 & 0 \\ 0 & 0 & 0 & 0 & 1 \end{pmatrix}$$

$$\mathbf{C}_4^3 = \begin{pmatrix} 1 & & & & \\ & 0 & 1 & & \\ & -1 & 0 & & \\ & & & -1 & \\ & & & & -1 \end{pmatrix}, \quad \sigma_{\mathbf{V}}^{(\mathbf{xz})} = \begin{pmatrix} 1 & & & & \\ & 1 & 0 & & \\ & 0 & -1 & & \\ & & & -1 & \\ & & & & 1 \end{pmatrix}$$

$$\sigma_{\mathbf{V}}^{(\mathbf{yz})} = \begin{pmatrix} 1 & & & & \\ & -1 & 0 & & \\ & 0 & 1 & & \\ & & & -1 & \\ & & & & 1 \end{pmatrix}, \quad \sigma'_{\mathbf{V}} = \begin{pmatrix} 1 & & & & \\ & 0 & 1 & & \\ & 1 & 0 & & \\ & & & 1 & \\ & & & & -1 \end{pmatrix}$$

$$\sigma_{\mathbf{V}''} = \begin{pmatrix} 1 & & & & \\ & 0 & -1 & & \\ & -1 & 0 & & \\ & & & 1 & \\ & & & & -1 \end{pmatrix}, \quad \mathbf{E} = \begin{pmatrix} 1 & & & & \\ & 1 & 0 & & \\ & 0 & 1 & & \\ & & & 1 & \\ & & & & 1 \end{pmatrix} \tag{4.28}$$

In the above matrices, the blanks indicate zero matrix elements. The representations occur in reduced forms and the different IR's numbered serially are:

IR	D(E)	D(C_4)	D(C_2)	D($C_4$3)	D($\sigma_V{}^{xz}$)	D($\sigma_V{}^{yz}$)	D($\sigma_{V'}$)	D($\sigma_{V''}$)
Γ_1	(1)	(1)	(1)	(1)	(1)	(1)	(1)	(1)
Γ_2	$\begin{pmatrix}1 & 0\\ 0 & 1\end{pmatrix}$	$\begin{pmatrix}0 & -1\\ 1 & 0\end{pmatrix}$	$\begin{pmatrix}-1 & 0\\ 0 & -1\end{pmatrix}$	$\begin{pmatrix}0 & 1\\ -1 & 0\end{pmatrix}$	$\begin{pmatrix}1 & 0\\ 0 & -1\end{pmatrix}$	$\begin{pmatrix}-1 & 0\\ 0 & 1\end{pmatrix}$	$\begin{pmatrix}0 & 1\\ 1 & 0\end{pmatrix}$	$\begin{pmatrix}0 & -1\\ -1 & 0\end{pmatrix}$
Γ_3	(1)	(–1)	(1)	(–1)	(–1)	(–1)	(1)	(1)
Γ_4	(1)	(–1)	(1)	(–1)	(1)	(1)	(–1)	(–1)

(4.29)

Character values

	$\chi_i(E)$	$\chi_i(C_4)$	$\chi_i(C_2)$	$\chi_i(C_{4^3})$	$\chi_i(\sigma_{v^{xz}})$	$\chi_i(\sigma_{v^{yz}})$	$\chi_i(\sigma_{v'})$	$\chi_i(\sigma_{v''})$
Γ_1	1	1	1	1	1	1	1	1
Γ_2	2	0	−2	0	0	0	0	0
Γ_3	1	−1	1	−1	−1	−1	1	1
Γ_4	1	−1	1	−1	1	1	−1	−1

(4.30)

We next turn our attention to $C_{3\hbar}$ the group elements of which are E, C_3, C_{3^2}, $\sigma_\hbar$, S_3 and S_{3^5}. An exercise for working out the relations similar to those of Eq. (4.27) results in the following:

$$\text{(1) } C_3^{-1}\mathbf{X}' \quad \left.\begin{aligned} x &= x'\cos\theta + y'\sin\theta \\ y &= -x'\sin\theta + y'\cos\theta \\ z &= -z' \\ \text{with } \theta &= \tfrac{2\pi}{3} \end{aligned}\right\} \qquad \text{(2) } (C_3^2)^{-1}\mathbf{X}' \quad \left.\begin{aligned} x &= x'\cos\theta + y'\sin\theta \\ y &= -x'\sin\theta + y'\cos\theta \\ z &= z' \\ \text{with } \theta &= 2\times\left(\tfrac{2\pi}{3}\right) \end{aligned}\right\}$$

$$\text{(3) } \sigma_{h^{-1}}\mathbf{X}' \left.\begin{aligned} x &= x' \\ y &= y' \\ z &= -z' \end{aligned}\right\} \quad \text{(4) } S_3^{-1}\mathbf{X}' \left.\begin{aligned} x &= x'\cos\theta + y'\sin\theta \\ y &= -x'\sin\theta + y'\cos\theta \\ z &= -z' \text{ with}\theta = \tfrac{2\pi}{3} \end{aligned}\right\} \quad \text{(5) } (S_3^2)^{-1}\mathbf{X}' \left.\begin{aligned} &\text{same as } S_{3^{-1}} \\ &\text{with} \\ &\theta = 2\times\left(\tfrac{2\pi}{3}\right) \end{aligned}\right\} \tag{4.31}$$

Use of Eq. (4.33) leads, on removing the primes, to

$$\begin{aligned} C_3 d_z{}^2 &= d_{z^2},\ C_3 d_{xz} = d_{xz}\cos\theta + d_{yz}\sin\theta, \\ C_3 d_{yz} &= -d_{xz}\sin\theta + d_{yz}\cos\theta \\ C_3 d_{xy} &= (x'\cos\theta + y'\sin\theta)(x'\sin\theta + y'\cos\theta)' \\ &= d_{xy}(\cos^2\theta - \sin^2\theta) - d_{(x^2-y^2)}\sin\theta\cos\theta \\ C_3 d_{(x^2-y^2)} &= (x'\cos\theta + y'\sin\theta)^2 - (x'\sin\theta + y'\cos\theta)^2 \\ &= d_{xy}4\sin\theta\cos\theta + d_{(x^2-y^2)}(\cos^2\theta - \sin^2\theta). \end{aligned}$$

The relations for the effect of C_3^2 are the same as those for C_3 with the difference that $\theta = 2\times\left(\frac{2\pi}{3}\right)$ and $\frac{2\pi}{3}$ for the two cases respectively.

Again,

$$\begin{aligned} \sigma_\hbar d_z{}^2 &= d_z{}^2,\ \sigma_\hbar d_{xz} = -d_{xz}\ ,\ \sigma_\hbar d_{yz} = -d_{yz}, \\ &\qquad\qquad \sigma_\hbar d_{xy} = d_{xy}\sigma_\hbar d_{(x^2-y^2)} = d_{(x^2-y^2)} \\ S_3 d_{z^2} &= d_{z^2},\ S_3 d_{xz} = -d_{xz}\cos\theta - d_{yz}\sin\theta, \\ S_3 d_{yz} &= d_{xz}\sin\theta - d_{yz}\cos\theta, \\ S_3 d_{xy} &= d_{xy}(\cos^2\theta - \sin^2\theta) - d_{(x^2-y^2)}\sin\theta\cos\theta \\ \text{and } S_3 d_{(x^2-y^2)} &= d_{xy}4\sin\theta\cos\theta + d_{(x^2-y^2)}(\cos^2\theta - \sin^2\theta). \end{aligned}$$

The S_3^5 effect on the basis functions is similar to that of S_3 except that θ in the former case is $2 \times \left(\frac{2\pi}{3}\right)$.

With the symmetry effect displayed by the foregoing relations, one can write out the corresponding matrices easily. Hereunder c and s represent $\cos\theta$ and $\sin\theta$ respectively with proper θ values as indicated.

$$\mathbf{D}(\mathrm{E}) = \begin{pmatrix} 1 & & & & \\ & 1 & 0 & & \\ & 0 & 1 & & \\ & & & 1 & 0 \\ & & & 0 & 1 \end{pmatrix} \quad \mathbf{D}(\mathrm{C}_3) = \begin{pmatrix} 1 & & & \\ & \mathrm{c} & -\mathrm{s} & \\ & \mathrm{s} & \mathrm{c} & \\ & & & (\mathrm{c}^2-\mathrm{s}^2)4\mathrm{cs} \\ & & & -\mathrm{cs}(\mathrm{c}^2-\mathrm{s}^2) \end{pmatrix} = \mathbf{D}(\mathrm{C}_3^2)$$

$$\left[\theta = \frac{2}{3}\pi \text{ for } \mathrm{C}_3 \text{ and } 2 \times \left(\frac{2}{3}\right)\pi \text{ for } \mathrm{C}_3^2\right]$$

$$\mathbf{D}(\sigma\hbar) = \begin{pmatrix} 1 & & & & \\ & -1 & 0 & & \\ & 0 & -1 & & \\ & & & 1 & 0 \\ & & & 0 & 1 \end{pmatrix} \quad \mathbf{D}(\mathrm{S}_3) = \begin{pmatrix} 1 & & & \\ & -\mathrm{c} & \mathrm{s} & \\ & -\mathrm{s} & -\mathrm{c} & \\ & & & (\mathrm{c}^2-\mathrm{s}^2)4\mathrm{cs} \\ & & & -\mathrm{cs}(\mathrm{c}^2-\mathrm{s}^2) \end{pmatrix}$$

$$= \mathbf{D}(\mathrm{S}_3^5)$$

$$\left[\theta = \frac{2\pi}{3} \text{ for } \mathrm{S}_3 \text{ and } 2 \times \left(\frac{2\pi}{3}\right) \text{ for } \mathrm{S}_3^5\right]$$

The matrices are thus seen to occur in block diagonal forms representing one-, two- and two-dimensional IR's with the functions d_{z^2}, $(d'_{xz}d_{yz})$ and $(d_{xy}, d_{(x^2-y^2)})$ serving as basis sets for the IR's. The table of the characters is compiled below.

Table 4.3

Character table-Group $C_{3\hbar}$

IR	$\chi_i(E)$	$\chi_i(C_3)$	$\chi_i(C_3^2)$	$\chi_i(\sigma_\hbar)$	$\chi_i(S_3)$	$\chi_i(S_3^5)$	Basis of representation
Γ_1	1	1	1	1	1	1	d_{z^2}
Γ_2	2	2c	2c	−2	−2c	−2c	(d_{xz}, d_{yz})
Γ_3	2	$2(c^2-s^2)$	$2(c^2-s^2)$	2	$2(c^2-s^2)$	$2(c^2-s^2)$	$(d_{xy}, d_{x^2-y^2})$

where proper θ values are to be used for actual evaluation of characters and c and s stand for $\cos\theta$ and $\sin\theta$ respectively.

Chapter 5

Reducible Representations, Irreducible Representations And Characters – Theorems And Properties

Some aspects of the representation theory of symmetry groups have been treated in the previous chapter. Representation theory is the key to the application of symmetry ideas in the solution of practical problems. The following shows the link.

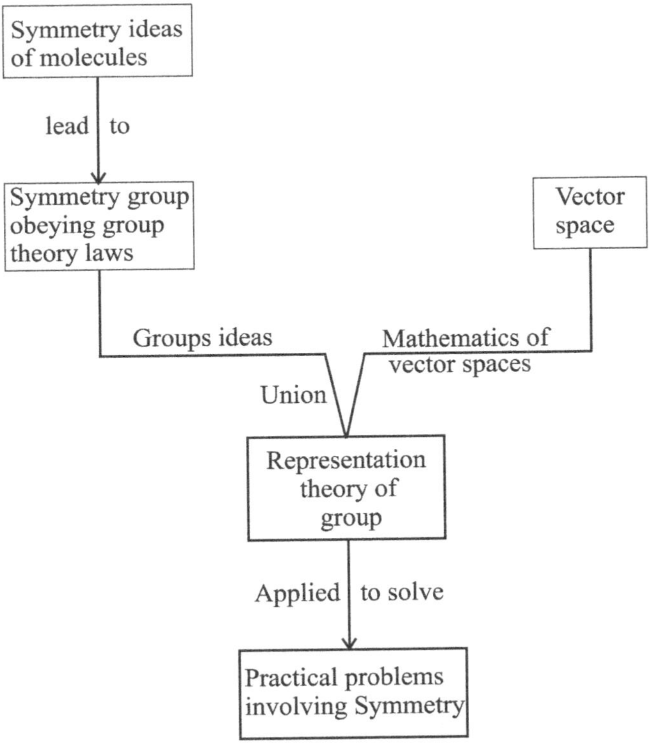

It is the suitable union of the ideas of vector space and group theory that generates the representation theory. The symmetry ideas carried indirectly by a representation, its IR's and characters are brought over to the realm of practical problems having similar symmetry. The solutions, however, depend on a number of generalisations which emerge from a few theorems on reducible representations, IR's and characters. In the present chapter, we have to wade through this marshy bog of theorems till we reach some point from where we can make further explorations.

5.1 Metrical Matrix–Positive Definiteness

The general scalar product of two vectors, $\overrightarrow{\mathrm{p}}$, $\overrightarrow{\mathrm{q}}$, has been defined as $\overrightarrow{P}^*\overrightarrow{Q}$ (Sec. 3.3) to make the product a real quantity. If a single vector $\overrightarrow{\mathrm{r}}$ be taken, the inner product with itself, viz., $\overrightarrow{r}^*\overrightarrow{r}$ should be real, positive and definite. The matrix equivalent of the latter is $\mathbf{r}^{\neq}\mathbf{r}$ wherein the elements are positive and definite.

Now suppose we have two basis sets **f** and **g** (not necessarily orthonormal) connected through the transformation matrix **A** and let us further suppose that the corresponding metrics are $\mathbf{M}_\mathrm{f}$ and $\mathbf{M}_\mathrm{g}$.

$$\mathbf{g} = \mathbf{fA} \qquad \therefore\ \mathbf{g}^{\neq}\mathbf{g} = (\mathbf{fA})^{\neq}\ (\mathbf{fA}) = \mathbf{A}^{\neq}\ \mathbf{f}^{\neq}\ \mathbf{fA}$$
$$\therefore\ \mathbf{M}_\mathrm{g} = \mathbf{A}^{\neq}\ \mathbf{M}_\mathrm{f}\mathbf{A} \tag{5.1}$$

Following the definitions of positive definiteness of scalar product of vectors and hence of the elements of the corresponding matrix product, it is seen that entities like $\mathbf{A}^{\neq}\mathbf{M}_\mathrm{f}\mathbf{A}$ or $\mathbf{r}^{\neq}\mathbf{M}_\mathrm{f}\mathbf{r}$ should be positive and definite, since $\mathbf{M}_\mathrm{g}$ and $\mathbf{M}_\mathrm{f}$ themselves (Eq. 5.1) are positive and definite.

5.2 Reducible Representations–Unitary Basis and Unitary Representation.

Earlier (Chap. 4) we have seen that one can change from one representation to another equivalent one by a mere change of basis set. We now establish the following two theorems the proofs of which may be skipped by readers in their first follow-up of the subject. All the same they ought to know the contents of the theorems before going over to Theorem 5.3.

Theorem 5.1. It is always possible to choose a unitary basis (Metric = unit matrix) for the representation of any symmetry group.

Let the group be G = {P, Q, R, S, ······} and the basis set **f** with the metric $\mathbf{M_f} = \mathbf{f}^{\neq}\,\mathbf{f}$. Since $\mathbf{f}^{\neq}\,\mathbf{f}$ is hermitian (Sec. 3.3), so is $\mathbf{M_f}$. Hence a unitary matrix **U** can diagonalise the hermitian matrix $\mathbf{M_f}$, to its diagonal eigenvalue matrix $\wedge$ (Theo. 3.2). One may write

$$\mathbf{U}^{\neq}\,\mathbf{M_f U} = \wedge \quad \text{i.e., } \mathbf{U} \wedge \mathbf{U}^{\neq} = \mathbf{M_f} \tag{5.2}$$

Since $\mathbf{U}^{\neq}\mathbf{M_f}\,\mathbf{U}$ should be positive definite (Sec. 5.1), the diagonal eigenvalue matrix $\wedge$ should consist of positive numbers.

Let us form a matrix $\mathbf{V} = \mathbf{U}\wedge^{\frac{1}{2}}$ were $\wedge^{\frac{1}{2}} = \begin{pmatrix} \lambda_1^{\frac{1}{2}} & & & \\ & \lambda_2^{\frac{1}{2}} & & \\ & & \lambda_3^{\frac{1}{2}} & \\ & & & \ddots \end{pmatrix}$

a diagonal matrix with its elements as the positive sq.roots of the eigenvalues of $\mathrm{M_f}$.

$$\mathbf{VV}^{\neq} = \mathbf{U}\wedge^{\frac{1}{2}}(\mathbf{U}\wedge^{\frac{1}{2}})^{\neq} = \mathbf{U}\wedge\mathbf{U}^{\neq}\left[\wedge^{\frac{1}{2}} = (\wedge^{\frac{1}{2}})^{\neq}\right] = \mathbf{M_f} \text{ (from Eq. 5.2)}$$

$$\begin{aligned} \mathbf{V}^{-1}\,\mathbf{M_f}\,\mathbf{V}^{\neq -1} &= \mathbf{1} \text{ (unit matrix)} \\ \text{or } \mathbf{V}^{-1}\,\mathbf{f}^{\neq}\,\mathbf{f}\mathbf{V}^{-1\neq} &= \mathbf{1} \text{ (unit matrix)} \end{aligned} \tag{5.3}$$

$$\text{Now}\,\mathbf{V}^{-1} = \left(\mathbf{U}\wedge^{\frac{1}{2}}\right)^{-1} = \mathbf{B}^{\neq}\text{(say), then } (\mathbf{V}^{-1})^{\neq} = (\mathbf{B}^{\neq})^{\neq} = \mathbf{B}.$$

We have, therefore, from (5.3)

$$\begin{aligned} \mathbf{B}^{\neq}\,\mathbf{f}^{\neq}\,\mathbf{fB} &= \mathbf{1} \\ \text{or } (\mathbf{fB})^{\neq}\,(\mathbf{fB}) &= \mathbf{1} \end{aligned}$$

Since **fB**=**h**, a new basis set, it follows $\mathbf{h}^{\neq}\mathbf{h}=\mathbf{1}$, i.e., a unitary basis set **h** can be found for the group G by transforming the old basis with the matrix $\mathbf{B}=\mathbf{V}^{-1\neq} = \mathbf{U}(\wedge^{\frac{1}{2}})^{-1}$, where **U** is the diagonalising matrix of the metric $\mathbf{M_f}$.

Theorem 5.2. Any symmetry group can always be represented by a unitary representation.

Consider a group G = {P, Q, R, S, ...} and let **P**, **Q**, **R**, **S**..... form a representation Γ of the group. Construct a matrix $\mathbf{H}=\sum_{\mathbf{P}}\mathbf{PP}^{\neq}$.

It is easy to establish the following three properties.

(i) $\mathbf{H}$ is hermitian, for its adjoint equals itself.
(ii) Any matrix of the type $\mathbf{B}^{\neq}\ \mathbf{HB}$ is positive definite.
(iii) A matrix $\mathbf{PHP}^{\neq} = \mathbf{H}$ and this is true for all the representative matrices $\mathbf{P}$, $\mathbf{Q}$, $\mathbf{R}$, $\mathbf{S}$.... of the representation Γ.

Following the arguments as in Theorem 5.1, properties (i) and (ii) enable us to write $\mathbf{W}^{\neq}\mathbf{HW}=\mathbf{D}$ where $\mathbf{W}$ is the unitary matrix diagonalising the hermitian matrix $\mathbf{H}$ to the diagonal eigenvalue matrix $\mathbf{D}$ consisting of positive definite numbers.

We may then write

$$\mathbf{H} = \mathbf{WDW}^{\neq} \tag{5.4}$$

Construct a matrix $\mathbf{X}=\mathbf{WD}^{\frac{1}{2}}$, where $\mathbf{D}^{\frac{1}{2}}$=a diagonal matrix of the positive square roots of the elements of $\mathbf{D}$.

$$\begin{aligned}\mathbf{XX}^{\neq} = \mathbf{WD}^{\frac{1}{2}}(\mathbf{WD}^{\frac{1}{2}})^{\neq} &= (\mathbf{WD}^{\frac{1}{2}})(\mathbf{D}^{\frac{1}{2}}W^{\neq}) = \mathbf{WDW}^{\neq}\\ &= \mathbf{H} \text{ (from 5.4)}\end{aligned} \tag{5.5}$$

$$\mathbf{X}^{-1}\ \mathbf{HX}^{\neq-1} = \mathbf{1} \text{ (unit matrix)} \tag{5.6}$$

Let $\mathbf{P}$', $\mathbf{Q}$', $\mathbf{R}$', $\mathbf{S}$'... be the similarity transforms of $\mathbf{P}$, $\mathbf{Q}$, $\mathbf{R}$, $\mathbf{S}$... with respect to $\mathbf{X}$ and the former set is, hence, an equivalent representation Γ of the group G (Sec. 4.3).

$$\begin{aligned}\text{Now } \mathbf{P}' &= \mathbf{X}^{-1}\mathbf{PX}\\ \mathbf{P}'\mathbf{P}'^{\neq} &= (\mathbf{X}^{-1}\mathbf{PX})(\mathbf{X}^{-1}\mathbf{PX})^{\neq} = (\mathbf{X}^{-1}\mathbf{PXX}^{\neq}\mathbf{P}^{\neq}\mathbf{X}^{-1\neq})\\ &= \mathbf{X}^{-1}(\mathbf{PHP}^{\neq})\mathbf{X}^{-1\neq} \quad \text{(from eq 5.5)}\\ &= \mathbf{X}^{-1}\mathbf{HX}^{-1\neq} \text{ (from property (iii))}\\ &= \mathbf{1} \text{ (unit matrix) (from Eq.5.6)}\end{aligned}$$

$\mathbf{P}'$ is a unitary matrix; similarly the other similarity transforms $\mathbf{Q}'$, $\mathbf{R}'$, $\mathbf{S}'$.... etc of Γ' are all unitary in character. Thus a unitary representation of the group is always possible by using $\mathbf{X}^{-1}$=$(\mathbf{WD}^{\frac{1}{2}})^{-1}$ where $\mathbf{W}$ diagonalises $\mathbf{H} = \sum_{\mathbf{P}} \mathbf{PP}^{\neq}$.

The application of Theo. 5.2 to our already familiar two dimensional representation Γ of the D_3 group in the basis set $\overrightarrow{l_1}$, $\overrightarrow{l_2}$ (cf. Eqs. 4.20, 4.13) will be instructive. Here the hermitian matrix

$$\mathbf{H} = \begin{pmatrix} 1 & 0 \\ 0 & 1 \end{pmatrix}\begin{pmatrix} 1 & 0 \\ 0 & 1 \end{pmatrix} + \begin{pmatrix} 0 & -1 \\ 1 & -1 \end{pmatrix}\begin{pmatrix} 0 & 1 \\ -1 & -1 \end{pmatrix} +$$

$$\begin{pmatrix} -1 & 1 \\ -1 & 0 \end{pmatrix}\begin{pmatrix} -1 & -1 \\ 1 & 0 \end{pmatrix} + \begin{pmatrix} 1 & -1 \\ 0 & -1 \end{pmatrix}\begin{pmatrix} 1 & 0 \\ -1 & -1 \end{pmatrix}$$
$$+ \begin{pmatrix} -1 & 0 \\ -1 & 1 \end{pmatrix}\begin{pmatrix} -1 & -1 \\ 0 & 1 \end{pmatrix} + \begin{pmatrix} 0 & 1 \\ 1 & 0 \end{pmatrix}\begin{pmatrix} 0 & 1 \\ 1 & 0 \end{pmatrix}$$
$$= \begin{pmatrix} 8 & 4 \\ 4 & 8 \end{pmatrix}$$

The unitary matrix **W** diagonalising **H** to its diagonal eigen value matrix **D** is found by first setting the determinant (Sec. 3.3)

$\begin{vmatrix} 8-\lambda & 4 \\ 4 & 8-\lambda \end{vmatrix} = 0$, whence the eigenvalues are evaluated to be 12 and 4. Hence the $\mathbf{D}^{\frac{1}{2}} = \begin{pmatrix} \sqrt{12} & 0 \\ 0 & \sqrt{4} \end{pmatrix}$ and **W**, the eigenvector matrix, $= \begin{pmatrix} \frac{1}{\sqrt{2}} & \frac{1}{\sqrt{2}} \\ \frac{1}{\sqrt{2}} & -\frac{1}{\sqrt{2}} \end{pmatrix}$. It follows, therefore, $\mathbf{X}=\mathbf{W}\mathbf{D}^{\frac{1}{2}} = \begin{pmatrix} \sqrt{6} & \sqrt{2} \\ \sqrt{6} & -\sqrt{2} \end{pmatrix}$

and its inverse, $\mathbf{X}^{-1} = \begin{pmatrix} \frac{1}{\sqrt{24}} & \frac{1}{\sqrt{24}} \\ \frac{1}{\sqrt{8}} & -\frac{1}{\sqrt{8}} \end{pmatrix}$

$$\mathbf{D}'(\mathrm{C}_3) = \mathbf{X}^{-1}\mathbf{D}(\mathrm{C}_3)\mathbf{X} = \begin{pmatrix} \frac{1}{\sqrt{24}} & \frac{1}{\sqrt{24}} \\ \frac{1}{\sqrt{8}} & -\frac{1}{\sqrt{8}} \end{pmatrix}\begin{pmatrix} 0 & -1 \\ 1 & -1 \end{pmatrix}\begin{pmatrix} \sqrt{6} & \sqrt{2} \\ \sqrt{6} & -\sqrt{2} \end{pmatrix}$$
$$\begin{pmatrix} -\frac{1}{2} & \frac{\sqrt{3}}{2} \\ -\frac{\sqrt{3}}{2} & -\frac{1}{2} \end{pmatrix}$$

which, on testing, is found to be unitary. Similarly the other matrices of the unitary representation can be found out. A discerning reader will certainly note that these new set of unitary matrices of the group elements are not the self same matrices of the *corresponding group elements* as found in the representation of the D_3 group in the basis set $\vec{e_1}$, $\vec{e_2}$ (Eq. 4.19, Fig. 4.1). These represent group elements in a new orthonormal basis $\vec{e_1}'$, $\vec{e_2}'$, where $(e_1'\ e_2')=(l_1\ l_2)\begin{pmatrix} \sqrt{6} & \sqrt{2} \\ \sqrt{6} & -\sqrt{2} \end{pmatrix}$ Alternatively, if we view from the same basis set $(e_1\ e_2)$ the newly generated matrices turn out to be conjugate elements. That is $\mathbf{D}(\mathrm{C}_3)$ leads to $\mathbf{D}'(\mathrm{C}_3^2)$, $\mathbf{D}(\mathrm{C}_3^2)$ to $\mathbf{D}'(\mathrm{C}_3)$, $\mathbf{D}(\mathrm{C}_2')$ to $\mathbf{D}'(\mathrm{C}_2''')$, $\mathbf{D}(\mathrm{C}_2'')$ to $\mathbf{D}'(\mathrm{C}_2'')$ and finally $\mathbf{D}(\mathrm{C}_2''')$ to $\mathbf{D}'(\mathrm{C}_2')$, where the primes on **D** represent the new unitary set, while the unprimed ones denote matrices in the $(\vec{l_1}\ \vec{l_2})$ basis.

5.3 Theorems – IR's And Characters.

We have already seen that there may be multitude of reducible representations of a symmetry group. Of these the equivalent representations in reducible or irreducible forms are distinguished by the sameness of the dimensionality and of the set of character values. Moreover it was also hinted before (Sec. 4.5) that the number of basic IR's of a group is also definite and limited. It is thus seen that the IR's and their characters bring in a lot of harmony in the representation theory in the midst of limitless choice of basis sets and numerous forms of representations. It is the intent of the present section to probe into the anatomy of the IR's and the properties of their characters in the form of theorems and their elucidations.

Theorem 5.3. This theorem, called the great orthogonality theorem, is the most basic of all the theorems in symmetry groups. The theorem is best stated mathematically followed by its logical interpretation. It states that for all nonequivalent unitary irreducible representations of a point group

$$\sum_{\mathrm{R}} \mathrm{D^i(R)^*_{mn} D^j(R)_{m'n'}} = \frac{\mathrm{g}}{\sqrt{l_i l_j}} \delta_{\mathrm{ij}} \delta_{\mathrm{mm'}} \delta_{\mathrm{nn'}} \tag{5.7}$$

where $\mathrm{D^i(R)_{mn}}$ $\mathrm{D^j(R)m'n'}$ denote the (mn)th and (m′n′)th matrix elements of the symmetry operation R in the Γ_i and Γ_j IR's of the group of order g. $\mathrm{l_i}$ and $\mathrm{l_j}$ represent the dimensionalities of the i'th and the j-th IR's respectively. The symbol δ_{ij} etc. denotes the Kronecker delta.

The theorem embodied in the relation (5.7) is best interpreted in piecemeal manner in three steps, each depending on vector analogies.

(i) The set of (mn)th elements of all the matrices of an IR, Γ_i and the set of (m′n′)th elements of all the matrices of a different IR, Γ_j of a symmetry group behave like components of two orthogonal vectors

$$\sum_{\mathrm{R}} \mathrm{D^i(R)^*_{mn} D^j(R)_{m'n'}} = 0 \quad \delta_{\mathrm{ij}} = 0 \tag{5.7a}$$

(ii) The set of (mn)th elements of the matrices of an IR, Γ_i, and the set of (m′n′)th elements of the same matrices of the same IR behave like components of two orthogonal vectors.

$$\sum_{R} D^i(R)^*_{mn} D^i(R)_{m'n'} = 0 \quad \text{if } m \neq m' \text{ and, or } n \neq n' \tag{5.7b}$$

(iii) The set of (mn)th elements of an IR, Γ_i, of a group behave like components of a vector normalised to $\frac{g}{l_i}$

$$\sum_{R} \{D^i(R)_{mn}\}^2 = \frac{g}{l_i} \tag{5.7c}$$

We shall adopt the orthogonality theorem (5.7) as true without trying to establish it, and for elucidation turn to the matrices of C_{4v} point group in the five d-orbital function space (Sec. 4:7). Consider the IR's Γ_2 and Γ_3 and the (12)th and (11)th matrix elements of the different group elements of C_{4v} having g equal to 8 as given in the relation (4.29). Use of (4.29) in Eq. (5.7a) with i=2, j=3, (mn)=(12) and (m′n′)=(ll) yields

$[0 \times 1] + [(-1) \times (-1)] + [0 \times 1] + [1 \times (-1)] + [0 \times (-1)] + [0 \times (-1)] + [1 \times 1] + [(-1) \times (-1)] = 0$ Eq.(5.7a). Even if m=m′=1 and n=n′=1, but i and j are 2 and 3 respectively, relation (4.29) leads to zero value as demanded by Eq. (5.7a). Sticking to Γ_2 we can easily verify Eq. (5.7b) by considering arbitrarily the matrix elements (11)th and (22)th of the different R's. Thus it is found
$[1 \times 1] + [0 \times 0] + [(-1) \times (-1)] + [0 \times 0] + [1 \times (-1)] + [(-1) \times 1] + [0 \times 0] + [0 \times 0] = 0$
as is to be expected from Eq. (5.7b). Once again choosing any IR, say, Γ_4 and employing the data of (4.29) in Eq, (5.7c), one finds $[1 \times 1] + [(-1) \times (-1)] + [1 \times 1] + [(-1) \times (-1)] + [1 \times 1] + [1 \times 1] + [(-1) \times (-1)] + [(-1) \times (-1)] = 8 = \frac{g}{l}$, where the value of l for this one dimensional representation is unity. In fact one may test the validity of Theo. 5.3 by examining the IR's of the other point groups already worked out such as those of D_3 and C_{3h} (Secs. 4.5, 4.7).

Corollary: Two basis functions ϕ_μ and ψ_ν, used in unitary representation of a group and belonging to IR's, Γ_i and Γ_j and to different rows, μ and ν, respectively of the transformation matrix are mutually orthogonal.

To establish this corollary, we shall see that this is a direct offshoot of the theorem 5.3 Let the group be of order g. Then remembering that the inner product of two functions of a vector space is conserved under symmetry operations by group operators, we may write

$<\phi_\mu|\psi_\nu> = <R\phi_\mu|R\psi_\nu> = <P\phi_\mu|P\psi_\nu> = \ldots$ (g such equalities on RHS)

$$\begin{aligned}
\text{i.e., } g <\phi_\mu \mid \psi_\nu> &= \sum_R <R\phi_\mu \mid R\psi_\nu> \\
&= \sum_R <\sum_\lambda D^i(R)_{\lambda\mu}\phi_\lambda \mid \sum_{\lambda'} D^j(R)_{\lambda'\nu'}\psi_{\lambda'}> \\
&= \sum_{\lambda\lambda'}\sum_R D^i(R)_{\lambda\mu} D^j(R)_{\lambda'\nu} <\phi_\lambda \mid \psi_{\lambda'}> \\
&= \sum_{\lambda\lambda'}\sum_R D^i(R)^*_{\mu\lambda} D^j(R)^*_{\nu\lambda'} <\phi_\lambda \mid \phi_{\lambda'}>
\end{aligned}$$

since the representations are unitary. on applying the great orthogonality theorem, the above can be written as

$$\begin{aligned}
g <\phi_\mu \mid \psi_\nu> &= \frac{g}{l_i}\sum_{\gamma\lambda'} \delta_{ij}\delta_{\lambda\lambda'}\delta_{\mu\nu} <\phi_\lambda \mid \psi_{\lambda'}> \\
\text{i.e., } <\phi_\mu \mid \psi_\nu> &= \frac{1}{l_i}\delta_{ij}\delta_{\mu\nu} \qquad (5.7\text{ d})
\end{aligned}$$

It means that unless the functions are of the same symmetry and have also the same set of transformation coefficients in the μ and the ν rows of the matrix for all operations R, the functions would be mutually orthogonal.

Theorem 5.4. The sum of the squares of the dimensions of all the distinct IR's of a group equals its order, i.e.,

$$\sum_i l_i^2 = l_1^2 + l_2^2 + \cdots\cdots + l_k^2 + \cdots\cdots + l_n^2 = g$$

It is already known from Theo. 5.3 that the (mn)th matrix elements of all operations, R, of any IR Γ_i behave as components of a vector normalised to $\frac{g}{l_i}$ and any two such vectors are mutually orthogonal. If l_i is the dimension of an IR, Γ_i, then there are altogether l_i^2 number of g-dimensional orthogonal vectors. These vectors will also be orthogonal to the other vectors given by the suitable sets of matrix elements of any ether IR, Γ_j.Hence the total number of g-dimensional mutually orthogonal vectors yielded by all the IR's of a group is $l_1^2 + l_2^2 + \ldots. + l_n^2 = \sum_i l_i^2$.

Since a g-dimensional group element vector space cannot have more than g number of linearly independent and mutually orthogonal vectors, it is apparent that

$$l_1^2 + l_2^2 + \cdots\cdots + l_n^2 = \sum_i l_i^2 \leqslant \mathrm{g} \tag{5.8}$$

We thus provide only a partial proof of the theorem in that the inequality sign is still present in Eq. (5.8). We shall attempt a complete proof after developing a few more propositions and concepts (Sec. 5.9). To test subjectively Eq. (5.8) with equality sign one may turn to different character tables (Appendix II) where the dimensions of the different IR's are identical with the corresponding character values of the identity element, E. One should guard against the temptation of testing Theo. 5.4 on any limited set of representations. For example, the theorem when tested with the four representations of C_{4v} in the basis set of the five d-orbitals (4.29) will lead to $\sum_i l_i^2 = 7$ against the expected value of 8. Such a wrong result emerges from the fact that the five basis functions partially span the total symmetry space of point group C_{4v}. Evidently there should be one more one-dimensional IR, the symmetry of which is not reflected in the symmetry behavior of the d-orbitals under C_{4v}, group operations.

Theorem 5.5.

(i) **The sets of characters of any two nonequivalent IR's of a point symmetry group behave like components of mutually orthogonal vectors.**

(ii) **The characters of any IR of a point symmetry group behave like components of a vector normalised to g, the order of the group.**

The two parts of the theorem, stated mathematically, are respectively.

$$\text{(i)} \quad \sum_{\mathrm{R}} \chi_{\mathrm{i}(R)}^{*} \chi_{\mathrm{j(R)}} = 0 \text{ when } \mathrm{i} \neq \mathrm{j} \tag{5.9a}$$

$$\text{(ii)} \quad \sum_{\mathrm{R}} \chi_{\mathrm{i}}(\mathrm{R})^{*} \chi_{\mathrm{i}}(\mathrm{R}) = \mathrm{g} \tag{5.9b}$$

To weave a common approach to prove the above relations, we start from the LHS of (5.9a).

$$\begin{aligned}\sum_R \chi_i(R)^*\chi_j(R) &= \sum_R \left\{\sum_m D^i(R)^*_{mm} \sum_n D^j(R)_{nn}\right\} \\ &= \sum_m\sum_n \left\{\sum_R D^i(R)^*_{mm} D^j(R)_{nn}\right\} \qquad (5.9) \\ &= \sum_m\sum_n \left\{\frac{g}{\sqrt{l_i l_j}}\,\delta_{ij}\delta_{mn}\right\}, \quad \text{using the orthogonality theorem.} \\ &= \frac{g}{\sqrt{l_i l_j}}\delta_{ij} \sum_m\sum_n \delta_{mn} \qquad (5.9c)\end{aligned}$$

Case I. If $i \neq j$, $\sum_R \chi_i(R)^*\chi_j(R) = 0$ as demanded by (5.9a). It means the two vectors which have the characters as the components in the character space are orthogonal.

Case II. If i=j, then (5.9c) yields

$$\begin{aligned}\sum_R \chi_i(R)^*\chi_i(R) &= \frac{g}{l_i}\sum_m\sum_n \delta_{mn}, \quad \text{where m and n can take up values from 1 to } l_i \\ &= \frac{g}{l_i}\,(1+1+1+\cdots\cdots \text{ terms}) = g.\end{aligned}$$

For concrete illustration, one may refer to the character values of the IR's of the C_{4v} point group recorded in (4.30) or to any other character table in the Appendix II. For Γ_1 and Γ_2 representations in the former case, one obtains

$$\begin{aligned}\sum_R \chi_1^*(R)\chi_2(R) &= [1\times 1]+[1\times 0]+[1\times(-2)] \\ &\quad +[1\times 0]+[1\times 0]+[1\times 0]+[1\times 0]+[1\times 0] \\ &= 0, \text{ satisfying (5.9a)}\end{aligned}$$

Similarly (5.9b) is obeyed when a single representation, say, Γ_2 is considered, i.e.,
$\sum_R [\chi_2(R)]^2 = 2^2+0+(-2)^2+0+0+0+0+0 = 8$=g of the C_{4v} point symmetry group.

Theorem 5.6. The characters of the conjugate elements of a class are the same.

We know that the group elements of a particular class (i.e., the conjugate elements) are related to each other through a similarity transform, i.e., $Q=T^{-1}PT$ (Sec. 2.2.6). Therefore, in matrix representation this relation becomes

$$\mathbf{Q} = \mathbf{T}^{-1}\mathbf{P}\ \mathbf{T}$$

Tr $\mathbf{Q}$=Tr($\mathbf{T}^{-1}\ \mathbf{P}\ \mathbf{T}$)=Tr $\mathbf{P}$, since the character of $\mathbf{P}$ is invariant under a similarity transformation (Theo. 4.2). This will apply to all the mutually conjugate elements and hence is the theorem. The latter is true both for reducible and irreducible representations. A confirmation of the theorem in terms of data is readily obtained from the character tables (Appendix II).

Theorem 5.7. The number of IR's of a point symmetry group is equal to the number of classes in the group.

The contents of Eqs. (5.9a and 5.9b) of Theo. 5.5. can be expressed as a summation over classes instead of operations, viz.,

$$\sum_{k} \chi_i(\varepsilon_k)^* \chi_j(\varepsilon_k) N_k = g\delta_{ij} \tag{5.10}$$

where ε_k is any typical group element of class k and N_k, the number of group elements in that class. One can regard (5.10) as representing orthogonality conditions where $\chi_i(\varepsilon_k)$, $\chi_j(\varepsilon_k)$ are the components of two mutually orthogonal vectors with a weighting factor N_k in the space in which the classes play the roles of coordinate axes. Alternatively, one may think of $\sqrt{N_k}\chi_i(\varepsilon_k)$ and $\sqrt{N_k}\chi_j(\varepsilon_k)$ stemming from two IR's, as the components of two orthogonal vectors in the space. Since the number of mutually orthogonal vectors in the space cannot exceed the number of classes and since each vector is built up with the components coming from the characters of one IR', it follows that the number of IR's cannot be more than the number of classes. Further follow up and extension[4] of the treatment show that these two are equal. (See Sec. 5.9).

Theorem 5.8. The number of times, n_i, that an IR, Γ_i, occurs in the reducible representation, $\Gamma^{(red)}$, is given by

$$n_i = \frac{1}{g} \sum_{R} \chi_i^*(R) \chi^{(red)}(R)$$

where $\chi^{(\mathrm{red})}(\mathrm{R})$ is the character of the R matrix in the reducible representation, $\Gamma^{(\mathrm{red})}$ and $\chi_i(\mathrm{R})$ that for the ith IR.

We know any reducible representation can be transformed to the reduced form by a suitable similarity transformation (Sec. 4.6) and under such a transformation, the trace remains unaltered (Theo. 4.2). Suppose the matrix $\mathbf{D}(\mathrm{R})$ of a reducible representation, $\Gamma^{(\mathrm{red})}$, for the group element R, is ultimately block diagonalised into a set of matrices $\mathbf{D}_1'(\mathrm{R})$, $\mathbf{D}_2'(\mathrm{R})$,....$\mathbf{D}_i'(\mathrm{R})$.... It should be noted that, depending on the basis set used, all the IR's of the group may or may not appear in the block diagonal reduced form of $\mathbf{D}(\mathrm{R})$. But whichever of these appear, the latter should satisfy for each R the relation

$$\chi^{(\mathrm{red})}(\mathrm{R}) = n_1\chi_1(\mathrm{R}) + n_2\chi_2(\mathrm{R}) + ... + n_i\chi_i(\mathrm{R}) + \cdots\cdots \tag{5.11}$$

where n_k's represent the number of kth IR occurring in the block diagonal form.

Multiplying by $\chi_i^*(\mathrm{R})$ and summing over all R's

$$\sum_{\mathrm{R}} \chi_i^*(\mathrm{R})\chi^{(\mathrm{red})}(\mathrm{R}) = n_1 \sum_{\mathrm{R}} \chi_i^*(\mathrm{R})\chi_1(\mathrm{R}) + n_2 \sum_{\mathrm{R}} \chi_i^*(\mathrm{R})\chi_2(\mathrm{R}) + \cdots + n_i \sum_{\mathrm{R}} \chi_i^*(\mathrm{R})\chi_i(\mathrm{R}) + \cdots \tag{5.12}$$

Using Theo. 5.5,

$$\sum_{\mathrm{R}} \chi_i^*(\mathrm{R})\chi^{(\mathrm{red})}(\mathrm{R}) = n_i g$$

$$n_i = \frac{1}{g} \sum_{\mathrm{R}} \chi_i^*(\mathrm{R})\chi^{(\mathrm{red})}(\mathrm{R}) \tag{5.13}$$

which proves the theorem.

The employment of this relation will frequently be made in the chemical applications of group theoretical principles. However, one or two such instances will be cited at the end of section 5.6 dealing with the reduction of a reducible representation.

5.4 Character Tables Principle Of Construction

Although the full matrices of the different IR's of a group provide more information on the symmetry properties of a system, the majority of the symmetry problems can be successfully handled with the help of the set of characters of the different IR's without any need for reference to the explicit forms of the matrices of the IR's. Interestingly enough, the setting up of the character tables is possible just with the help of the Theorems 5.1 to 5.7. Of special importance in this connection are the following:

(i) $\sum_i l_i^2 = g$ (Theo. 5.4).

(ii) Orthogonality relation of characters, viz.,
$\sum_k \chi_i^*(\varepsilon_k)\chi_j(\varepsilon_k)N_k = g\ \delta_{ij}$ (Eq. 5.10 & Theo. 5.5)

(iii) Number of IR's=number of classes, k, in the group. (Theo. 5.7).

(iv) A second orthogonality relation, which we have not proved but shall accept as true, viz.,
$\sum_i \chi_i^*(\varepsilon_k)\chi_i(\varepsilon_l) = \frac{g}{\sqrt{N_k N_l}}\ \delta_{kl}$

(v) Using a basis $e^{i\frac{2\pi}{n}}$ (see the conclusion of this chapter)

Example 5.1

Let us try to set up the character table of C_{4v}, point group of which the group elements, distributed classwise, are E, $2C_4$, C_2, $2\sigma_v$, $2\sigma_d$.

(a) As there are five classes, there should be five IR's.

(b) The dimensions of the IR's should be $l_i^2 + l_j^2 + l_k^2 + l_m^2 + l_n^2 = g = 8$. The unique way in which this can hold good is for $l_i = l_j = l_k = l_m = 1$ and $l_n = 2$. Hence the IR's Γ_1, Γ_2, Γ_3, Γ_4 are all one-dimensional and Γ_5; is two-dimensional.

(c) One IR of all groups is a totally symmetric representation since the matrix (1) satisfies all the conditions necessary for defining a group. Hence

	χ				
	E	$2C_4$	C_2	$2\sigma_v$	$2\sigma_d$
Γ_1	1	1	1	1	1

(d) The characters of Γ_2 can be built up from Γ_1 by remembering the orthogonality condition. Since the orthogonality condition can be achieved in more than one way from Γ_2, Γ_3 and Γ_4 can also be simultaneously written down.

	χ				
	E	$2C_4$	C_2	$2\sigma_v$	$2\sigma_d$
Γ_2	1	1	1	−1	−1
Γ_3	1	−1	1	−1	1
Γ_4	1	−1	1	1	−1

(e) The remaining IR, Γ_5, should have $\chi(E) = 2$ since it is two-dimensional. The only single way the characters can be assigned satisfying both the orthogonality conditions is

	χ				
	E	$2C_4$	C_2	$2\sigma_v$	$2\sigma_d$
Γ_5	2	0	−2	0	0

There is no other way to construct Γ_5 that will conform to the rules (i) to (iv) stated earlier in this section. It may, incidentally, be noted that the characters of four representations agree with the ones obtained after the detailed tabulations of the IR's in the d-orbital function space (Eq. 4.30, Sec. 4.7). One representation is lacking in the latter because of the fact that the five d-orbitals span a subspace of the total symmetry space of C_{4v} symmetry group.

A more mechanical way is to solve a set of simultaneous equations. For example, the character table of C_{2v}, consisting of four group elements, has four classes and hence four one dimensional iR's. Since Γ_1 is totally symmetric and orthogoanal to Γ_2, we can write the characters of these two specifically and of the remaining two generally.

	χ			
	E	C_2	$\sigma_v^{(xz)}$	$\sigma_v^{(yz)}$
Γ_1	1	1	1	1
Γ_2	1	1	−1	−1
Γ_3	1	a	b	c
Γ_4	1	d	e	f

Application of rule (iv) in stages to the first and second columns, the first and third columns and to the third and the fourth columns and utilization of the results obtained in each stage enable the determination of the unknowns a, b, c, d, e and f.

First and second columns : Two relations accrue:
$1^2 + 1^2 + a^2 + d^2 = 4$ i.e., $a^2 + d^2 = 2$
and 1.1+1.1+1.a+1.d=0, i.e., a+d=−2
whence a=d=−1.

First and third columns : The pair of relations are
$1^2 + 1^2 + b^2 + e^2 = 4$, i.e., $b^2 + e^2 = 2$

and $1-1+b+e=0$, i.e., $b=-e$
whence $b=1$ and $e=-1$ (or vice versa)

Third and fourth columns : The relations are
$1+1+c-f=0$ i.e., $f=c+2$
and $1+1+c^2+f^2=4$ i.e., $c^2+f^2=2$
whence $c=-1$ and $f=1$.

The point group C_{2v}, has, therefore, the following set of characters,

	χ			
	E	C_2	$\sigma_v^{(xz)}$	$\sigma_v^{(yz)}$
Γ_1	1	1	1	1
Γ_2	1	1	−1	−1
Γ_3	1	−1	1	−1
Γ_4	1	−1	−1	1

For bigger groups, however, such a compilation process for a full character table is laborious; but this is no problem with the availability of programmes based on this principle and the facility of computers[6]. Some years ago, a process has been developed for generating the IR's of a group from the set of characters[7].

The character tables of pure rotational point groups (C_n) with complex characters (Appendix II) can easily be compiled by finding the matrix representation of the group operations using the spherical harmonics (m=0, 1, −1, 2, −2, etc.) as the base function [cf. also Sec. 4,4]. In principle, the character tables of all axial point groups can be worked out from those of the basic C_n point groups and the spherical harmonics basis functions.

5.5 Character Tables-Description; Notations For Irreducible Representations.

In the foregoing section, theorems of finite groups pertaining to dimensionality and characters of representation have been formally used to illustrate the construction of the character tables of symmetry groups. Conventionally, however, a character table contains more information than just a mere exhibition of the characters of the IR's. There are, in general, four sectors in a character table; the first segment contains the different symbols of the IR's of the group. These symbols are generally given in Mulliken's notations. The second segment contains the characters of the types of symmetry operations falling under different classes of the symmetry group. The third and the fourth segments exhibit the various vectors and functions that form the basis of the respective IR's under the group operations. The reader should check all these by scrutiny of the character tables given in the Appendix II.

Notations for IR's – There are two conventions for assignment of notations to IR's of finite groups. The first is to use the symbols Γ_1, Γ_2, ...Γ_μ serially. The second and the more conventional one is the set of Mulliken's symbols which are given below. It should be borne in mind that if $\chi(R)=+1$ the operation is termed symmetric and if it is -1, the operation is antisymmetric (These are $+2$ and -2 respectively for two dimensional representations).

	Criterion/Property	Nature of Property	Notations in the form of subscripts/ superscripts	IR Notations (examples)
		1		A or B
		2		E
1.	Dimensions of IR	3		T
		4		U
		5		V
2.	(a) Axial groups (i) Behavior of basis functions with respect to the principal axis			
	(ii) Symmetry behavior with respect to S_{2n} axis in groups where the highest fold proper axis is C_n and where i or σ_n do not occur	Symmetric		A

	Criterion/Property	Nature of Property	Notations in the form of subscripts/ superscripts	IR Notations (examples)
	(b) Cubic groups Behaviour of basis functions with respect to the generator axis C_3	Anti-symmetric		B
	(c) groups with no proper axis C_1, C_2, C_3			A
3.	Symmetry behaviour of basis functions with respect to inversion (C_{nh}, D_{nh} with even n and D_{nd} with odd n)	Symmetric	Subscript, g	A_g, B_g, E_g, T_g
		Anti-symmetric	Subscript, u	A_u, B_u, E_u, T_u
4.	when i is lacking symmetry behaviour of basis function with respect to σ_h(c_{nh},D_{nh} with odd n	Symmetric	superscript$'$	A',B',E'
		Anti-Symmetric	superscript$''$	A'',B'',E''
5.	Behaviour of the basis functions with respect to one or more C_2's perpendicular to the principal axis (or with respect to σ_v's where C_2's do not occur)	Symmetric or mostly Symmetric	Subscript,1	A_1,B_1
		Anti-symmetric	Subscript, 1,2..	A_2, B_2, B_3

	Criterion/Property	Nature of Property	Notations in the form of subscripts/ superscripts	IR Notations (examples)
6.	Two dimensional IR's E. Behaviour of basis functions comprising pairs of spherical harmonics	Basis Spherical harmonics with	4	5
	such as (Y_{11}, Y_{1-1}) (Y_{22},	$\vert m\vert=1$	subscript, 1	E_1
	Y_{2-2}) and pair of product of	$\vert m\vert=2$	subscript=2	E_2
	spherical harmonics (Y_{11} Y_{10},$Y_{1-1}Y_{10}$) ($Y_{22}.Y_{10}$, Y_{2-2},Y_{10}) or ($Y_{2,1}$ Y_{2-1}) etc. One is thus to start with low order harmonics.	$\vert m\vert=3$	subscript=3	E_3 There may occur certain departure also from this
7.	For triply degenerate representations T_1, T_2 etc. Behaviour of triplets of	Basis spherical harmonics with		
	spherical harmonics for	l=1	Subscript 1	T_1
	different specific values of *l* in $Y_{L,M}$, when i is present (as in O_h) g or u subscripts should be property assigned	l=2	Subscript 2	T_2 T_{1g}, T_{1u}, T_{2g}, T_{2u}.

If triplets of higher order spherical harmonics form basis functions of representations of molecular or crystallographic point groups, their transformation properties will be similar to those of T_1 or of T_2.

Where conflicts are likely to arise, a decision is made by the precedence to be assigned in the order (a) g, u subscripts (b) prime, double primes superscripts (c) numerical subscripts. The following additional points may be noted (compare character tables-Appendix II).

[1] If u and g subscripts are adequate to distinguish the representations (two A's, two B's or two T's) the numerical subscripts are omitted. For example in D_{3d}, these are E_g and E_u.

[2] If the inversion operation i is lacking (hence ruling out the occurrence of u and g subscripts) but $\sigma_\hbar$ occurs, primes and double primes are attributed. If these alone are adequate to distinguish the representations, numerical subscripts are not added. If inadequate, numerical subscripts have to be added. Primes and double primes do not bear company with u, g subscripts.

[3] Lack of i or $\sigma_\hbar$, makes it necessary to use the numerical subscripts to distinguish the plurality of a particular representation, e.g., B_1, B_2 of C_{4v} and E_1, E_2, E_3 of D_{4d} and T_1, T_2 of T_d.

[4] In case g, u notations (or primes and double primes) are not by themselves enough, suitable numerical subscripts are also used in conformity with the principle stated earlier, e.g., E_{1u}, E_{1g}, E_{2g}, E_{2u} or A_1' A_2', A_1'', A_2'' etc.

Linear Point Groups – Characters And Notations:

As already explained in Sec. 2.3.5, the groups of linear molecules are $C_{\propto v}$, and $D_{\propto \hbar}$, and each contains an infinite number of group elements. The theorems of finite groups are not all applicable here. Hence their character tables have to be built up in a different way. Taking $C_{\propto v}$ as an illustration, it is to be noted that any rotation ϕ or $(-\phi)$ ranging from infinitesimal to 2π constitutes a symmetry operation and each such pair of rotations, ϕ and $(-\phi)$, belongs to a single class. Remembering the forms of matrices in the basis sets (e_3) and (e_1, e_2) for rotations about $\vec{e_3}$, it is at once evident that the character for rotation in $C_{\propto v}$ will be 1 and $2\cos\phi$ for one-and two dimensional representations in the reduced forms. When special methods are applied to find the IR's of the group $C_{\propto v}$ it is found that the latter consists of two one-dimensional (A_1, A_2) and an infinite number of two-dimensional (E_1, E_2, E_3......) representations. Some times instead of Mulliken's notations, the greek letters $\sum$, Π, Δ, Φ etc. are used. The characters of $C_{\propto v}$ are

			χ		
			E	$2C_{\propto}(\phi)$	$\propto \sigma_v$
A_1	=	$\sum^+$	1	1	1
A_2	=	$\sum^-$	1	1	−1
E_1	=	π	2	$2\cos\phi$	0
E_2	=	Δ	2	$2\cos 2\phi$	0
E_3	=	Φ	2	$2\cos 3\phi$	0
⋮			⋮		⋮

The group $D_{\propto\hbar}$ is a direct product group, viz., $D_{\propto\hbar} = C^{-}_{\propto v} \otimes C_i$. Hence, the IR's in $D_{\propto\hbar}$ will be those of $C_{\propto v}$ with g and u subscripts such as $\sum_g^+$, $\sum_g^-$, $\sum_u^+$, $\sum_u^-$, Π_g, Π_u and so on.

We have, in the previous and the present sections, confined ourselves to a discussion of the first two segments of the character table. The final segments containing the basis vectors and functions, having the transformation properties of the different IR's, will be taken up now in a general manner. To this end, the idea of projection operator is invaluable and the concept is developed and employed in the next section.

5.6 Projection Operators, Basis Functions And Reduction Of Representations.

The twin problems of finding basis functions of desired symmetry (i.e., a specific IR) and the quantitative technique of the reduction of the reducible representation of a group, qualitatively discussed earlier in Sec. 4.6, are resolved through the use of projection operators.

In physical science, depending on the nature of the problems, one uses different types of projection operators such as operators for projecting a component vector, eigenvector projection operator and group theoretical projection operator. It is the latter that we shall be concerned with in this section. To start with, we shall proceed with the proofs of three propositions.

Proposition One :

A knowledge of one base function of a set of n_μ functions forming a basis for the IR, Γ_μ, enables, with the help of the suitable projection operator to be constructed, the discovery of the other functions of the basis set.

Let the basis set $(f_1^\mu, f_2^\mu ... f_{n\mu}^\mu)$ span the n_μ dimensional IR, Γ_μ, and we just happen to know the functional form of just f_j^μ. If R be an element of the group G of which Γ_μ is an IR, then

$$\hat{R}f_j{}^\mu = \sum_i f_i{}^\mu D^\mu(R)_{ij} \tag{5.14}$$

where $D^\mu(R)_{ij}$ is the ij-the element of the representative matrix $\mathbf{D}^\mu(R)$ in IR, $\Gamma\mu$. Multiplying both sides of Eq. (5.14) by $\sum_R D^\nu(R)^*_{kl}$ one obtains

$$\begin{aligned}\sum_R D^\nu(R)^*_{kl}\hat{R}f_j{}^\mu &= \sum_R \sum_i f_i{}^\mu D^\nu(R)^*_{kl} D^\mu(R)_{ij} \\ &= \sum_i f_i{}^\mu \left\{ \sum_R D^v(R)^*_{kl} D^\mu(R)_{ij} \right\} \qquad (5.15)\end{aligned}$$

Employing Theo. 5,3 in Eq. (5.15), the latter is recast as

$$\sum_R D^\nu(R)^*_{kl}\hat{R}f_j{}^\mu = \sum_i \frac{g}{\sqrt{n_\mu\ n_\nu}} f_i{}^\mu \delta_{ik}\delta_{jl}\delta_{\mu\nu} \tag{5.16}$$

where g is the order of the group G.
Writing P^ν_{kl} for $\frac{n_v}{g}\sum_R D^\nu(R)^*_{kl}\hat{R}$, it follows from Eq. (5.16)

$$P_{kl}{}^\nu f_j{}^\mu = \sum_i f_i{}^\mu \delta_{ik}\delta_{jl}\ \delta_{\mu\nu} \tag{5.16a}$$

It is to be noted that the operator P^ν_{kl} is a linear combination of symmetry operators in G and the coefficients in this linear combination are $\frac{n_\nu}{g}$ times the complex conjugates of the kl-th matrix elements occurring in the IR, Γ_ν. Eq. (5.16a) shows that such an operator, called *projection operator* and built with features of the IR, Γ_ν, when allowed to act upon $f_j{}^\mu$, a base function of another symmetry Γ_μ, annihilates the function. It is also clear that for n_ν-dimensional IR, Γ_ν. one can build n_ν^2 number of projection operators using the different n_ν^2 matrix elements of each operator matrix.

If we now form the projection operator $P_{kj}{}^{\mu} = \frac{n_\mu}{g}\sum_{R} D^{\mu}(R)^{*}_{kj}\hat{R}$ characteristic of IR, Γ_μ, Eq. (5.16a) enables us to write, with $\nu = \mu$ and l=j,

$$P_{kj}{}^{\mu} f_j{}^{\mu} = f_k{}^{\mu} \tag{5.17}$$

Hence, the projection operator, characteristic of Γ_μ symmetry, operating upon the base function $f_j{}^{\mu}$ of the subspace projects another base function $f_k{}^{\mu}$ belonging to the same subspace. Continuing in this way, the total basis set of functions transforming as Γ_μ can be generated.

Proposition Two :

If the projection operator $P_{kl}{}^{\mu}$ operates on a general function $\Psi = \sum_{i=1}^{n_\mu} C_i f_i{}^{\mu}$ belonging to subspace characterized by Γ_μ, it projects only a multiple of the component f_k^{μ} and **annihilates the rest.**

$$P_{kj}{}^{\mu}\psi = P_{kj}{}^{\mu}\sum_{i=1}^{n} C_i f_i{}^{\mu} = P_{kj}{}^{\mu}\,(C_1 f_1{}^{\mu} + C_2 f_2{}^{\mu} + \cdots + C_k f_k{}^{\mu} + \cdots + C_n f_n{}^{\mu})$$

Using (5.17), $P_{kj}{}^{\mu}(C_j f_j{}^{\mu}) = C_j P_{kj}{}^{\mu} f_j{}^{\mu} = C_j f_k{}^{\mu}$ (5.18)

A corollary to the above is the general validity of (5.18) even when a general function ϕ belongs to the entire symmetry space (i.e., representation space) of the group G.

We are still not yet out of the woods. While Eqs. (5.17) and (5.18) are by themselves important, their usefulness in practice is very limited. This is because the employment of each equation demands a preknowledge of the actual matrices of the IR's which are often not available readily. Instead, one has characters ready on hand. Can we thus switch over to a new formulation of a projection operator which utilises characters in lieu of the matrix elements of the IR's? The answer is in the affirmative and we now set to build it up.

The operator, $P_{kj}{}^{\mu} = \frac{n_\mu}{g}\sum_{R} D^{\mu}(R)^{*}_{kj}\hat{R}$, for a diagonal matrix element (k=j) takes the form

$\mathrm{P_{kk}}^{\mu} = \frac{n_\mu}{g} \sum_{R} D^{\mu}(R)_{kk^*}\hat{R}$, which on summing over k, becomes

$$\begin{aligned}\mathcal{P}^{\mu} &= \sum_{k} \mathrm{P_{kk}}^{\mu} = \frac{n_\mu}{g} \sum_{R} \left\{ \sum_{k} D^{\mu}(R)_{kk^*} \right\} \hat{R} \\ &= \frac{n_\mu}{g} \sum_{R} \chi_\mu(R)^* \hat{R} \end{aligned} \tag{5.19}$$

Proposition Three :

The operator, $\mathcal{P}^{\mu}$, projects from a general function F, which is a linear combination of the basis functions of the IR's, Γ_μ, Γ_ν, Γ_η etc. a linear combination of basis functions transforming as the IR, Γ_μ. Such a symmetry adapted linear combination of basis functions is called a 'SALC'.

$$\begin{aligned}\text{Let F} &= \sum_{i=1}^{n_\mu} C_i f_i{}^{\mu} + \sum_{j=1}^{n_\nu} b_j f_j^{v} + \sum_{l=1}^{n_\eta} d_l f_l{}^{\eta} + \cdots \\ \therefore \mathcal{P}^{\mu}F &= \mathcal{P}^{\mu}\left(C_1 f_1{}^{\mu} + C_2 f_2{}^{\mu} + \cdots + C_n f^{\mu}_{n\mu}\right) + \mathcal{P}^{\mu}\left(b_1 f_1^{v} + b_2 f_2^{v} + \cdots \right. \\ &\quad \left. + b_n f^{v}_{nv}\right) + \mathcal{P}^{\mu}\left(d_1 f_1{}^{\eta} + d_2 f_2{}^{\eta} + \cdots + d_n f^{\eta}_{n\eta}\right) + \cdots \\ &= \sum_{k} P_{kk}{}^{\mu}\left(C_1 f_1{}^{\mu} + C_2 f_2{}^{\mu} + \cdots + C_n f_{n\mu^{\mu}}\right) + \sum P_{kk}{}^{\mu}\left(b_1 f_1{}^{\gamma}\right. \\ &\quad \left. + b_2 f_2^{v} + \cdots + b_n f_{nv^{v}}\right) + \sum_{k} P_{kk^{\mu}}\left(d_1 f_1^{\eta} + \cdots\cdots + d_n f_{n\eta^{\eta}}\right)\end{aligned}$$

Now using Eqs. (5.16) and (5.17), it is found that

$$\mathcal{P}^{\mu}F = C_1 f_1{}^{\mu} + C_2 f_2{}^{\mu} + \cdots + C_k f_k{}^{\mu} + \cdots + C_n f^{\mu}_{n\mu} \tag{5.20}$$

which is the required SALC having the symmetry of the IR, Γ_μ. Terms relating to functions of Γ_ν, $\Gamma_\eta \cdots\cdots$ symmetries vanish due to Eq. (5.16).

We have thus achieved our first goal of obtaining a basis set for an IR by projection from an arbitrary function. Our treatment has so far been bristled with sigmas, suffixes and superscripts. It is time we seek temporary relief from these and turn to applications illustrating the use of Eq. (5.20).

Example 5.2

Form SALC's from d-orbital space that can form basis of irreducible representations of the C_{4v} point group.

A general function F in the d-orbital space may be represented as $F = d_z^2 + d_{xz} + d_{yz} + d_{xy} + d(x^2 - y^2)$ and the IR's of C_{4v} point group are A_1 A_2, B_1. B_2 and E. Hence the projection operators that are to be employed on F are $\mathcal{P}A_1$, $\mathcal{P}A_2$, $\mathcal{P}B_1$, and $\mathcal{P}E$. Since all these operators involve the effect of R running over the group elements of C_{4v} (Eq. 5.19), we need to know the transformation properties of the d-orbitals under R, Utilizing the results following Eq. (4.27) in Sec. 4.7 and the group elements of C_{4v}, the following table can be drawn up.

	E	C_4	C_2	C_4^3	$\sigma_v^{(xz)}$	$\sigma_v^{(yz)}$	$\sigma_{v'}$	$\sigma_{v''}$
d_{z^2}	d_{z^2}	d_{z^2}	d_{z^2}	d_{z^2}	d_{z^2}	d_{z^2}	d_{z^2}	d_{z^2}
d_{xz}	d_{xz}	d_{yz}	$-d_{zz}$	$-d_{yz}$	d_{xz}	$-d_{xz}$	d_{yz}	$-d_{yz}$
d_{yz}	d_{yz}	$-d_{xz}$	$-d_{yz}$	d_{xz}	$-d_{yz}$	d_{yz}	d_{xz}	$-d_{xz}$
d_{xy}	d_{xy}	$-d_{xy}$	d_{xy}	$-d_{xy}$	$-d_{xy}$	$-d_{xy}$	d_{xy}	d_{xy}
$d_{(x^2-y^2)}$	$d_{(x^2-y^2)}$	$-d_{(x^2-y^2)}$	$d_{(x^2-y^2)}$	$-d_{(x^2-y^2)}$	$d_{(x^2-y^2)}$	$d_{(x^2-y^2)}$	$-d_{(x^2-y^2)}$	$-d_{(x^2-y^2)}$

(5.21)

Remembering, $\mathcal{P}^\mu = \frac{n_\mu}{g} \sum_R \chi_\mu^*(R)\hat{R}$, the projection operators are

$$\mathcal{P}^{A_1} = \frac{1}{8}\left[\hat{E} + \hat{C}_4 + \hat{C}_2 + \hat{C}_4^3 + \hat{\sigma}_{v(xz)} + \hat{\sigma}_{v(yz)} + \hat{\sigma}_{v'} + \hat{\sigma}_{v''}\right]$$
$$\mathcal{P}^{A_2} = \frac{1}{8}\left[\hat{E} + \hat{C}_4 + \hat{C}_2 + \hat{C}_4^3 - \hat{\sigma}_{v(xz)} - \hat{\sigma}_{v(yz)} - \hat{\sigma}_{v'} - \hat{\sigma}_{v''}\right]$$
$$\mathcal{P}^{B_1} = \frac{1}{8}\left[\hat{E} - \hat{C}_4 + \hat{C}_2 - \hat{C}_4^3 + \hat{\sigma}_{v(xz)} + \hat{\sigma}_{v(yz)} - \hat{\sigma}_{v'} - \hat{\sigma}_{v''}\right]$$
$$\mathcal{P}^{B_2} = \frac{1}{8}\left[\hat{E} + \hat{C}_4 + \hat{C}_2 - \hat{C}_4^3 - \hat{\sigma}_{v(xz)} - \hat{\sigma}_{v(yz)} + \hat{\sigma}_{v'} + \hat{\sigma}_{v''}\right]$$
$$\mathcal{P}^{E} = \frac{2}{8}\left[2\hat{E} - 2\hat{C}_2\right]$$

Utilizing these operators and the results of transformations (5.21) we readily find $\mathcal{P}^\mu F$'s which are
$\mathcal{P}^{A_1}F = {d_z}^2$, $\mathcal{P}^{A_2}F = 0$, $\mathcal{P}^{B_1}F = d(x^2 - y^2)$, $\mathcal{P}^{B_2}F = d_{xy}$
and $\mathcal{P}^{E}F = d_{xz} + d_{yz}$

It is thus concluded that A_1, B_1 and B_2 are spanned separately by ${d_z}^2$, $d(x^2 - y^2$ and d_{xy} respectively. The two dimensional IR, E, has the basis (d_{xz}, d_{yz}). No SALC can be formed from d-orbital space that has the symmetry of A_2 representation.

Example 5.3

Construct the projection operators of the different IR's of the C_3 point group and form, with the d orbitals, the basis of E representation.

First Part. The C_3 point group consists of the IR's A and E. Using the character values, one can write

$$\begin{aligned}\mathcal{P}^{A} &= \frac{1}{3}\left(1.\hat{E}+1.\hat{C}_3+1.\hat{C}_3^2\right)=\frac{1}{3}\left(\hat{E}+\hat{C}_3+\hat{C}_3^2\right)\\ \mathcal{P}^{E} &= \frac{2}{3}\left(2\hat{E}-1.\hat{C}_3-1.\hat{C}_3^2\right)=\frac{2}{3}\left(2\hat{E}-\hat{C}_3-\hat{C}_3^2\right)\end{aligned}$$

Second Part. We now turn to the transformation properties of the d-orbitals under the group operations of C_3 point group. These can be picked up from those appearing under $C_{3\hbar}$, of Sec. 4.7. Using these, we may write

$$\begin{aligned}\mathcal{P}^{E}F &= \frac{2}{3}(2\hat{E}-\hat{C}_3-\hat{C}_{3^2})F,\ \text{where } F=d_z{}^2+d_{xz}+d_{yz}+d_{xy}+d_{(x^2-y^2)}\\ \frac{2}{3}(2\hat{E}F) &= \frac{2}{3}\left(2d_{z^2}+2d_{xz}+2d_{yz}+2d_{xy}+2d_{(x^2-y^2)}\right)\\ \frac{2}{3}C_3F &= \frac{2}{3}\left\{d_z{}^2-\frac{1}{2}d_{xz}+\frac{\sqrt{3}}{2}d_{yz}-\frac{\sqrt{3}}{2}d_{xz}-\frac{1}{2}d_{yz}+\frac{1}{4}d_{xy}-\frac{3}{4}d_{xy}\right.\\ &+ \left.\frac{\sqrt{3}}{4}d_{x^2-y^2}-\sqrt{3}d_{xy}+\frac{1}{4}d_{(x^2-y^2)}\frac{3}{4}d_{(x^2-y^2)}\right\}\left[\because\theta=\frac{2\pi}{3}\right]\\ \frac{2}{3}C_3^2F &= \frac{2}{3}\left\{d_z{}^2-\frac{1}{2}d_{xz}-\frac{\sqrt{3}}{2}d_{yz}+\frac{\sqrt{3}d}{2}xz-\frac{1}{2}d_{yz}+\frac{1}{4}d_{xy}-\frac{3}{4}d_{xy}\right.\\ &\left.-\frac{\sqrt{3}}{4}d_{(x^2-y^2)}+\sqrt{3}d_{xy}+\frac{1}{4}d_{(x^2-y^2)}-\frac{3}{4}d_{(x^2-y^2)}\right\}\\ \therefore\ \mathcal{P}^{E}F &= \frac{2}{3}\left(3d_{xz}+3d_{yz}+3d_{xy}+3d_{(x^2-y^2)}\right)\\ &= 2\left(d_{xz}+d_{yz}+d_{xy}+d_{x^2-y^2}\right)\end{aligned}$$

The resultant combination appears somewhat intriguing since these four d-orbitals should span a four dimensional function space whereas the E representation is indicative of a two dimensional subspace. The paradox is lifted when we closely look at the transformation properties. It is seen that under C_3 and C_3^2 operations, the orbitals d_{xy} and d_{yz} get mixed into linear combinations and similarly d_{xy} and $d_{(x^2-y^2)}$ (cf. transformations under $C_{3\hbar}$ sec. 4.7). But none of the former set gets mixed with any or both of the latter. Our conclusion, therefore, is that the pair of orbitals (d_{xz}, d_{yz}) and also (d_{xy}, $d_{(x^2-y^2)}$) form two separate sets of basis functions for the E-representation:

It is not essential that one should always stick to the formal methods of projection operator in making linear combination of functions of desired symmetries. As often happens for bigger systems the formal technique becomes a lengthy process and may be substituted, either partially, mostly or in full, by the method of inspection and mental reckoning, if possible. One example is provided below to typify the method of inspection.

Example 5.4

For the square planar $[PtCl_4]^{2-}$ ion, construct four linear combinations of the p_x orbital on the four chlorine atoms having the symmetries of A_{lg}, B_{lg} and E_{2u}.

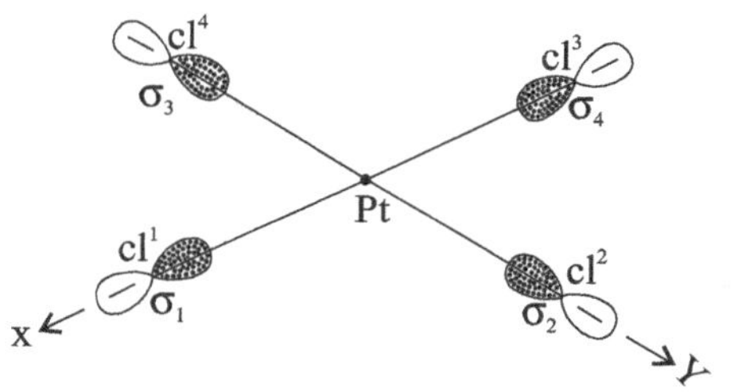

Fig. 5.1: The four p_x- orbitals of Cl-atoms direct to Pt.

The fig. 5.1 depicts the x, y axes with the chlorine atoms located on the axes in such a way that the positive lobes of the p_{x-} orbitals point to the central metal ion. These are indicated by σ_1, σ_2, σ_3 and σ_4. This planar complex ion belongs to the point group $D_{4\hbar}$.

The character table of $D_{4\hbar}$ for the relevant IR's is

	E	$2C_4$	C_2	$2C_{2'}$	$2C_{2''}$	i	$2S_4$	$\sigma_\hbar$	$2\sigma_v$	$2\sigma_d$
A_{1g}	1	1	1	1	1	1	1	1	1	1
B_{1g}	1	−1	1	1	−1	1	−1	1	1	−1
E_{2u}	2	0	−2	0	0	−2	0	2	0	0

Linear combination for a_{1g} : Such a combination is $\phi(a_{1g}) = \frac{1}{2}(\sigma_1+\sigma_2+\sigma_3+\sigma_4)$ as it evidently satisfies the relevant portion of the character table. The factor $\frac{1}{2}$ is the normalising factor.

Combination for b_{1g} : Since $\chi(C_4) = -1$, the linear combination should change into its negative under C_4 operation. An apparent combination is $\phi(b_{1g}) = \frac{1}{2}(\sigma_1 - \sigma_2 + \sigma_3 - \sigma_4)$, where C_4 turns σ_1 into the negative of σ_2 and similarly, σ_3 into the negative of σ_4. The resultant orbital distribution appears

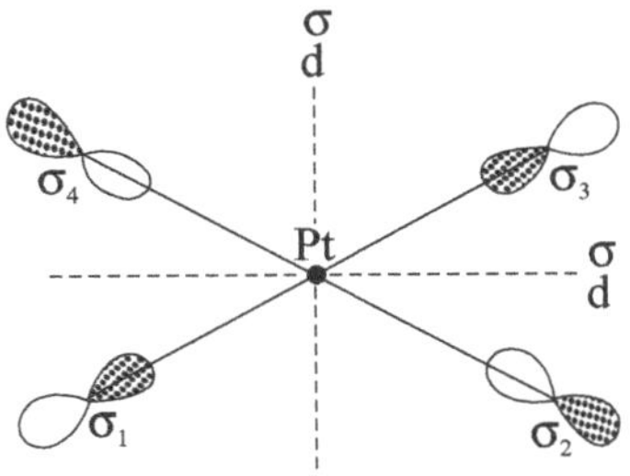

Fig. 5.2: Orientation of the p_x orbitals of b_{lg} symmetry.

in Fig. 5.2. That this satisfies the character table criterion may be tested by noting the transformations under randomly selected symmetry elements, say, $2\sigma_\mathrm{v}$, and $2\sigma_\mathrm{d}$.

Linear combination for e_{2u} : This is a pair of degenerate combinations. Since $\chi(C_2) = -2$ and $\chi(C_4) = 0$, the latter implies that the linear combination should not reproduce itself under C_4 and the former indicates that under C_2, a negative of the combination should result. The pair of combinations satisfying these conditions are (Fig. 5.3)

$$\phi_1(\mathrm{e_u}) = \frac{1}{\sqrt{2}}(\sigma_1 - \sigma_3)$$

$$\phi_2(\mathrm{e_u}) = \frac{1}{\sqrt{2}}(\sigma_2 - \sigma_4)$$

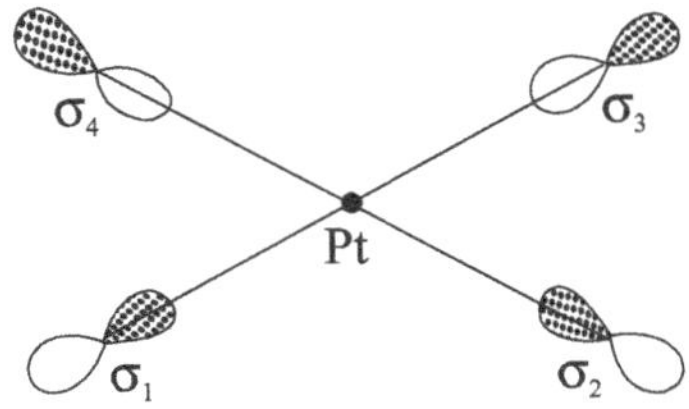

Fig. 5.3: Orientation of the p_x-orbitals of Cl atoms of e_u symmetry.

Reduction of Reducible Representation:

Attention is now paid to the second objective set in the caption of this section. The reduction of a reducible representation of a group in a certain basis set involves *three distinct stages:*

(a) To ascertain, by applying relation of Theo. 5.8, the number and kinds of IR's contained in the reducible representation Γ^{red}.

(b) To form by projection operator technique new basis sets for the IR's obtained in step (a) through suitable linear combinations of the base vectors of the reducible representation.

(c) Of many such projected linear combinations obtainable in step (b), a requisite number of linearly independent combinations are chosen which are finally normalized.

These linear combinations, if used as a basis set, will give rise to a reduced form of representation. Alternatively, from (b) and (c), the transformation matrix between the old and the new basis sets can be written and with this matrix a similarity transformation may be effected on the reducible representation to yield the reduced form of the representation as referred to in Sec. 4.6.

Principle:

We shall start with the assumption that step (a) has already been completed and one knows the specific IR's contained in $\Gamma^{red} = \Gamma_1 + \Gamma_2 + ... + \Gamma_v$. Two cases may arise, viz., (i) no IR in Γ^{red} occurs more than once, (ii) one or more IR's in Γ^{red} occur a number of times.

Case 1. No IR occurs more than once

Let, as found from (a), $\Gamma^{red} = \Gamma_1 + \Gamma_2 + \cdots\cdots + \Gamma_v$ with $g_1, g_2, ... g_j ..... g_n$ serving as base vectors for the n-dimensional reducible representation. The total representation space of Γ^{red} when reduced, will yield invariant subspaces $V(1), V(2), ... V(\gamma)$ of the IR's $\Gamma_1, \Gamma_2, ... \Gamma_v$. Let us suppose that these subspaces, in the order mentioned, be spanned by the basis sets

$f_1(^1), f_2(^1) .... f_{n1}(^1), f_1(^2), f_2(^2), ... f_{n2}(^2) ... f_{1(v)}, f_{2(v)}, f_{n(\nu)}$

$$\text{Any base vector } g_j = \sum_i f_i(1)\alpha_{ij} + \sum_i f_i(2)\beta_{ij} + ... + \sum_i f_{i(v)}\xi_{ij}$$

$$= F_j(1) + F_j(2) + ... + F_{j(v)} \quad (5.22)$$

where $F_j(1)$ is a vector in $V(1)$ subspace of symmetry Γ_1, $F_j(2)$ a vector in $V(2)$ subspace of symmetry Γ_2 and so on.

$$\therefore g_j = \sum_{\mu=1}^{v} F_j(\mu) \quad (5.23)$$

Constructing the projection operator $\mathcal{P}^{\mu}$ and allowing it to act upon g_j

(Eq. 5.23), we get.

$$\mathcal{P}^{\mu} g_j = F_j^{(\mu)} \tag{5.24}$$

Since $\mathcal{P}^{\mu}$ is a linear combination of operators, its action of gj will project a linear combination of the g-vectors all in $V^{(\mu)}$ subspace and having Γ_{μ}. symmetry. Allowing j to run from 1 to n, (i.e., g_1 to g_n) we can thus accumulate many linear combinations of g-vectors with the help of Eq. (5.24) using the same operator $\mathcal{P}^{\mu}$. Not all these are linearly independent. We can choose $n\mu$ number of linearly independent combinations and normalise the set to span the IR, $\Gamma\mu$. In this way, all the basis sets for Γ_1, Γ_2.....Γ_v may be found by using the projection operators $\mathcal{P}^1$, $\mathcal{P}^2$, ...$\mathcal{P}^v$.

Example 5.5.

We now illustrate this principle by a concrete case. Let us consider the (d_1, d_2, d_3) orthogonal basis set and the reducible representation of the point group D_3 (Sec. 4.3, relation 4.15). The IR's of D_3, are A_1, A_2 and E.

$$\begin{aligned} n_{A1} &= \frac{1}{6}[1.3 + 1.0 + 1.0 - 1.1 - 1.1 - 1.1] = 0 \\ n_{A2} &= [1.3 + 1.0 + 1.0 + (-1).(-1) + (-1).(-1) + (-1).(-1)] = 1 \\ n_E &= \frac{1}{6}[2.3 + 0.(-1) + 0.(-1) + (-1).0 + (-1).0 + (-1).0] = 1 \end{aligned}$$

Hence $\Gamma^{red} = A_2 \oplus E$ and the stage (a) is completed. We now turn to stage (b) to find $\mathcal{P}^{A_2 d_1}$, $\mathcal{P}^{A_2 d_2}$, $\mathcal{P}^{A_2 d_3}$ first and then $p^E d_1$, $p^E d_2$ and $p^E d_3$,. Application of projection operator presupposes a knowledge of the transformation properties of d_1, d_2, d_3 which are tabulated below [use is made of the matrices (Eq. 4.15) Sec. 4-3]

	E	C_3	C_3^2	$C_{2'}$	$C_{2''}$	C_2'''
d_1	d_1	d_2	d_3	$\frac{1}{3}(d_1\text{-}2d_2\text{-}2d_3)$	$\frac{1}{3}(\text{-}2d_1\text{-}2d_2\text{+}d_3)$	$\frac{1}{3}(\text{-}2d_1\text{+}d_2\text{-}2d_3)$
d_2	d_2	d_3	d_1	$\frac{1}{3}(\text{-}2d_1\text{-}2d_2\text{+}d_3)$	$\frac{1}{3}(\text{-}2d_1\text{+}d_2\text{-}2d_3)$	$\frac{1}{3}(d_1\text{-}2d_2\text{-}2d_3)$
d_3	d_3	d_1	d_2	$\frac{1}{3}(\text{-}2d_1\text{+}d_2\text{-}2d_3)$	$\frac{1}{3}(d_1\text{-}2d_2\text{-}2d_3)$	$\frac{1}{3}(\text{-}2d_1\text{-}2d_2\text{+}d_3)$

The projection operators are

$$\mathcal{P}^{A_2} = \frac{1}{6}\left[\hat{E} + C_3 + C_3^2 - C_{2'} - C_{2''} - C_{2'''}\right] \tag{5.25}$$

$$\text{and } \mathcal{P}E = \frac{2}{6}\left[2\hat{E} - C_3 - C_3^2\right]$$

Now $$\mathcal{P}^{A_2} d_1 = \frac{1}{6}\left(d_1 + d_2 + d_3 - \frac{1}{3}d_1 + \frac{2}{3}d_2 + \frac{2}{3}d_3 + \frac{2}{3}d_1 + \frac{2}{3}d_2 - \frac{1}{3}d_3 + \frac{2}{3}d_1 - \frac{1}{3}d_2 + \frac{2}{3}d_3\right) = \frac{1}{3}(d_1 + d_2 + d_3)$$

It is also found $\mathcal{P}^{A_2}d_1 = \mathcal{P}^{A_2}d_2 = \mathcal{P}^{A_2}d_3 = \frac{1}{3}(d_1 + d_2 + d_3)$
Hence the linear combination, which is symmetry adapted for A_2 and is in normalized form, is $\frac{1}{\sqrt{3}}(d_1 + d_2 + d_3)$.

$$\begin{aligned}
\text{Similarly}\quad \mathcal{P}^{E}d_1 &= \frac{2}{6}\left(2\hat{E} - \hat{C}_3 - \hat{C}_3^2\right)d_1 = \frac{2}{6}(2d_1 - d_2 - d_3) \\
\mathcal{P}^{E}d_2 &= \frac{2}{6}(2d_2 - d_3 - d_1) \\
\text{and}\quad \mathcal{P}^{E}d_3 &= (2d_3 - d_1 - d_2).
\end{aligned}$$

But these three linear combinations are not linearly independent since $\mathcal{P}^{E}d_1 + \mathcal{P}^{E}d_2 + \mathcal{P}^{E}d_3 = 0$ [Recall stage(c)].

The two orthonormal base vectors may be obtained by retaining $\mathcal{P}^{E}d_1$ and taking the difference of $\mathcal{P}^{E}d_2$ and $\mathcal{P}^{E}d_3$. Written in normalized forms, these final vectors are

$$\begin{aligned}
F_1 &= \frac{1}{\sqrt{3}}(d_1 + d_2 + d_3) \\
F_2 &= \frac{1}{\sqrt{6}}(2d_1 - d_2 - d_3) \text{ and } F_3 = \frac{1}{\sqrt{2}}(d_2 - d_3).
\end{aligned}$$

We thus find the orthonormal basis functions for A_2 and E representations of the D_3 group. If these basis functions are used, Γ^{red} will appear in block diagonal forms. To show the link between our findings here with the principle of similarity transformation in the qualitative discussion of block digitalization of reducible representation (Sec. 4.6) we find

$$(F_1\ F_2\ F_3) = (d_1\ d_2\ d_3)\begin{pmatrix} \frac{1}{\sqrt{3}} & \frac{2}{\sqrt{6}} & 0 \\ \frac{1}{\sqrt{3}} & -\frac{1}{\sqrt{6}} & \frac{1}{\sqrt{2}} \\ \frac{1}{\sqrt{3}} & -\frac{1}{\sqrt{6}} & -\frac{1}{\sqrt{2}} \end{pmatrix}$$

i.e., $\mathbf{F} = \mathbf{dB}$

If now a similarity transformation be effected with the **B** matrix upon the reducible matrices in the basis set ($d_1\ d_2\ d_3$) of Sec. 4.3, the latter will be thrown into reduced form.

Example 5.6

Obtain the reduced form of the matrices in Eq. (4.15).
We shall illustrate here the reduction of $\mathbf{D^d}(C_3)$ matrix only leaving the remainder as exercises for the interested readers.

$$\mathbf{D^d}(C_3) = \begin{pmatrix} 0 & 0 & 1 \\ 1 & 0 & 0 \\ 0 & 1 & 0 \end{pmatrix}$$

The **B** matrix is orthogonal (since it is the transformation matrix from one orthogonal basis set $\vec{d_1}$, $\vec{d_2}$, $\vec{d_3}$ to another $\vec{F_1}$, $\vec{F_2}$, $\vec{F_3}$, hence

$$\mathbf{B}^{-1} = \widetilde{\mathbf{B}} = \begin{pmatrix} \frac{1}{\sqrt{3}} & \frac{1}{\sqrt{3}} & \frac{1}{\sqrt{3}} \\ \sqrt{\frac{2}{3}} & -\frac{1}{\sqrt{6}} & -\frac{1}{\sqrt{6}} \\ 0 & \frac{1}{\sqrt{2}} & -\frac{1}{\sqrt{2}} \end{pmatrix}$$

$$\mathbf{D}^{F}(C_3) = \mathbf{B}^{-1}\mathbf{D}^{d}(C_3)\mathbf{B} = \begin{pmatrix} 1 & 0 & 0 \\ 0 & -\frac{1}{2} & -\frac{1+\sqrt{4}}{\sqrt{12}} \\ 0 & \frac{1+\sqrt{4}}{\sqrt{12}} & -\frac{1}{2} \end{pmatrix}$$

Case II. An IR occurs more than once in Γ^{red}

The alternative case of the occurrence of any IR more than once in the reduction of a reducible representation presents a difficult problem.

Let us suppose that an IR, Γ_μ, occurs twice in Γ^{red}. The relation (5.24) will then comprise a combination of functions belonging to the two subspaces of symmetry Γ_μ. That is, $\mathcal{P}^{\mu}g_j = F_{j^\mu} + F_{j'^\mu}$. This will also happen for other base vectors of the basis set, g. There is no general way of segregating the linear combination into two parts, each of symmetry $\Gamma\mu$, without considering closely the forms of the matrices of the reducible representation.

A somewhat analogous problem was encountered by us in Example 5.3 where we sought to project a linear combination of d-orbitals of IR E of the D_3 point group. A linear combination of d_{xz}, d_{yz}, d_{xy}, and $d(x^2-y^2)$ was projected which was ultimately argued out as two different linear combinations both belonging to E representation.

5.7 Direct Product Representation:(Tensor Product Representation)

The concept of linear direct product space was elucidated in Sec. 3.2. Each basis function of a direct product space is a product of two basis functions of another two linear spaces. Working upon this, it can be shown that the product functions of the direct product space can serve as the basis for the representation of the group.

Consider two basis sets **f**$= (f_1\ f_2 \cdots\cdots f_m)$ and **g**$= (g_1\ g_2 \cdots\cdots g_n)$ spanning the IR's Γ_μ and Γ_ν, of a group. If R be any group element, then $Rf_i = \sum_j f_j D_j{}^{\mu}(R)_{ji}$ and $Rg_k = \sum_l g_l D^{\nu}{}_g\,(R)_{lk}$ where $D_f{}^{\mu}(R)_{ji}$ is the ji-th element of the matrix representing R in the Γ_μ representation in the basis set **f**. A similar interpretation is valid for $D^{\nu}{}_g(R)_{lk}$. The effect of R on the product function $(f_i g_k)$ is

$$\begin{aligned} R(f_i g_k) &= (Rf_i)(Rg_k) \\ &= \sum_j^m \sum_l^n f_j g_l D_f{}^{\mu}(R)_{ji} D_{g^v}(R)_{lk} \end{aligned} \tag{5.26}$$

Let the series of product functions, like $(f_i g_k)$, be symbolized by $h_1, h_2, \cdots h_r, h_s, \cdots h_v$ where each new subscript, r,s,v, etc., stand for a couple of subscripts associated with the component base functions. Thus $f_i g_k = h_r$, $f_j g_l = h_s$, $f_m\ g_n = h_v$ and so on. With such substitution the relation (5.26) can be put in the form

$$Rh_r = \sum_s^{mn=v} h_s D^{(\mu\times\nu)}(R)_{sr} \tag{5.27}$$

where the matrix element $D^{(\mu\times\nu)}\ (R)_{sr}$. is written for $D_f{}^{\mu}(R)_{ji} D_g{}^{v}(R)_{lk}$ and the summation over j and l substituted by one over s ranging from 1 to mn. Eq. (5.27) permits us to conclude:

(i) The product functions $h_1, h_2 \ldots h_r \ldots h_v$ form a basis of representation, called the direct product representation of the group and the corresponding representation space is termed a direct product space $\mathbf{f}\otimes\mathbf{g}$

(ii) The matrix representing R in the new space is related to those of the component spaces **f** and **g**

$$\mathbf{D}^{(\mu\times\nu)}(R) = \mathbf{D}_f{}^{\mu}(R) \otimes \mathbf{D}_g^{v}(R) \tag{5.28}$$

and thus the direct product representation is conventionally symbolized as

$$\Gamma^{(\mu\times\nu)} \text{ or } \Gamma_\mu \otimes \Gamma_v \tag{5.29}$$

It will be useful to remember the following:

a. The direct product representation $\Gamma^{(\mu\times\nu)} = \Gamma_\mu \otimes \Gamma_v$ is valid whether or not Γ_μ and Γ_v are reucible or irreducible.

b. $\Gamma^{(\mu\times\nu)}$ is usually a reducible representation.

c. The actual representative matrices for different R's given by Eq. (5.26), though somewhat difficult to work out (cf. Sec. 3.2), are scarcely required in practical applications.

d. The direct product representation is of extreme importance in representation theory because of some inherent properties associated with its character.

Theorem 5.9.

The character of any group element, R, in a direct product representation $\Gamma\mu \otimes \Gamma\nu$ is a product of its characters in $\Gamma\mu$ and $\Gamma\nu$.

Exploiting the elementwise relation given previously, viz., $\mathbf{D}^{(\mu\times\nu)}(\mathrm{R})_{\mathrm{sr}} = \mathrm{D_f}^{\mu}(\mathrm{R})_{\mathrm{ji}}\mathrm{D_g}^{\mathrm{v}}(\mathrm{R})_{lk}$ where s stands for the couple of subscripts jl and r for ik, we can write

$$\begin{aligned}\chi_{(\mu\times\nu)}(\mathrm{R}) &= \sum_{\mathrm{s}} \mathrm{D}^{(\mu\times\nu)}(\mathrm{R})_{\mathrm{ss}} \\ &= \sum_{\mathrm{j}} \mathrm{D_f}^{\mu}(\mathrm{R})_{\mathrm{jj}} \sum_{\mathrm{l}} \mathrm{D_g}^{\mathrm{v}}(\mathrm{R})_{\mathrm{ll}} = \chi_{\mu}(\mathrm{R})\chi_{\mathrm{v}}(\mathrm{R})\end{aligned}$$

which proves the theorem.

Keeping in view the earlier theorems and the properties of a direct product representation, the reader will readily appreciate the validity of the following generalizations.

A. The principle of direct product representation may be extended to cover the products of more than two basis set functions. Accordingly, the theorem concerning its character (cf. Theo. 5.9) is similarly extensible, i.e.,

$$\Gamma^{(\mu\times\nu\times\eta)} = \Gamma^{(\mu\times\nu)} \otimes \Gamma_{\eta} = \Gamma_{\mu} \otimes \Gamma_{\nu} \otimes \Gamma_{\eta}$$

In terms of characters, it means for the group element (R)

$$\chi_{(\mu\times\nu\times\eta)}(\mathrm{R}) = \chi_{\mu}(\mathrm{R}).\ \chi_{\nu}(\mathrm{R})\chi_{\eta}(\mathrm{R}) \tag{5.30}$$

B. Since the RHS of Eq. (5.30) involves the product of scalar quantities (i.e., the characters), it is independent of the sequence of the occurrence of the representations in the direct product.

$$\begin{aligned}\chi_{(\mu\times\nu\times\eta)}(\mathrm{R}) &= \chi_{(\mu\times\eta\times\nu)}(\mathrm{R}) = \chi_{(\eta\times\mu\times\nu)}(\mathrm{R}) \\ &= \chi_{\mu}(\mathrm{R})\chi_{\nu}(\mathrm{R})\chi_{\eta}(\mathrm{R})\end{aligned}$$

C. Any direct product representation, if a reducible one, is ultimately transformable into block diagonal form. Hence Theo. 5.8 is applicable to find n_i, the number of times the IR, Γ_i, occurs in the direct product representation $\Gamma^{(\mu\times\nu)}$, i.e.,

$$n_i = \frac{1}{g}\sum_R \chi_i(R)^* \chi_{(\mu\times\nu)}(R) = \frac{1}{g}\sum_R \chi_i(R)^* \chi_\mu(R)\chi_v(R) \quad (5.31)$$

The contraction of symbolism for product of pair of functions (such as using h_r for $f_i g_k$ etc.) and of suffixes (r for ik s for lk) was aimed at lessening confusion in the proper placement of double suffixes. But most authors use the convention of retaining the pair of suffixes in preference to a single substitute subscript. Thus the product of individual matrix elements $D_f^\mu(R)_{ji}$ and $D_g^\mu(R)_{lk}$ are retained in double subscripted form in the corresponding direct product matrix element as

$$D_f{}^\mu R_{ji}.D_g^\nu(R)_{lk} = D^{(\mu\times\nu)}(R)_{jlik}$$

and also like

$$\begin{aligned} R(f_i g_k) &= \sum_{j,\,l} (f_j g_l) D_f{}^\mu(R)_{ji} D_g(R)_{lk} \\ &= \sum_{jl} f_j g_l \; D^{(\mu\times\gamma)}(R)_{jlik} \end{aligned}$$

The readers need to cultivate this practice, noting specially the sequence of placement of suffixes in the last factor. Finally many authors use the expression "tensor product representation" in preference to"direct product representation" mainly for two reasons:

(a) the product of basis functions or kets is a tensor. Kets of Quantum mechanics are tensors.

(b) to avoid the apparent but mistaken notion that the direct product representation is the representation of the direct product group.

Operational Significance Of The Direct Product Symbol $\otimes$ As Distinct From Matrix Multiplication:

The matrix for direct product representation of R, expessed by Eq. (5.28), has been obtained through the intermediacy of the h_r, functions. A reflection on Eq. (5.26) indicates that for the symmetry operation R on the product function $(f_i g_k)$ expressible in the product basis set

$\{f_j g_l\}$, i.e., **h** the coefficients of the linear combination are given by all possible twin products of the elements occurring in the i-th and the k-th columns of the matrices $\mathbf{D}^{\mu}_{f}(R)$ and $\mathbf{D}^{\nu}_{g}(R)$ respectively. Therefore, the effect of R on $(f_i g_k)$ for all i's and k's, expressed compactly in the direct product relation of Eq. (5.28), will involve all the possible mutual twin products of all the elements occurring in the matrices $\mathbf{D}^{\mu}_{f}(R)$ and $\mathbf{D}^{\nu}_{g}(R)$ and is not merely a multiplication of the two matrices. This important aspect is sought to be expressed through the symbol $\otimes$, indicating direct product. Thus the direct product.

$$\mathbf{A} \otimes \mathbf{B} = \begin{pmatrix} a_{11}\mathbf{B} & a_{12}\mathbf{B} & \cdots\cdots a_{1m}\mathbf{B} \\ a_{21}\mathbf{B} & a_{22}\mathbf{B} & \cdots\cdots a_{2m}\mathbf{B} \\ \vdots & & \vdots \\ a_{m1}\mathbf{B} & a_{m2}\mathbf{B} & \cdots\cdots a_{mm}\mathbf{B} \end{pmatrix}$$

The above direct product matrix can thus be partitioned into m^2 submatrices. In actual practice one will scarcely need the detailed forms of the direct product matrices although some examples aimed at clarifying the concepts are provided here.

Example 5.7.

The matrix representations of C_3 of the point group $D_{3\hbar}$ in the basis set of components (x, y) and of (z) are given by

$$\mathbf{D}(C_3)\begin{pmatrix} x \\ y \end{pmatrix} = \begin{pmatrix} \cos\theta & -\sin\theta \\ \sin\theta & \cos\theta \end{pmatrix}\begin{pmatrix} x \\ y \end{pmatrix} \text{ and } \mathbf{D}(C_3)z = (1)z,$$

where $\theta = \frac{2\pi}{3}$. Find the C_3 representation in the direct product space (xz, yz).

It may be noted that although x, y and z are components of a vector, xz and yz are to be treated as basis functions.

$$\begin{aligned} \mathbf{D}^{xz,\ yz}(C_3) &= \mathbf{D}^{x,y}(C_3) \otimes \mathbf{D}^{z}(C_3) \\ &= \begin{pmatrix} -\frac{1}{2}\mathbf{D}^{z} & -\frac{\sqrt{3}}{2}\mathbf{D}^{z} \\ \frac{\sqrt{3}}{2}\mathbf{D}^{z} & -\frac{1}{2}\mathbf{D}^{z} \end{pmatrix} = \begin{pmatrix} -\frac{1}{2} & -\frac{\sqrt{3}}{2} \\ \frac{\sqrt{3}}{2} & -\frac{1}{2} \end{pmatrix} \end{aligned}$$

which is in conformity with what has been found in Sec. 4.7.

Example 5.8.

Find the matrix representation of C_3 of the $D_{3\hbar}$ point group in the direct product space of two basis sets (e_1 e_2) and (l_1 l_2) where the orientations of the vectors are the same as described in Fig. 4.2

The respective matrices scooped from Sec. 4.2 and 4.3 are

$$\mathbf{D}^{\mathrm{e}}(\mathrm{C}_3) = \begin{pmatrix} -\frac{1}{2} & -\frac{\sqrt{3}}{2} \\ \frac{\sqrt{3}}{2} & -\frac{1}{2} \end{pmatrix} \text{ and}$$

$$\mathbf{D}^{\mathrm{l}}(\mathrm{C}_3) = \begin{pmatrix} 0 & -1 \\ 1 & -1 \end{pmatrix}$$

The direct product space consists of the basis functions (e_1 l_1, e_1 l_2, e_2 l_1, e_2 l_2). The direct product matrix for C_3 is given by

$$\begin{aligned} \mathbf{D}^{(\mathrm{e}\times\mathrm{l})}(\mathrm{C}_3) &= \mathbf{D}^{\mathrm{e}}(\mathrm{C}_3) \otimes \mathbf{D}^{\mathrm{l}}(\mathrm{C}_3) \\ &= \begin{pmatrix} 0\begin{pmatrix} -\frac{1}{2} & -\frac{\sqrt{3}}{2} \\ \frac{\sqrt{3}}{2} & -\frac{1}{2} \end{pmatrix} & -1\begin{pmatrix} -\frac{1}{2} & -\frac{\sqrt{3}}{2} \\ \frac{\sqrt{3}}{2} & -\frac{1}{2} \end{pmatrix} \\ 1\begin{pmatrix} -\frac{1}{2} & -\frac{\sqrt{3}}{2} \\ \frac{\sqrt{3}}{2} & -\frac{1}{2} \end{pmatrix} & -1\begin{pmatrix} -\frac{1}{2} & -\frac{\sqrt{3}}{2} \\ \frac{\sqrt{3}}{2} & -\frac{1}{2} \end{pmatrix} \end{pmatrix} \\ &= \begin{pmatrix} 0 & 0 & \frac{1}{2} & \frac{\sqrt{3}}{2} \\ 0 & 0 & -\frac{\sqrt{3}}{2} & \frac{1}{2} \\ -\frac{1}{2} & -\frac{\sqrt{3}}{2} & \frac{1}{2} & \frac{\sqrt{3}}{2} \\ \frac{\sqrt{3}}{2} & -\frac{1}{2} & -\frac{\sqrt{3}}{2} & \frac{1}{2} \end{pmatrix} \end{aligned}$$

The foregoing example is a vindication of the Theo. 5.9, viz., the character of a symmetry element in the direct product representation equals the product of the characters of the symmetry elements in the component representations.

$$\chi(\mathrm{e} \times \mathrm{l})(\mathrm{C}_3) = 1 = \chi_{\mathrm{e}}(\mathrm{C}_3).\ \chi_{\mathrm{l}}(\mathrm{C}_3) = (-1).(-1).$$

Example 5.9

Set up the character table of the direct product representations of all possible two different IR's of the point group C_{3v}.

The character table of C_{3v}, is

	χ		
IR's	E	$2C_3$	$3_{\sigma v}$
A_1	1	1	1
A_2	1	1	−1
E	2	−1	0

Using Theo 5.9, we can construct the characters of the necessary direct product representations.

	χ		
	E	$2C_3$	$3_{\sigma v}$
$\Gamma(A_1 \times A_2)$	1	1	−1
$\Gamma(A_1 \times E)$	2	−1	0
$\Gamma(A_2 \times E)$	2	−1	0

Example 5.10

A d^2 transition metal ion in infinitely strong octahedral crystal field has a configuration $(t_{2g})^2$ and is characterized by a direct product representation $T^{(T_{2g} \times T_{2g})}$. What IR's are involved?

From $O_\hbar$−character table (Appendix II) we have

	E	$6C_4$	$3C_2$	$6S_4$	$8C_3$	$8S_6$	$3_{\sigma\hbar}$	i	6^{σ}_{d}	$6C'_2$
T_{2g}	3	−1	−1	−1	0	0	−1	3	1	1
$T_{2g} \times T_{2g}$	9	1	1	1	0	0	1	9	1	1

We may now apply Eq. 5.31 to find the IR's contained in it.

$$\begin{aligned}
n_{T2g} &= \frac{1}{48}[3.9 - 6.1.1 - 3.1.1 - 6.1.1 - 3.1.1 + 3.9 + 6.1.1 + 6.1.1] = 1 \\
n_{T1g} &= \frac{1}{48}[3.9 + 6.1.1 - 3.1.1 + 6.1.1 - 3.1.1 + 3.9 - 6.1.1 - 6.1.1] = 1 \\
nE_g &= \frac{1}{48}[2 \times 9 + 3.2.1 + 3.2.1 + 9.2] = 1 \\
nA_{1g} &= \frac{1}{48}[9.1 + 6.1.1 + 3.1.1 + 6.1.1 + 3.1.1 + 1.9 + 6.1.1 + 6.1.1] = 1
\end{aligned}$$

Application of Eq. (5.31) to the other IR's of $O_\hbar$ point group results in their nonoccurrence in $\Gamma\,(T_{2g} \times T_{2g})$

$$\text{Hence} \quad \Gamma\left(T_{2_g} \times T_{2_g}\right) = A_{1g} + E_g + T_{1g} + T_{2g}.$$

5.8 Some General Remarks-Transformations, Bases And Characters.

Upto this point whatever transformation properties we have considered, these are mostly with respect to basis sets of configuration space vectors, translational vectors, rotational vectors and functions all of which had their origins (or origin of the variables in case of functions) embedded in the stationary point of the symmetry group. In all such cases it was invariably found that each base vector either transformed as an IR of the point group or transformed, in association with one or more necessary base vectors, as a degenerate two or three dimensional IR. The question is what will happen if one utilises, as the basis of representation of a group, a number of vectors having their origins at different points and away from the stationary point of the symmetry group of the molecule. *The answer is that in general none of the vectors individually will transform as an IR, but these will collectively constitute the basis of a reducible representation which can finally be split into IR's. The same remark also applies to the basis set of functions defined in terms of variables measured from different origins the locations of which do not coincide with the stationary point chosen as origin of the coordinate axes for symmetry operations.*

One usually avoids detailed algebraic procedure of transformation (cf. Sec. 3.5) to pile up the representative matrices if the results of symmetry transformations are such as to permit of ready inference. In such cases of avoidance, one can still find the characters of the respective symmetry elements. To do this it is only necessary to remember that when any basis function, say f_k, is operated upon by R, then a linear combination results, i.e.,

$$\hat{R}f_k = \sum_l f_l D(R)_{lk} \tag{5.32}$$

Of the different coefficients, the diagonal one, viz., $D(R)_{kk}$ of f_k in the RHS of (5.32) contributes to the character value $\mathbf{D}(R)$. Hence $\chi(R) = \sum_k D(R)_{kk}$. In cases where transformation results are mentally inferred, it pays to remember the following rules for computation of $\chi(R)$, the rules in ultimate analysis being just equivalent to finding $\sum_k D(R)_{kk}$.

(i) Each vector or basis function that is unshifted in position or sign under symmetry operation R contributes +1 to the value of $\chi(R)$.

(ii) Vectors or basis functions that shift completely under the effect of R from their original locations contribute nil to $\chi(R)$ value.

(iii) Vectors or functions, changing directions or signs only but remaining unshifted from their original locations, contribute each, to the net value of $\chi(R)$, a quantity commensurate with the change in direction. The contribution is -1 for each vector or function changing its direction by 180° (i.e.π).

Once the characters of the reducible representation are. obtained, Theo. 5.8 is then invoked to help disclose the IR's spanned by the basis vectors or functions. A number of examples will now be given which will bear out clearly all the concepts presented in this section.

Example 5.11.

(a) Find the transformation properties of typical elements of each class of the O subgroup of a molecule with octahedral symmetry with complex d-wavefunctions as the basis set.

(b) Make a list of the corresponding characters. What IR's are involved?

[The variables of the d-wavefunctions are defined with respect to the central point of the octahedral molecule as the origin, the C_4-axis being the quantising axis].

(a) The five typical group elements of the O-subgroup are E, $C_4(z)$, $C_2(z)$, C_3 and $C_{2'}(xy)$. The complex *d*-wavefunction (dropping the radial f(r) part and the common multiplicative factor $\left(\sqrt{\frac{5}{4\pi}}\right)$ are

$$\begin{aligned} d_{\pm 2} &= \sqrt{\frac{3}{8}}(x \pm iy)^2,\ d_{+1} = -\sqrt{\frac{3}{2}}(x+iy)z,\ d_{-1} = \sqrt{\frac{3}{2}}(x-iy)z \\ \text{and}\quad d_0 &= \frac{1}{2}(3z^2 - r^2). \end{aligned}$$

The functions are complicated enough and do not permit a ready mental reckoning of the transformation properties. So following the conventional technique detailed in Sec. 3.5, the transformations of the basis set of functions under E, $C_4(z)$, $C_2(z)$, $C_{2'}(xy)$ are readily found. These are given in the table. The transformations under C_3 are a little hard to work out and will engage our attention here. We note that under C_3, x changes to y, y to z and z to x in a cubic symmetry where e_1 e_2 and e_3 axes point to the midpoints of the faces. Consider the functional parts of d-functions without the numerical normalising factors and designate these functions as d'_2, d'_{-2}, d'_1, d'_{-1} and d'_0 stated above.

$$C_3 d_2 = C_3\sqrt{\frac{3}{8}}(x+iy)^2 = \sqrt{\frac{3}{8}}(y+iz)^2 = \sqrt{\frac{3}{8}}(y^2 - z^2 + 2iyz)$$

$\therefore$ $C_3 d'_2 = (y^2 - z^2 + 2iyz)$ and this is expected to be a linear combination of d'_2, d'_{-2} (since y^2 occurs),d'_1, d'_{-1}, (since 2iyz occurs) and d'_0 (since z^2 occurs). Let us assume

$$C_3 d'_2 = (y^2 - z^2 + 2iyz) = td'_0 + pd'_2 + qd'_{-2} + rd'_1 + sd'_{-1} \tag{5.33}$$

where p, q, r, s and t are to be so chosen as to make the RHS equal the left.

$$\begin{aligned} C_3 d'_2 &= y^2 - z^2 + 2iyz = t(3z^2 - r^2) + p(x^2 - y^2 + 2ixy) \\ &\quad + q(x^2 - y^2 - 2ixy) - r(x+iy)z + s(x-iy)z \end{aligned} \tag{5.34}$$

A little trial reckoning enables the evaluation of p, q, r, s and t. Noting the signs of y^2, z^2 and yz on the LHS of Eq.(5.34), it is evident that, p, q, r, s and t should all be negative. Again since xy, x^2 should disappear from the RHS of (5.34), p=q=$\frac{1}{2}$t. For similar reasons involving the nonappearance of x on the LHS of (5.34), r should equal s to cause the former (viz. x) to disappear from RHS. The occurrence of one 2iyz makes it imperative that r=s=−1 on the RHS of (5.34). Similarly one y^2 on the left demands p=q=$-\frac{1}{4}$ and t=$-\frac{1}{2}$ in conformity with what has been stated before.

$$\begin{aligned}
&\text{Thus } c_3d_2' = y^2 - z^2 + 2iyz = \\
&-\frac{1}{2}(3z^2 - r^2) - \frac{1}{4}(x^2 - y^2 + 2ixy) \\
&-\frac{1}{4}(x^2 - y^2 - 2ixy) - 1(-1)(x+iy) - 1(x-iy)z \\
&= -\frac{1}{2}d_0' - \frac{1}{4}d_2' - \frac{1}{4}d_{-2}' - d_1' - d_{-1}' \\
&= -\frac{1}{2}.2d_0 - \frac{1}{4}\sqrt{\frac{8}{3}}d_2 - \frac{1}{4}.\sqrt{\frac{8}{3}}d_{-2} - \sqrt{\frac{2}{3}}d_1 - \sqrt{\frac{2}{3}}d_{-1} \\
\text{or } C_3d_2 &= \sqrt{\frac{3}{8}}(y^2 - z^2 + 2iyz) \\
&= -\sqrt{\frac{3}{8}}d_0 - \frac{1}{4}d_2 - \frac{1}{4}d_{-2} - \sqrt{\frac{2}{3}}.\sqrt{\frac{3}{8}}d_1 - \sqrt{\frac{2}{3}}.\sqrt{\frac{3}{8}}d_{-1} \\
&= -\sqrt{\frac{6}{16}}d_0 - \sqrt{\frac{1}{16}}d_2 - \sqrt{\frac{1}{16}}d_{-2} - \sqrt{\frac{4}{16}}d_1 - \sqrt{\frac{4}{16}}d_{-1}
\end{aligned} \tag{5.35}$$

With just a trivial extension of the arguments, it is readily seen that

$$\hat{C}_3d_{-2} = -\sqrt{\frac{6}{16}}d_0 - \sqrt{\frac{1}{16}}d_2 - \sqrt{\frac{1}{16}}d_{-2} + \sqrt{\frac{4}{16}}d_1 + \sqrt{\frac{4}{16}}d_{-1} \tag{5.36}$$

Let us now turn to $\hat{C}_3d_1$

$$\begin{aligned}
\hat{C}_3d_1 = \hat{C}_3\left\{-\sqrt{\frac{3}{2}}(x+iy)z\right\} &= -\sqrt{\frac{3}{2}}(y+iz)x \\
&= \sqrt{\frac{3}{2}}[-(xy)+izx]
\end{aligned}$$

$\hat{C}_3d_1' = -(xy + izx)$ is expected to be a linear combination of d_2', d_{-2}' (since xy occurs in them), d_1' and d_{-1}' (for zx occurs in both). d_0' should not occur as z^2 does not figure in the rotated function.

$$\begin{aligned}
\hat{C}_3d_1' &= -(xy + izx) = pd_2' + qd_{-2}' + rd_1' + sd_{-1}' \\
&= p(x^2 - y^2 + 2ixy) + q(x^2 - y^2 - 2ixy) + \\
&\quad r[-(x+iy)z] + s(x-iy)z
\end{aligned}$$

Noting the concurrent presence of terms on the LHS and the RHS of the foregoing expression and also the due disappearances of some terms of the RHS one readily settles the values of p, q, r and s, viz.,

$$p = -q = \frac{i}{4} \text{ and } r = \frac{i}{2} = -s.$$

$$\begin{aligned}
\text{Hence } \hat{C}_3 d_1' &= \frac{i}{4}d_2' - \frac{i}{4}d'_{-2} + \frac{i}{2}d_1' - \frac{i}{2}d'_{-1} \\
&= \frac{i}{4}\sqrt{\frac{8}{3}}d_2 - \frac{i}{4}\sqrt{\frac{8}{3}}d_{-2} + \frac{i}{2}\sqrt{\frac{2}{3}}d_1 - \frac{i}{2}\sqrt{\frac{2}{3}}d_{-1} \\
\therefore \hat{C}_3 d_1 &= \sqrt{\frac{3}{2}}[(-1)(y+iz)x] \\
&= \frac{i}{4}\sqrt{\frac{8}{3}}\sqrt{\frac{3}{2}}d_2 - \frac{i}{4}\sqrt{\frac{8}{3}}\sqrt{\frac{3}{2}}d_{-2} + \frac{i}{2}\sqrt{\frac{2}{3}}\sqrt{\frac{3}{2}}d_1 - \frac{i}{2}\sqrt{\frac{2}{3}}\sqrt{\frac{3}{2}}d_{-1} \\
&= \frac{i}{2}d_2 - \frac{i}{2}d_{-2} + \frac{i}{2}d_1 - \frac{i}{2}d_{-1} \qquad (5.37)
\end{aligned}$$

We can immediately write down (noting the differences in the signs of $\hat{C}_3 d_1$ and $\hat{C}_3 d_{-1}$)

$$\hat{C}_3 d_{-1} = -\frac{i}{2}d_2 + \frac{i}{2}d_{-2} + \frac{i}{2}d_1 - \frac{i}{2}d_{-1} \qquad (5.38)$$

The last transformation still to be worked out is $\hat{C}_3 d_0$

$$\hat{C}_3 d_0 = \hat{C}_3\left[\frac{1}{2}(3z^2 - r^2)\right] = \frac{1}{2}(3x^2 - r^2)$$

$\therefore \hat{C}_3 d_0' = 3x^2 - r^2$ should merely be a linear combination of d_0, d_2 and d_{-2} since the rotated function contains x^2, y^2 and z^2 only.

$$\hat{C}_3 d_0' = (3x^2 - r^2) = t(3z^2 - r^2) + p(x^2 - y^2 + 2ixy) + q(x^2 - y^2 - 2ixy)$$

Since ixy does not occur in the LHS it is due for disappearance from the RHS which demands p be equal to q. The plus and minus signs of r^2 in the LHS and RHS mean that t should be negative. A little trial shows that with $p=q=\frac{3}{4}$ and $t=-\frac{1}{2}$ the equality of LHS and RHS is ensured.

$$\begin{aligned}
\hat{C}_3 d_0' &= (3x^2 - r^2) = \frac{3}{4}(x^2 - y^2 + 2ixy) + \frac{3}{4}(x^2 - y^2 - 2ixy) \\
&\qquad - \frac{1}{2}(3z^2 - r^2) \\
&= \frac{3}{4}(x^2 - y^2 + 2ixy) + \frac{3}{4}(x^2 - y^2 - 2ixy) \\
&\qquad - \frac{1}{2}(3r^2 - 3x^2 - 3y^2 - r^2)
\end{aligned}$$

$$
\begin{aligned}
&= \frac{3}{4}d_2' + \frac{3}{4}d_{-2}' - \frac{1}{2}d_0' \\
&= \frac{3}{4}\sqrt{\frac{8}{3}}d_2 + \frac{3}{4}\sqrt{\frac{8}{3}}d_{-2} - \frac{1}{2}.2d_0 \\
\text{or } \hat{C}_3 d_0 &= \frac{1}{2}(3x^2 - r^2) = \sqrt{\frac{3}{8}}d_2 + \sqrt{\frac{3}{8}}d_{-2} - \sqrt{\frac{2}{8}}d_0 \\
&= \sqrt{\frac{6}{16}}d_2 + \sqrt{\frac{6}{16}}d_{-2} - \sqrt{\frac{4}{16}}d_0 \qquad (5.39)
\end{aligned}
$$

The transformation results of the symmetry operators E $C_4{}^z C_2{}^z$ $C_2'(xy)$ and C_3 (Eqns. 5.35-5.39) are tabulated below.

	E	$C_4{}^z$	$C_2{}^z$	$C_2{}'^{xy}$	C_3
d_0	d_0	d_0	d_0	d_0	$-\sqrt{\frac{4}{16}}d_0+\sqrt{\frac{6}{16}}d_2+\sqrt{\frac{6}{16}}d_{-2}$
d_2	d_2	$-d_2$	d_2	$-d_{-2}$	$-\sqrt{\frac{6}{16}}d_0-\sqrt{\frac{1}{16}}d_2-\sqrt{\frac{1}{16}}d_{-2}-\sqrt{\frac{4}{16}}d_1-\sqrt{\frac{4}{16}}d_{-1}$
d_{-2}	d_{-2}	$-d_{-2}$	d_{-2}	$-d_2$	$-\sqrt{\frac{6}{16}}d_0-\sqrt{\frac{1}{16}}d_2-\sqrt{\frac{1}{16}}d_{-2}+\sqrt{\frac{4}{16}}d_1+\sqrt{\frac{4}{16}}d_{-1}$
d_1	d_1	$-id_1$	$-d_1$	id_{-1}	$\frac{i}{2}d_2-\frac{i}{2}d_{-2}+\frac{i}{2}d_1-\frac{i}{2}d_{-1}$
d_{-1}	d_{-1}	id_{-1}	$-d_{-1}$	id_1	$-\frac{i}{2}d_2+\frac{i}{2}d_{-2}+\frac{i}{2}d_1-\frac{i}{2}d_{-1}$

This completes the answer to part (a) of the problem.
(b) We now turn to part (b) of the problem. Summing the coefficients of the diagonal terms under E, C_4^z, C_2^z, $C_2{}'^{xy}$ and C_3 as given in the above table we find

$$\chi(E) = 5,\ \chi(C_4^z) = -1,\ \chi(C_2{}^z) = 1,\ \chi(C_2{}'^{xy}) = 1,\ \chi(C_3) = -1$$

These values are also obtained by applying the rules for shifted and unshifted functions. We can now use the character table of O appearing in the table of O_h to find IR's using Theo. 5.8 [Eqn. 5.13].

$$
\begin{aligned}
n_E &= \frac{1}{24}[5 \times 2 + 6.0(-1) + 3.1.2 + 6.0.1 + 8x(-1).(-1)] = 1 \\
n_{T2} &= \frac{1}{24}[5 \times 3 + 6.(-1).(-1) + 3.1(-1) + 6.1.1 + 8.0.(-1)] = 1
\end{aligned}
$$

This basis set of d-wavefunctions span the IR's E and T_2 of the point group O.

Example 5.12

What basis sets of the foregoing example belong to the IR's T_2 and E?

To find the basis we can use any one of the following two methods.

First method: Use of projection operators $\mathcal{P}^E$ and $\mathcal{P}^{T_2}$ for the O-subgroup can be,made (cf. Sec. 5.6) to project functions from the d-wavefunction space which will have symmetries of the IR's E and T_2. In following this method one can profitably use the transformation properties as tabulated in the foregoing example.

Second method : This method depends on a close scrutiny of the transformation properties supplemented by piecemeal arguments and their follow-up. It is seen from the table of the previous example that (i) d_2 and d_{-2} and (ii) d_1 and d_{-1} both mix together under the operation of $C_2'^{xy}$. Let us construct two pairs of orthogonal functions, viz.
(1) $\phi_1 = \frac{1}{\sqrt{2}}(d_{-2}+d_2)$; $\phi_2 = \frac{1}{\sqrt{2}}(d_{-2}-d_2)$ and
(2) $\psi_1 = \frac{1}{\sqrt{2}}(d_{-1}+d_1)$; $\psi_2 = \frac{1}{\sqrt{2}}(d_{-1}-d_1)$
and investigate the transformation properties of d_0, ϕ_1, ϕ_2, ψ_1 and ψ_2 under the operations of the typical elements of the O-subgroup. These are very easy to work out in view of the existing table of the previous example. The tabulation appears below.

	E	C_4^z	C_2^z	$C_2'^{xz}$	C_3
d_0	d_0	d_0	d_0	d_0	$-\sqrt{\frac{4}{16}}d_0+\sqrt{\frac{12}{16}}\phi_1$
$\frac{1}{\sqrt{2}}(d_{-2}+d_2)=\phi_1$	ϕ_1	$-\phi_1$	ϕ_1	$-\phi_1$	$-\sqrt{\frac{12}{16}}d_0 - \sqrt{\frac{4}{16}}\phi_1$
$\frac{1}{\sqrt{2}}(d_{-2}-d_2)=\phi_2$	ϕ_2	$-\phi_2$	ϕ_2	ϕ_2	ψ_1
$\frac{1}{\sqrt{2}}(d_{-1}+d_1)=\psi_1$	ψ_1	$i\psi_2$	$-\psi_1$	$i\psi_2$	$-i\psi_2$
$\frac{1}{\sqrt{2}}(d_{-1}-d_1)=\psi_2$	ψ_2	$i\psi_1$	$-\psi_2$	$-i\psi_1$	$i\phi_2$

It will he observed from the table that d_0 and ϕ_1 mix forming a basis for E representation. The other three functions ϕ_2, ψ_1, and ψ_2 also get mixed up and thus constitute the basis functions for the triply degenerate T_2. All these functions as described here are found to be symmetric with respect to inversion operation of the O_h group. Hence these are all gerade functions. We can thus conclude:

$$\left\{d_0,\ \frac{1}{\sqrt{2}}(d_{-2}+d_2)\right\} \text{ basis for } E_g$$

$$\left\{\frac{1}{\sqrt{2}}(d_{-2}-d_2),\ \frac{1}{\sqrt{2}}(d_{-1}+d_1),\ \frac{1}{\sqrt{2}}(d_{-1}-d_1)\right\} \text{ basis for } T_{2g}$$

It may be additionally remarked that the transformation properties of d_1, d_{-1} and $\frac{1}{\sqrt{2}}(d_{-2}-d_2)$ are such that these three may also form the basis for the T_{2g} representation. This may be verified by the interested readers. One may also scoop, by using the rules, the-character values of the different symmetry elements from the transformations of $\{\phi_2, \psi_1, \psi_2\}$ and of $\{d_0, \phi_1\}$ shown in the table of the foregoing example and confirm that the sets do really form bases for T_{2g} and E_g.

As examples of basis sets where the functions are defined with respect to origins not coincident with the stationary point of the symmetry group, two molecules are considered below to show the transformations of the basis set, the evaluation of characters, determination of the IR's and the linear combinations of the basis functions transforming as the IR's.

Example 5.13.

Which IR's are spanned by the four 1s orbitals of the H-atoms in ethylene molecule?

To solve the problem we have to find successively the transformations of the s-orbitals, evaluation of characters and then the IR's. Ethylene belongs to the point group $D_{2\hbar}$ (Fig. 5.4) where the four s-orbital lobes centred on the

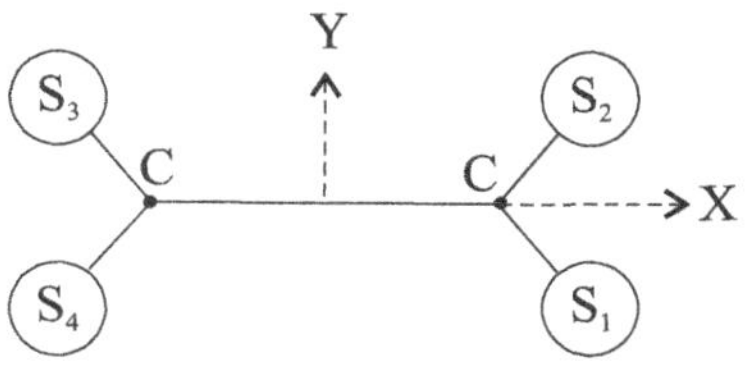

Fig. 5.4: Four S-orbitals on the H-atoms of ethylene. Z-axis is perp. to the plane

four hydrogens are marked S_1, S_2, S_3 and S_4. Our objective will be achieved by first considering the operations of the symmetry elements of $D_{2\hbar}$ upon these four functions. A mental reckoning of the transformation of the basis set is here feasible and the tabulation appears below.

	E	$C_2{}^z$	$C_2{}^x$	$C_2{}^y$	$\sigma_\hbar$	$i=S_2{}^z$	$S_2{}^x$	$S_2{}^y$
S_1	S_1	S_3	S_2	S_4	S_1	S_3	S_3	S_3
S_2	S_2	S_4	S_1	S_3	S_2	S_4	S_4	S_4
S_3	S_3	S_1	S_4	S_2	S_3	S_1	S_1	S_1
S_4	S_4	S_2	S_3	S_1	S_4	S_2	S_2	S_2

The characters of the reducible representation are found to be (on using the rules of shifting or nonshifting of functions)

	E	$C_2{}^z$	$C_2{}^y$	$C_2{}^x$	$\sigma_\hbar$	i	$S_2{}^x$	$S_2{}^y$
χ^{red}	4	0	0	0	4	0	0	0

The IR's of $D_{2\hbar}$ contained in the representation in the given basis set are found out with the help of the well known relation
$n_i = \frac{1}{g}\sum_R \chi^{red}(R)\chi i^*(R)$. It is found to contain A_g, B_{1g}, B_{2u} B_{3u}. One may

then utilise the appropriate projection operators to find the four suitable linear combinations of S_1, S_2, S_3, and S_4 which would transform as the above four IR's respectively.

It thus turns out that functions, defined with respect to origins away from the stationary point of the symmetry group, do not individually form bases of the one dimensional representations, but do so only in combinations.

Example. 5.14.

Find the IR's spanned by the p_z orbitals centred on the four carbon atoms of cyclobutadiene and also the basis functions belonging to the IR's.

Cyclobutadiene molecule (Fig. 5.5) belongs to the point group $D_{4\hbar}$. It is planar and the p_z-orbitals, perpendicular to the molecular plane and with their positive lobes pointing upwards, are indicated in the figure by Z_1, Z_2, Z_3 and Z_4. The transformation properties under D_4 (for similicity D_4 subgroup of $D_{4\hbar}$

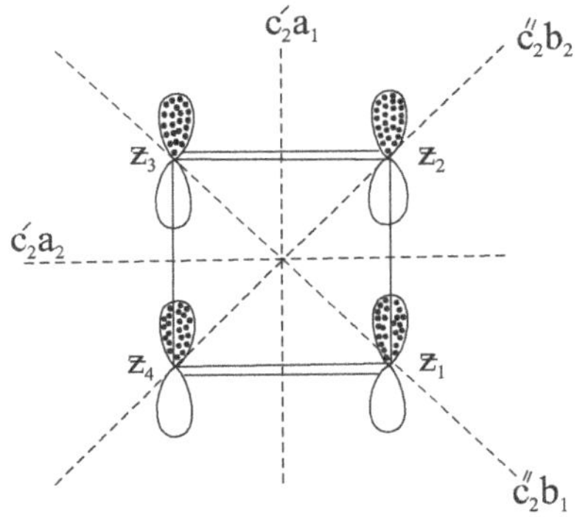

Fig. 5.5: Cyclobutadiene with four p_z-orbitals.

will suffice) can clearly be written down by inspection.

	E	C_4	C_2	C_4^3	$C_{2a_1}{}'$	$C_{2a_2}{}'$	$C_{2b_1}{}''$	$C_{2b_2}{}''$
Z_1	Z_1	Z_2	Z_3	Z_4	$-Z_4$	$-Z_2$	$-Z_1$	$-Z_3$
Z_2	Z_2	Z_3	Z_4	Z_1	$-Z_3$	$-Z_1$	$-Z_4$	$-Z_2$
Z_3	Z_3	Z_4	Z_1	Z_2	$-Z_2$	$-Z_4$	$-Z_3$	$-Z_1$
Z_4	Z_4	Z_1	Z_2	Z_3	$-Z_1$	$-Z_3$	$-Z_2$	$-Z_4$
χ^{red}	4	0	0	0	0	0	-2	-2

These characters of the reducible representation yield, with the help of the character table of D_4 and the relation $n_i = \frac{1}{g}\sum_R \chi^{red}(R)\chi_i^*(R)$, the IR's A_2, B_1 and E.

Constructing the projection operators $\mathcal{P}^{A_2}$, $\mathcal{P}^{B_1}$ and $\mathcal{P}^{E}$ of the D_4 point group and applying these to the original basis functions (Z_1, Z_2, Z_3, Z_4), the symmetry adapted functions that result are

$$\phi(a_2) = \frac{1}{2}(Z_1 + Z_2 + Z_3 + Z_4)$$
$$\phi(b_1) = \frac{1}{2}(Z_1 - Z_2 + Z_3 - Z_4)$$
$$\begin{cases} \phi_1(e) &= \frac{1}{\sqrt{2}}(Z_1 - Z_3) \\ \phi_2(e) &= \frac{1}{\sqrt{2}}(Z_2 - Z_4). \end{cases}$$

For the sake of completness, g and u symbols should be ascribed and this can be done easily.

5.9 Regular Representation.

The group multiplication table itself bares the fact that the possible result of operations of any group element P on each element in turn must be a linear combination of all the group elements. Thus referring to sec. 2.1, it is evident

$$PQ = O.E + O.P + O.Q + 1.R + O.S + \ldots \ldots \ldots$$

and this similar for others replacing Q in LHS with the operator P remaining the same this is true where we choose the group elements as the basis set and also as operators. The representation D(P) is a (g × g) dimensional matrix (g = order of the group) with zeros and ones appearing as elements of the representative matrix. This will show up in all the representative matrices. We thus find that the group elements, if chosen as basis set, will behave as orthogonal basis set of kets. The representation obtained in this way is called a regular representation. A regular representation is always feasible for any finite group.

We shall now consider a regular representation, for which we turn to the group multiplication table written in such a way that the vertical array of operators are the inverses, appearing sequentially, of the group elements in the horizontal row.

Thus for the group with g elements $\{E, A, B, C, \ldots P, Q, R, \ldots\} \in G$ the peripheral structure of the multiplication table is

$$
\begin{array}{cc|l}
 & & \overbrace{\mathrm{E\ \ B\ \ C.....P\ \ Q\ \ R....}}^{\mathrm{A}_\mu(\mu=1 \text{ to g})} \\
\hline
\mathrm{A}_{\mu^{-1}} & \left\{ \begin{array}{c} \mathrm{E}^{-1} \\ \mathrm{B}^{-1} \\ \mathrm{C}^{-1} \\ \vdots \\ \mathrm{P}^{-1} \\ \mathrm{Q}^{-1} \\ \mathrm{R}^{-1} \\ \vdots \end{array} \right. &
\end{array}
$$

The complete table drawn up in this fashion evidently contains E (the identity element) along the diagonal. To be specific, let us rewrite the group multiplication table of $\mathrm{C_{3v}}$ following the above framework (see page 26):

$\mathrm{C_{3v}}$	E	$\mathrm{C_3}$	$\mathrm{C_3}^2$	σ_{v1}	σ_{v2}	σ_{v3}
E^{-1}	E	$\mathrm{C_3}$	$\mathrm{C_3}^2$	σ_{v1}	σ_{v2}	σ_{v3}
$\mathrm{C_3}^{-1}$	C_3^2	E	$\mathrm{C_3}$	σ_{v3}	σ_{v1}	σ_{v2}
$\mathrm{C_3}^{2-1}$	$\mathrm{C_3}$	$\mathrm{C_3}^2$	E	σ_{v2}	σ_{v3}	σ_{v1}
$\sigma_{v1}{}^{-1}$	σ_{v1}	σ_{v3}	σ_{v2}	E	$\mathrm{C_3}^2$	$\mathrm{C_3}$
$\sigma_{v2}{}^{-1}$	σ_{v2}	σ_{v1}	σ_{v3}	$\mathrm{C_3}$	E	$\mathrm{C_3}^2$
$\sigma_{v3}{}^{-1}$	σ_{v3}	σ_{v2}	σ_{v1}	$\mathrm{C_3}^2$	$\mathrm{C_3}$	E

Using this multiplication table repeatedly, the representation matrices for the group elements $\mathrm{C_{3v}}$ are now written in the form of six dimensional matrices consisting merely of elements 0 and 1. The element 1 occurs once in each row of the matrix only at the location where the corresponding group element appears as the product in the group multiplication table.

Thus

$$
\mathbf{D}^{\mathrm{reg}}(\mathrm{E}) = \begin{pmatrix} 1 & 0 & 0 & 0 & 0 & 0 \\ 0 & 1 & 0 & 0 & 0 & 0 \\ 0 & 0 & 1 & 0 & 0 & 0 \\ 0 & 0 & 0 & 1 & 0 & 0 \\ 0 & 0 & 0 & 0 & 1 & 0 \\ 0 & 0 & 0 & 0 & 0 & 1 \end{pmatrix} \quad \mathbf{D}^{\mathrm{reg}}(\mathrm{C_3}^2) = \begin{pmatrix} 0 & 0 & 1 & 0 & 0 & 0 \\ 1 & 0 & 0 & 0 & 0 & 0 \\ 0 & 1 & 0 & 0 & 0 & 0 \\ 0 & 0 & 0 & 0 & 1 & 0 \\ 0 & 0 & 0 & 0 & 0 & 1 \\ 0 & 0 & 0 & 1 & 0 & 0 \end{pmatrix}
$$

and similarly for the other group elements. These six matrices (each six dimensional) form a regular representation of the point group C_{3v}. We take note of certain general features of the so called representative matrices before turning to prove the authenticity of the representation.

The characteristics of the matrices are

(i) $\chi(E)=g$ (order of the group)

(ii) $\chi(A)=\chi(B)=...=\chi(R)=0$, for all R's$\neq$E

(iii) Only the $\mathbf{D}^{reg}(E)$ matrix is diagonal, the rest are nondiagonal.

(iv) If $A\mu$'s (μ=1 to g) represent the pseudonames of the group elements E, A, B,....P, Q... in the horizontal row and if the μth and that νth elements (i.e., $A\mu$ and $A\nu$) in the sequential counting, be P and Q, then the corresponding μth and the νth elements ($A\mu^{-1}$ and $A\nu^{-1}$) in the vertical column are P^{-1} and Q^{-1} respectively in the peripheral frame of the group multiplication table.

(v) The general structure of any element, say, the ($\mu\nu$)th matrix element of $\mathbf{D}^{reg}(B)$ is

$$D(B)_{\mu\nu} = D(B)_{A\mu^{-1}A_\nu} = 1 \text{ or } 0 \tag{5.40}$$

It is 1 if $A_\mu^{-1}A_\nu=B$, i.e., if $A_\mu B=A_\nu$ otherwise, it is zero. Thus we may write formally

$$D(B)_{\mu\nu} = D(B)_{A\mu^{-1}A\nu} = \delta_{A\mu B,A_\nu} \tag{5.41}$$

In the foregoing expressions $A\mu^{-1}$, $A\nu$ represent both group elements and also proxies for the row and column numbers μ and ν of the ($\mu\nu$)th element.

We have now to prove that the matrices form a representation of the group G. In other words the product law, say, F=JK of the group elements should be obeyed by the corresponding matrices, i.e., we have to show

$$\mathbf{D}^{reg}(F) = \mathbf{D}^{reg}(JK) = \mathbf{D}^{reg}(J)\mathbf{D}^{reg}(K) \tag{5.42}$$

Now, using (5.41), we have

$$\begin{aligned} D(F)_{\mu\gamma} &= D(F)_{A\mu^{-1}A\nu} \\ &= \delta_{A\mu F,A\nu} = \delta_{PJK,Q} \end{aligned} \tag{5.43 a}$$

The $(\mu\nu)$ th element of the RHS of the product matrix

$$\{\mathbf{D}^{\mathrm{reg}}(\mathrm{J})\mathbf{D}^{\mathrm{reg}}(\mathrm{K})\}_{\mathrm{A}\mu^{-1}\mathrm{A}_\gamma} = \sum_{\lambda} \mathrm{D}(\mathrm{J})_{\mathrm{A}\mu^{-1}\mathrm{A}_\lambda} \mathrm{D}(\mathrm{K})_{\mathrm{A}_\lambda^{-1}\mathrm{A}_\gamma}$$

$$\left.\begin{aligned} &= \sum_{\lambda} \delta_{\mathrm{A}\mu\mathrm{J},\mathrm{A}_\lambda} \cdot \quad \delta_{\mathrm{A}\lambda\mathrm{K},\mathrm{A}_\gamma} \\ &= \sum_{\lambda} \delta_{\mathrm{A}\mu\mathrm{JK},\mathrm{A}_\lambda\mathrm{K}} \cdot \quad \delta_{\mathrm{A}\lambda\mathrm{K},\mathrm{A}_\gamma} \\ &= \delta_{\mathrm{A}\mu\mathrm{JK},\mathrm{A}_\lambda\mathrm{K}} \cdot \quad \delta_{\mathrm{A}\lambda\mathrm{K},\mathrm{A}_\gamma} \end{aligned}\right\} \begin{array}{l}\text{These are true both}\\ \text{suffix}-\text{wise and group}\\ \text{element product}-\text{wise}\end{array} \qquad (5.43\mathrm{b})$$

The summation sign in (5.43b) disappears since $\mathrm{A}_\mu\mathrm{JK}$ is a specific group element that is identical with $\mathrm{A}_\lambda\mathrm{K}$ only for a specific value of λ and hence A_λ. The relation (5.43b) can now be expressed in one δ notation, viz.,

$$\{\mathbf{D}^{\mathrm{reg}}(\mathrm{J})\mathbf{D}^{\mathrm{reg}}(\mathrm{K})\}_{\mu\gamma} = \delta_{\mathrm{A}\mu\mathrm{JK},\mathrm{A}_\gamma} = \delta_{\mathrm{PJK,Q}} \qquad (5.43\ \mathrm{c})$$

which is the same as (5.43 a). This establishes the validity of the representation.

Before we discuss the direct utility of a regular representation, it is necessary to prove a theorem concerning the frequency of occurrence of the IR's of the point group in a regular representation.

Theorem 5.10: All the IR's of a point group occur in its regular representation with a frequency each equal to the dimensionality of the corresponding IR.

Let G be the point group of order g with the IR's, Γ_i(i=1 to n) of dimensionalities l_i's (i=1 to n). If n_i, be the number of times that Γ_i occurs in the block diagonal form of Γ^{reg}, then applying relation (5.13) one can write

$$\left.\begin{aligned} n_i &= \frac{1}{g}\sum_{R} \chi_i{}^*(R)\chi^{\mathrm{reg}}(R) \\ &= \frac{1}{g}\chi_i{}^*(E)\chi^{\mathrm{reg}}(E) \\ &= \frac{1}{g} l_i g = l_i \end{aligned}\right\} \left.\begin{array}{l} \because \chi^{\mathrm{reg}}(R) = 0 \\ \text{for all R's} \neq E \\ \text{and} \\ \chi^{\mathrm{reg}}(E) = g \\ \chi_i(E) = l_i \end{array}\right\}$$

This proves the celebrated theorem.

The utility of a regular representation lies in validating the equality signs of theorems 5.4 and 5.7. Consider the block diagonal form of

$\mathbf{D}^{reg}(E)$. Since $\chi^{reg}(E)$ remains unaltered during blocking and since all the IR's (in their E representations) occur in it

(Theorem 5.10) leads to

$$\chi^{reg}(E) = \sum_{i} l_i \chi_i(E)$$

i.e., $g=\sum_{i} l_i^2$. which establishes the equality sign of Theorem 5.4.

Let us now consider Theorem 5.7. When all the IR's consistent with the condition $\sum_{i} l_i^2$ have been found out, all the possible distinct character sets which are mutually orthogonal in the group element space are thus known. These form a complete set. When the dimensionality of the group element space is totally shrunk to that of the class space (viz., k dimension), the latter number of the independent mutually orthogonal character vector sets with reduced number of components (viz k, since the group elements are lumped into classes) still form the complete set in the class space and there cannot be anymore independent set in that dimensional space. The number of distinct IR's thus equals the number of classes in the point group (Theo. 5.7).

We include an additional theorem in the pack of this chapter

Theorem 5.10: In an abelian group all the irreducible representations are each one dimensional.

It is readily understood that the application of the conjugacy relation $g_i^{-1} g_j g_i = g_j$ does not lead to any new group element of the group G of order g. Due to mutual commutation of the elements in the abelian group, each group element q_i forms a class by itself the number of group elements = g = no. of classes = no. of distinct IR's of G.

Now applying the dimensionality theorem (theo. 5.4)

$\sum l_i^2 = g =$ Sum of the squares of dimensions of g no. of IR's

On analysing $\sum l_i^2$, it is found that each l_i is one dimensional.

Hence all the IR's of the abelian group are each one dimensional.

To conclude the chapter we now respond to the suggestion (v) of sec. 5.4. The character sets of the pseudo degenerate pairs of IR's of the abelian cyclic groups, C_n's can be easily be found out. The C_n groups, being abelian, can affard to have suitable numerical basis, also incidentally serving as characters, to satisfy the group multiplication law like what unity (1) does for the completely symmetric IR's of all the symmetric groups.

Any of the n roots of unity $e^{i2\pi}$ i.e. $e^{(i\frac{2\pi}{n})}K$ where k ranges from 1 to n, can serve as a basis for the group elements of C_n The results for the remaining elements $C_n^2, C_n^3 ... c_n^n$ may their be easily worked out. Each exponential and its complex conjingate can independently serve as separate bases for C_n and others for the generation of a pair of degenerate representations We illustrate the cases for C_3, C_4 and C_5

The group C_3

basis	C_3	C_3^2	C_3^3=E		
$e^{i\frac{2\pi}{3}}$	$\in$	$\in^2$	$\in^3$		
$=\in$	$\in$	$\in^*$	1	$\rightarrow$	$\chi(C_3)$
$e^{-i\frac{2\pi}{3}}$	$\in^*$	$\in^{*2}$	$\in^{*3}$		
$=\in^*$	$\in^*$	$\in$	1	$\rightarrow$	$\chi(C_3)$

The group C_4 The group C_3

basis	C_4	$C_4^2 = C_2$	C_4^3	$C_4^4 = E$
$e^{i\frac{2\pi}{4}} =\in$	$\in$	$\in^2$	$\in^3$	$\in^4$
	i	-1	$-i$	$1 \rightarrow \chi(C_4)$
basis				
$e^{-i\frac{2\pi}{4}} =\in^*$	$\in^*$	$\in^{*2}$	$\in^{*3}$	$\in^{*4}$
	$-i$	-1	i	$1 \rightarrow \chi(C_4)$

Similar procedure can be applied to the group C_5 We, however, have two pairs of pseudo degenerate representations and have to work out separately with four basis exponentials (a) $e^{i2\pi/5} = \epsilon$ (b) $e^{-i2\pi/5} = \epsilon*$ (c) $e^{(i2\pi/5)^2} = \epsilon^2$ (d) $e^{(-i2\pi/5)^2} = \epsilon^2*$

All the C_n groups have A_1 IR if the initial root ϵ be chosen as $e^{(}i2\pi/n)^n$ for basis representing also the primary operation C_n. The C_n. groups, with n = even, have B_1 IR wherein the primary operation is the root $e^{(}i2\pi/n)^{n/2}$ i.e. $k = \frac{n}{2}$.

Chapter 6

Representation Theory And Quantum Mechanics

In this chapter an attempt will be made to forge a link between the symmetry properties reflected in the representations and certain properties of wave functions, i.e., the solutions of the Schrodinger equation.

6.1 Symmetry Operators, Hamiltonian Operator And Wave Functions.

Consider a molecule of any point group G with elements P, Q, R, S, T,...... and its wave equation $H\psi=E\psi$. Let us now make a note of some properties of the symmetry operators and some of the terms occurring in the wave equation. As to the former, the following are worthy of attention:

(i) Each symmetry operator R in G will commute with the unperturbed hamiltonian H of the molecule. It means that H is invariant under all symmetry operations of the group, a property that can be established quantitatively (Appendix I). Qualitatively speaking, if the molecule be subjected to any symmetry operation of the relevant group, the requisite transformation will not cause any distortion of the molecule. There would thus be no change in the energy eigenvalue of the molecule implying, thereby, that the hamiltonian H will remain unaffected before or after the symmetry operation. In technical language it means that the commutator [H, R] vanishes or that HR=RH.

(ii) For any IR of the point group G, symmetry adapted functions in the corresponding subspace can be built up (Sec. 5.6). Each such one dimensional IR symmetry adapted function is an eigenfunction of the operators P, Q, R......with the corresponding characters of the one dimensional representations as the eigenvalues.

(iii) If the molecule be subjected to an external' perturbation affecting its point group symmetry (i.e., a descent in symmetry), the hamiltonian will also be transformed into some new perturbed hamiltonian. Consequences resulting from this will be taken up later (Sec. 6.3).

We now enlist the following relevant properties of the solutions of the *Schrodinger equation.*

(a) Non degenerate solutions, say ψ_i corresponding to energy eigenvalue E_1, may result on solving the wave equation.

(b) The wave equation may also yield n number of linearly independent solutions ψ_1, ψ_2...... ψ_j..... ψ_n corresponding to the n-fold degenerate energy eigenvalue En.

(c) These eigenfunctions of H can always be obtained, if desired, in orthonormal forms.

(d) Any linear combination of n degenerate wavefunctions $\psi' = \sum_j^n c_j\psi_j$ is also an eigenfunction with the same degenerate eigenvalue En.

(e) Molecular wavcfuuctions are some suitable antisymmetrised combinations of the spin orbitals, the latter being the Hartree product of the molecular orbitals and spin wavefunctions.

$$\psi = \sum_P (-1)^P P\left\{S_1(1)S_2(2)\cdots S_n(n)\right\}, \text{ where } \sum_P (-1)^P P$$

is the antisymmetrising operator and S_k(k) is the spin orbital occupied by the k-th electron. Each such $S_k = \phi_k(k)\alpha(k)$ or $\phi_k(k)\beta(k)$ where ϕ_k's are the molecular orbitals.

(f) For an unperturbed system and also for a π system the basic wave equation can be split (after elimination or proper sizing of the internal perturbations) into a number of component wave equations $H\phi_k = \in_k \phi_k$, where ϕ's are the molecular orbitals and $\in_k$'s the energy eigenvalues. The basic properties listed under (a) to (d) are thus also applicable to the molecular orbitals and their energies. Implications that ensue from combination of these two sets of properties [(i) to (iii) and a to f] will be discussed in the next section.

6.2 Representations And Molecular Orbitals As Basis Set.

(1) Consider the nondegenerate eigenvalue $\in_i$ and the eigenfunction, ϕ_i and apply the R operator to the Schrodinger equation.

$$\begin{aligned} \widehat{R}H\phi_i &= \widehat{R} \in_i \phi_i \\ \text{or } H(\widehat{R}\phi_i) &= \in_i (\widehat{R}\phi_i) \cdots\cdots \text{from property (i)} \cdots\cdots \quad (6.1) \end{aligned}$$

$\therefore \widehat{R}\phi_i$ is also an eigenfunction of H with the eigenvalue $\in_i$, which, however, is nondegenerate. Hence it follows that ϕ_i and $\widehat{R}\phi_i$ should be linearly dependent, i.e.,

$\widehat{R}\phi_i = b\phi_i$, where b is a numerical constant. Again, from property (c) and from the fact that the inner product is preserved, we can write,

$$\int (\widehat{R}\phi_i)^*(\widehat{R}\phi_i)^*(\widehat{R}\phi_i)d\tau = 1 = bb^* \int \phi_{i*}\phi_i d\tau$$

$$\text{which means } \widehat{R}\phi_i = \pm 1\ \phi_i \text{ or } e^{j\theta}\phi_i \qquad (6.2)$$

ϕ_i, provided that it is real, is thus an eigenfunction of R with eigenvalue +1 or −1. It is thus seen from Eq. (6.2) that the effect of each symmetry operator on a nondegenerate molecular orbital ϕ_i is to reproduce ϕ_i multiplied by +1 or −1. Hence each such molecular orbital serves as a basis for one-dimensional representation of G with the matrix elements (+1) or (−1). Again, since the representation Γ is one-dimensional, it must be some irreducible representation of A or B symmetry. We can thus conclude **"Each nondegenerate molecular orbital of a molecule belonging to a particular point group serves as a basis for a one-dimensional IR of the point group."**

(2) The degenerate Case. For an n-fold degenerate eigenvalue $\in_n$, there occur ϕ_1, $\phi_2 \ldots \phi_k \ldots \phi_n$ degenerate molecular orbitals which span an n-dimensional linear function space. Applying any symmetry operator R on any degenerate eigenfunction ϕ_j of the set leads to another eigenfunction. This can be easily shown.

$$\begin{aligned} R \in_n \phi_j &= RH\phi_j = HR\phi_j \\ \text{i.e., } \in_n (R\phi_j) &= H(R\phi_j) \end{aligned}$$

$\widehat{R}\phi_j$ = an eigenfunction of H with the same degenerate energy $\in_n$ ······ [property (i) and cf. Eq. (6.1)]

= a linear combination of ϕ's... [property (d)]
This conclusion is true when we allow j's. to run from 1 to n and also when R is replaced by any other group element E, P, Q... etc. Hence the degenerate set of molecular orbitals forms a basis, set spanning an n−dimensional representation space of the group.

$$\therefore \mathrm{R}\phi_{\mathrm{j}} = \sum_{\mathrm{i}} \phi_{\mathrm{i}}\mathrm{D(R)_{ij}}, \text{ a specific linear combination} \cdots\cdots \quad (6.3)$$

=one general function in the n-dimensional function space spanned by the degenerate set.
Allowing j to run from 1 to n, we have from Eq. (6.3)

$$\widehat{\mathrm{R}}\Phi = \Phi\mathbf{D}(\mathrm{R}) \quad (6.4)$$

where $\mathbf{D}(\mathrm{R})$ an n×n matrix, represents R in the basis set $\mathbf{\Phi} = (\phi_1\ \phi_2.......\phi_{\mathrm{n}})$. The other symmetry operators P, Q, S. T..... belonging to G similarly generate the matrices $\mathbf{D}(\mathrm{P})$, $\mathbf{D}(\mathrm{Q})$ $\mathbf{D}(\mathrm{S})$, $\mathbf{D}(\mathrm{T})$ ... respectively in the basis set Φ. That these matrices retain the group property of G can be tested by examining, for example, the product relation.

Let us suppose that for G, P$\widehat{\mathrm{R}}$=T

$$\begin{aligned}
\text{Now } \mathrm{P}\widehat{\mathrm{R}}\phi_{\mathrm{j}} = \mathrm{P}(\widehat{\mathrm{R}}\phi_{\mathrm{j}}) &= \mathrm{P}\sum_{\mathrm{i}}\phi_{\mathrm{i}}\mathrm{D(R)_{ij}} = \sum_{\mathrm{i}}\mathrm{P}\phi_{\mathrm{i}}\mathrm{D(R)_{ij}} \\
&= \sum_{\mathrm{i}}\sum_{\mathrm{l}}\phi_{\mathrm{l}}\mathrm{D(P)_{li}D(R)_{ij}} \\
&= \sum_{\mathrm{l}}\phi_{\mathrm{l}}\left(\mathbf{D}(\mathrm{P})\mathbf{D}(\mathrm{R})\right)_{\mathrm{lj}}
\end{aligned}$$

$$\therefore \mathrm{P}\widehat{\mathrm{R}}\Phi = \Phi\mathbf{D}(\mathrm{P})\mathbf{D}(\mathrm{R}) \quad (6.5)$$

$$\begin{aligned}
\text{Again } \mathrm{P}\widehat{\mathrm{R}}\phi_{\mathrm{j}} &= \mathrm{T}\phi_{\mathrm{j}} = \sum_{\mathrm{l}}\phi_{\mathrm{l}}\mathrm{D(T)_{lj}} \\
\therefore \mathrm{P}\widehat{\mathrm{R}}\Phi &= \mathrm{T}(\Phi) = \Phi\mathbf{D}(\mathrm{T}) \quad (6.6)
\end{aligned}$$

Eqs. (6.5) and (6.6) lead to

$$\mathbf{D}(\mathrm{T}) = \mathbf{D}(\mathrm{P})\mathbf{D}(\mathrm{R}) \quad (6.7)$$

The discussion so far proves two things viz., Φ, the degenerate set, serves as a basis and the transformation matrices, retaining the group property, form a representation of the group. But what representation-reducible or irreducible? One makes the assumption that it is an irreducible one. We can thus generalise that **the set of n degenerate molecular orbilals forms a basis for the representation of an n-dimensional IR of the point group to which the molecule belongs.** This set spans an n-dimensional invariant subspace representing the IR.

(3) Since the molecular orbitals can be obtained, if desired, in orthonormal forms, it is possible to choose (Theo. 5.1) such a degenerate set of molecular orbitals as will form a unitary basis for the representation of an IR of the group.

Having thus brought the molecular orbitals within the ambit of the representation theory, one can anticipate that it should also be possible to assign a symmetry tag (in the form of an IR) to a given molecular orbital of a molecule belonging to a certain symmetry group. Practical illustration of this will be provided in Chapter 7.

(4) The foregoing descriptions of (1), (2) and (3) are just illustrative of the general concepts associated with the Hilbert space and its block diagonal form under group operations (sec 3.6). In fact the concepts are so important that a theorem has been framed.

Theorem 6.1: If a hermitian operator H (Such as the Hamiltonian) commutes with all the elements $\mathrm{D}(R)$ of the representation of a group G, then one can choose the eigenstates of H to transform according to the irreversible representations of the group. If an irreducible representation appears only once in the Hilbert space, every state in the IR is an eigenstate of H with the same eigenvalue.

It is however, to be remembered (see corollary theo 5.3) that the eigenstates belonging to different IR's (and even the component functions transforming differently) are mutually orthogonal.

6.3 Perturbations And Symmetry

The exact solutions of Schrodinger equation for many electron systems are not possible to obtain and in such cases it is usual to bank on approximate solutions and approximate energies. The methods employed are, amongst others, those of perturbations and variations. In the perturbative method the actual hamiltonian H is broken up into an idealised unperturbed hamiltonian H_0 and a perturbation hamiltonian H', i.e.,

$$H = H_0 + H'$$

Let us suppose that the molecule in the unperturbed state belongs to the point group G_0. When the perturbation H' is applied, two things may happen depending on the nature of the perturbation:

(i) the symmetry group of the perturbed molecule remains the same, viz., G_0

(ii) the symmetry group of the perturbed molecule becomes G and G is just a subgroup of G_0.

In the light of discussions in Sec. 6.2 we can thus conclude that

(a) for case (i), the set of degenerate molecular orbitals, corresponding to an energy eigenvalue $\in$ and forming a basis of an IR of G_0 is replaced by a radially different but symmetrically equivalent new degenerate set of molecular orbitals. This new set as also the former can still constitute a basis for the same IR in the perturbed state and the degeneracy is not lifted. H_0 as well as H' (and hence H) remain invariant under all symmetry operations of the group G_0.

(b) for the case (ii), while H_0 will be invariant under all the symmetry operations of G_0, H' (and H) will be so only under those of G, a subgroup of G_0. The energy levels of the unperturbed system will be classified according to IR's of G_0, while those of the perturbed one, with $H=H_0+H'$, according to those of G. Hence an energy $\in_k$ (k-fold degenerate) of the unperturbed state, classified by an IR. Γ of G_0, will split into levels $\in_1$, $\in_2$ ......in the perturbed state, i.e., the IR Γ of G_0 transforming into a reducible representation Γ^{red} in the perturbed state. The set of degenerate wave functions spanning the IR Γ of G_0 divides itself, under the influence of perturbation, into groups spanning the IR's Γ_1, Γ_2......of G such that $\Gamma^{red} = \Gamma_1 + \Gamma_2 + \cdots\cdots$characterised by energies $\in_1$ $\in_2$ $\cdots\cdots$respectively. There is thus a complete or partial lifting of degeneracy. This whole process is termed "branching rule" which gives an idea of the IR's stemming from descent in symmetry.

Trekking the reverse direction, i.e., from the perturbed to the unperturbed state with the gradual withdrawal of the perturbative forces, will result in the coalescence of energies ϵ_1, ϵ_2 ...... of the IR's Γ_1, Γ_2 ......into a single degenerate energy level ϵ_k, the degenerate wave functions of which span an IR in the unperturbed state.

It is necessary that we now cite examples to make the foregoing ideas of the influence of perturbation on symmetry more distinct and communicative. An atom belongs to a continuous rotational symmetry group and is characterised by the hamiltonian (neglecting magnetic interaction terms).

$$\begin{aligned} H &= -\frac{\hbar^2}{2m}\sum_i \nabla_i^2 - \sum_i \frac{Ze^2}{r_i} + \sum_{i<j} \frac{e^2}{r_{ij}} \\ &= H_0 + \sum_{i<j} \frac{e^2}{r_{ij}} = H_0 + H' \end{aligned} \qquad (6.8)$$

Here H_0 is the idealised unperturbed hamiltonian of the hypothetical atom without the electron-electron repulsion, viz., $\sum_{i<j} \frac{e^2}{r_{ij}}$. H' is the perturbation hamiltonian comprising the repulsion term in the real atom. It can be shown that the commutator $[H_0, H']$ vanishes, i.e., the symmetry group G_0 of the unperturbed atom and of the perturbed atom (i.e., the real atom) remains the same. But in this case, the representation Γ in the basis set of degenerate wave functions of the hypothetical unperturbed atom is reducible and is a direct sum of number of IR's of G_0. On introducing H', the same IR's of G_0 are obtained from Γ but only with different energy eigenvalues. Thus the point group remaining the same, the perturbation may still cause a partial or even a total lift of degeneracy.

To illustrate the second case, vtz., (b) let us continue with the previous example where the atom (or a transitional metal ion) is placed at the centre of an octahedral crystal field emanating from ligands (say CN groups) at the six vertices of an octahedron. The hamiltonian of the atom (or ion) is

$$H_1 = H_0 + \sum_{i<j} \frac{e^2}{r_{ij}} + H_{crys} = H + H_{crys} \qquad (6.9)$$

where H, given by (6.8), represents now the unperturbed atom (ion) and H_{crys} the perturbation hamiltonian.

It can be shown that H and H_{crys} no longer commute. The symmetry group of the unperturbed atom is thus lowered from one of complete rotational symmetry to $O_\hbar$ point group on introducing the perturbation hamiltonian H_{crys}. The five degenerate d-atomic orbitals, which span the IR, Γ_1, of the unperturbed atom (ion) of spherical symmetry with energy ϵ_1', now carry two IR's T_{2g}, and E_g, with different energies in the perturbed state. The t_{2g} state is spanned by $(d_{xz}d_{yz}d_{xy})$ and the e_g by $\left(d_{z^2}d(x^2 - y^2)\right)$. [One should note that in the present case the two new sets of degenerate d-orbitals chracteristic of the atom (ion) in the presence of H_{crys}, may also be made the basis functions in investigating the symmetry behaviour of the perturbed state.] That is the IR, Γ_1, carried by the five d-orbitals becomes reducible in the perturbed state, viz., $\Gamma_1 = T_{2g} \oplus E_g$. The degeneracy is thus partially lifted and the five d-orbitals are accidentally degenerate in the unperturbed (initial) state.

The above example shows that an IR, Γ_1, of the initial state branches into two IR's, T_{2g} and E_g, accompanied by a descent in symmetry in the subsequent perturbed state when H_{crys} is switchetd on. Two further step-downs in symmetry are discussed below.

(i) The transition metal ion in the octahedral crystal field of six CN ligands now constitutes the initial state and let our attention be fixed on the IR's, T_{2g} and E_g. The substitution of two diametrically opposite trans CN units by two H_2O units $[M(CN)_4(H_2O)_2]$ will effectively mean a perturbation imposed upon the initial state. This will lead to a tetragonal distortion and the symmetry of the system (i.e., the ion and the environmental field) will be lowered from $O_\hbar$ to $D_{4\hbar}$. The IR, T_{2g}, carried by $(d_{xz}\ d_{yz}\ d_{xy})$ will bifurcate into one dimensional B_{2g} sustained by d_{xy} and E_g by $(d_{xz}\ d_{yz})$ in the new situation. Similarly the original IR, E_g, of $O_\hbar$, will branch into A_{1g}, and B_{1g} spanned by the oribitals $(d_{z^2},\ d_{x^2-y^2})$ respectively.

(ii) As a further step down in symmetry, let us consider a situation where one H_2O ligand of the previous example is substituted by Cl, $[M(CN)_4Cl\ H_2O]$. This gives rise to a new perturbed system comprising the ion accompanied by a different distribution of the environmental crystal field. The system thus acquires a symmetry of the point group C_{4v} which is lower than that of the initial state of $D_{4\hbar}$. With this new descent in symmetry, there however occurs no further splitting of the one dimensional IR's of $D_{4\hbar}$ as is to be naturally expected. Thus, in the perturbed state C_{4v} the orbital, a_{1g} is transformed to a_1, b_{1g} to b_1 and b_{2g} to b_2. The degeneracy of

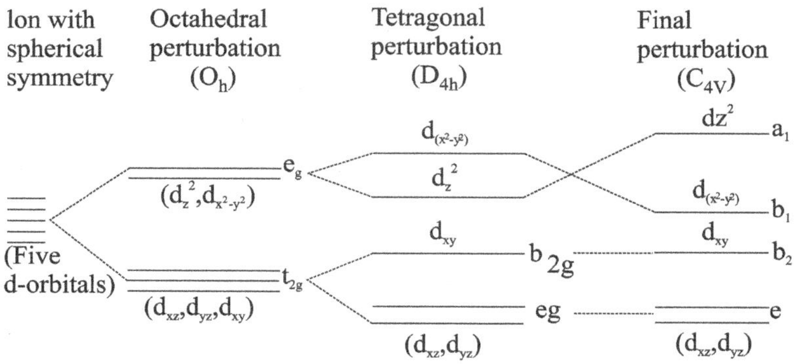

Fig. 6.1: Branching of IR's with descent in symmetry. Orbitals resulting from perturbation).

e_g carried by $(d_{xz}\ d_{yz})$ under $D_{4\hbar}$ of the initial state is not lifted. It is merely changed to e. The whole branching rule is schematically shown above.

In order to understand the branching rule of the IR's quantitatively, elaboration of the method with some typical examples will be made before concluding this chapter. This quantitative information needs three stages of data handling.

(i) Correspondence of the group elements of the lower group with those of the higher.
(ii) Tabulation of characters of the relevant group elements (higher group) under the corresponding elements of the lower group.
(iii) Use of Theo. (5.8) in conjunction with (ii) to ascertain the IR's of the lower group.

Of course, there exists an alternative but somewhat tortuous process of finding out the IR's when branching occurs. One starts from the set of basis functions spanning a particular IR of the higher symmetry group. The transformation behaviour of this basis under the symmetry operations of the lower group (the perturbed state) gives rise to matrices from which the IR's of the lower group can be spotted. One then concludes how an IR of the higher group splits into IR's of the lower group under perturbation. This method, employed in Examples 6.3 and 6.4, is not anything new and has been amply used in many examples of chapter 5.

Example 6.1.

How will the IR, T_{2g} of $O_\hbar$ group branch when tetragonal distortion is applied?

On application of the perturbation, the perturbed system will belong to $D_{4\hbar}$. The transformation from $O_\hbar$ to $D_{4\hbar}$ can be understood by considering the groups O and D_4, the effect of the inversion element i (or alternatively of $\sigma_\hbar$) may subsequently by indicated by g and u symbols as necessary. We now turn to the three successive stages.

(i)	D_4 Elements	C_2(about	$2C_{2'}$(about x	$2C_{2''}$
	E $2C_4$	z−axis)	and y axes)	
	O Elements	C_2	$2C_{2'}$(,,)	$2C_{2''}$
	E $2C_4$			
(ii)	Character tabulation			
	Elements of			

D_4	E	$2C_4$	C_2	$2C_{2'}$	$2C_{2''}$
Corresponding characters of O group under T_2 symmetry	3	−1	−1	−1	1

(iii) The IR's of D_4 are A_1, A_2, B_1, B_2 and E. Using the character table of D_4 and applying the Theo. (5.8)

$$n_{A_1} = \frac{1}{g}\sum_R \chi^{red}(R)\chi_{A_1^*}(R), \text{ we have}$$

$$n_{A_1} = \frac{1}{8}[3.1 + 2.(-1)(1) + (-1).1 + 2(-1)(1) + 2.(1).(1)] = 0$$

Similarly n_{A_2} and n_{B_1} are found to be zero.

$$n_{B_2} = \frac{1}{8}[3.1 + 2.(-1)(-1) + (-1).1 + 2.(-1)(-1) + 2.1.1] = 1$$

$$\text{and } n_E = \frac{1}{8}[3.2 + 2.(-1).0 + (-1)(-2) + 2.(-1).0 + 2.1.0] = 1$$

Therefore T_{2g} of $O_{\hbar}$ foliates into B_{2g} and E_g under tetragonal distortion.

Example 6.2.

What will be the effect of perturbation on the degenerate symmetries of BF_3, when one F atom is substituted by Cl atom, assuming the perturbation to be small?

BF_3 belongs to $D_{3\hbar}$ point group of which the degenerate IR's are E′ and E″. When one F is substituted by Cl the symmetry of planar molecule (Fig. 6.2) is lowered to C2v,. It is required to find the nature of splittings of E′ and E″ under C_{2v}.

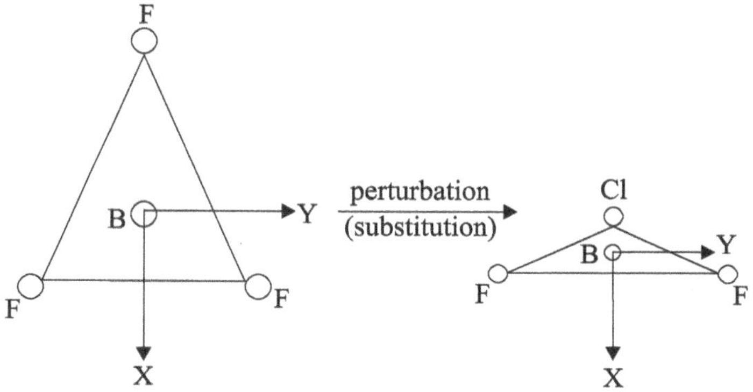

(Fig. 6.2: Perturbation effect on BF_3 molecule caused by substitution).

(i) Correspondence

C_{2v} group elements	E	C_2(about x-axis)	$\sigma_{v^{xy}}$	$\sigma_{v^{xz}}$
D_{3h} group elements	E	$C_{2'}$(about x-axis)	σ_h	$\sigma_{v'}$

(ii)

Character tabulation Elements of C_{2v}	E	C_2	$\sigma_{v^{xy}}$	$\sigma_{v^{xz}}$
Corresponding characters of D_{3h} under E′	2	0	2	0
E″	2	0	−2	0

(iii) The IR's under C_{2v} are A_1, A_2, B_1 and B_2 splitting of E′:

$$\begin{aligned}
nA_1 &= \frac{1}{4}(2.1 + 0.1 + 2.1+;0.1) = 1 \\
nA_2 &= \frac{1}{4}[2.1 + 0.1 + 2.(-1) + 0.(-1)] = 0 \\
nB_1 &= \frac{1}{4}[2.1 + 0.(-1) + 2.(-1) + 0.1] = 0 \\
nB_2 &= \frac{1}{4}[2.1 + 0.(-1) + 2.1 + 0.(-1)] = 1 \\
E' &= A_1 + B_2
\end{aligned}$$

Splitting of E″: Following the technique as shown above it is found $E''=A_2+B_1$

N.B. In writing the elements of C_{2v} in (ii) above the x-axis serves as the C_2 axis whereas the character table contains data with the z-axis as the C_2-axis. Hence any base function. say d_{xy} of $D_{3\hbar}$ should be read as d_{zy} (i.e., d_{yz}) under such a perturbative transformation. Similarly d_{yz} basis function of $D_{3\hbar}$ should be read as d_{yx} (i.e., d_{xy}. Which transforms as A_2 under C_{2v}. Such a caution for change of symbols in the basis function is called for only when the principal axis changes from the usual z-direction to x or y direction in the perturbed field as in the present instance.

The foregoing illustrations for quantitative determination of the reduction of symmetries are independent of any use of basis set. If one uses the alternative process, referred to earlier, of using a basis set, the latter may be wave functions, molecular orbitals or atomic orbitals spanning a certain IR of the initial state which may be regarded as the unperturbed state. These functions of the unperturbed state IR are used as basis functions for the representation of the group in perturbed slate. The representation in general will turn out to be reducible, the characters of which may be utilised to find the IR's involved. This will indicate the descent in symmetry from the initial slate. Examples are provided below.

Example 6.3.

The five d-orbitais $(d_{z^2},\ d_{(x^2-y^2)},\ d_{xz},\ d_{yz},\ d_{xy})$ span a five dimensional IR of a transition metal ion of complete spherical symmetry. How will the symmetry be affected if the ion be placed centrally in an octahedral electrostatic field of point charges.

The analytical forms of the five d-orbitals (real function) are obtained from suitable combination of the complex d-wavefunctions. The functions are (dropping the radial part f(r) and the common multiplicative factor)

$$
\begin{aligned}
d_{z^2} &= d_0 = \frac{1}{2}(3z^2 - r^2)\\
d_{x^2-y^2} = \frac{1}{\sqrt{2}}(d_{-2} + d_2) &= \frac{1}{\sqrt{2}}\sqrt{\frac{3}{8}}\left[(x - iy)^2 + (x + iy)^2\right]\\
&= \frac{\sqrt{3}}{2}(x^2 - y^2)\\
d_{xy} = \frac{\sqrt{i}}{2}(d_{-2} - d_2) &= \sqrt{\frac{3}{8}}\frac{i}{\sqrt{2}}\left[(x - iy)^2 - (x + iy)^2\right]\\
&= \frac{1}{\sqrt{2}}\sqrt{\frac{3}{8}}4xy = \sqrt{3}xy\\
d_{yz} = \frac{1}{\sqrt{2}}(d_{-1} + d_1) &= \frac{1}{\sqrt{2}}\sqrt{\frac{3}{2}}\left[i(x - iy)z - i(x + iy)z\right] = \sqrt{3}yz\\
d_{xz} = \frac{1}{\sqrt{2}}(d_{-1} - d_1) &= \frac{1}{\sqrt{2}}\sqrt{\frac{3}{2}}\left[(x - iy)z + (x + iy)z\right] = \sqrt{3}xz.
\end{aligned}
$$

Method I.

Let us now compile the transformation results of the typical symmetry operations (representing each class) of the O-group upon the d-orbitals. We can chalk out the transformations keeping in view the nature of changes of x, y and z under the symmetry operations. The tabulation appers below.

	E	C_{4^z}	C_{2^z}	$C_{2^{xy}}$	$C_3 \neq$
Changes of x, y and $z \rightarrow$	$x \rightarrow x$ $y \rightarrow y$ $z \rightarrow z$	$x \rightarrow y$ $y \rightarrow -x$ $z \rightarrow z$	$x \rightarrow -x$ $y \rightarrow -y$ $z \rightarrow z$	$x \rightarrow y$ $y \rightarrow x$ $z \rightarrow -z$	$x \rightarrow y$ $y \rightarrow z$ $z \rightarrow x$
$d_{z^2} = \frac{1}{2}$ $(3z^2 - r^2)$	d_{z^2}	d_{z^2}	d_{z^2}	d_{z^2}	$-\frac{1}{2}d_{z^2}$ $+\frac{\sqrt{3}}{2}d_{x^2-y^2}$
$d_{(x^2-y^2)}$ $= \frac{\sqrt{3}}{2}(x^2 - y^2)$	$d_{(x^2-y^2)}$	$-d_{(x^2-y^2)}$	$d_{(x^2-y^2)}$	$-d_{(x^2-y^2)}$	$-\frac{1}{2}d_{(x^2-y^2)}$ $-\frac{\sqrt{3}}{2}d_{z^2}$
$d_{xy} = \sqrt{3}xy$ $d_{xz} = \sqrt{3}xz$ $d_{yz} = \sqrt{3}yz$	d_{xy} d_{xz} d_{yz}	$-d_{xy}$ d_{yz} $-d_{xz}$	d_{xy} $-d_{xz}$ $-d_{xz}$	d_{xy} $-d_{yz}$ $-d_{xz}$	d_{yz} d_{xy} d_{xz}
χ	5	-1	1	1	-1

The last row is an array of the character values obtained by adding the diagonal elements of the corresponding five dimensional matrices. The IR's contained in this representation can be obtained by employing the relation of Theo. 5.8. As in the case of the complex d-wavefunction (example 5.11) the IR's are found to be E and T_2 spanned by $(d_z^2,\ d(x^2 - y^2))$ and $(d_{xy},\ d_{xz},\ d_{zy})$ respectively which is evident from the transformation table. Moreover, since all the d-orbitals are symmetric with respect to inversion, the IR's involved are E_g and T_{2g} of the $O_\hbar$ symmetry group.

Method II.

Often the formal group theoretical methods are substituted, if possible, by the quicker qualitative methods of inspection and mental reflection (cf. examples 5.12, 5.14). This may be done in the present case. Conider x, y and z axes (e_1 e_2 and e_3) of the transition metal octahedral complex as pointing towards the ligands. Since the polar graphs of d_z^2 and $d_{(x^2-y^2)}$ have their axes of symmetries lying respectively along the z axis and x, y axes, their orbital lobes point directly towards the ligands. These orbitals are thus subjected to stronger crystal field effect than d_{xy}, d_{xz}, d_{yz} orbitals the relative dispositions of which are similar with respect to the field. Consequently from the symmetry viewpoint, $(d_z^2,\ d_{(x^2-y^2)})$ should span a two-dimensional function space of $O_\hbar$, point group

∗ The first two entries in the last column are made following the arguments similar to those in Example 5.11.

(or O-group) and (d_{xy}, d_{xz}, d_{yz}) ought to from bases for a triply degenerate IR. One may now easily construct the partial representations of the typical elements of the O-point group with $(d_{z^2},\ d_(x^2 - y^2))$ and (d_{xy}, d_{xz}, d_{yz}) as basis sets and confirm from their character values that these belong to the IR's E and T_2 respectively.

Example 6.4.

What IR's will be spanned by the d-orbitals when a transition metal ion of complete spherical symmetry is placed in the tetrahedral field of four ligands?

The axial system in the present instant is so chosen that these point to the mid points of the faces of the cube in which the tetrahedron may be imagined to be inscribed. Since there are five classes in the T_d point group, we consider the transformation properties under one typical element of each class and then compute the character values from the diagonal terms. The transformation table is

	E	C_2^z	S_4^z	C_3	σ_d(accom modating z−axis
Changes of x, y and $z \rightarrow$	$x \rightarrow x$ $y \rightarrow y$ $z \rightarrow z$	$x \rightarrow -x$ $y \rightarrow -y$ $z \rightarrow z$	$x \rightarrow y$ $y \rightarrow -x$ $z \rightarrow -z$	$x \rightarrow y$ $y \rightarrow z$ $z \rightarrow x$	$x \rightarrow y$ $y \rightarrow x$ $z \rightarrow z$
$d_{z^2} = \frac{1}{2}$ $(3z^2 - r^2)$	d_{z^2}	d_{z^2}	d_{z^2}	$-\frac{1}{2}d_{z^2}+$ $\frac{\sqrt{3}}{2}d_{x^2-y^2}$	d_{z^2}
$d_{(x^2-y^2)}$ $= \frac{\sqrt{3}}{2}(x^2 - y^2)$	$d_{(x^2-y^2)}$	$d_{(x^2-y^2)}$	$-d_{(x^2-y^2)}$	$-\frac{1}{2}d_{(x^2-y^2)}$ $-\frac{\sqrt{3}}{2}d_{z^2}$	$-d_{x^2-y^2}$
$d_{xy} = \sqrt{3}xy$ $d_{xz} = \sqrt{3}xz$ $d_{yz} = \sqrt{3}yz$	d_{xy} d_{xz} d_{yz}	d_{xy} $-d_{xz}$ $-d_{yz}$	$-d_{xy}$ $-d_{yz}$ d_{xz}	d_{yz} d_{xy} d_{xz}	d_{xy} d_{yz} d_{xz}
χ	5	1	−1	−1	1

It is evident from the table that d_{z^2} and $d_{(x^2-y^2)}$ mix and so also do the three functions d_{xy}, d_{xz} and d_{yz}. Using the character table of T_d and Theo. 5.8, it is easily found that the IR's involved are E and T_2 as in the previous example. Qualitative considerations of the polar graphs of the d-orbitals and the locations of the ligands at the opposite corners of the faces of the cube (cf. Fig. 2.5) in tetrahedral symmetry would convince one that the triple set is in the more intense part of the crystal field than the twin set of d_{z^2}, $d_{(x^2-y^2)}$. Energetically, therefore, the doubly degenerate state represented by e will lie lower than the triply degenerate state t_2. The situation is thus the reverse of what obtains in octahedral crystal field.

6.4 Direct Product And Quantum Mechanical Integrals

Very often in quantum mechanics we come across numerous integrals containing products of functions, wave functions or molecular oribtals of the types $\int \phi_\mu \; F\phi_\nu \; d\tau$ or $\int \phi_\mu \; \phi_\nu \; d\tau$. These integrals, for their evaluation, require a considerable input of mental labour. However, application of symmetry arguments often decides whether some of the integrals are going to vanish or not. Although the non-zero integrals are not evaluated by symmetry arguments, the detection of the vanishing integrals is a great relief.

Consider the integral of the product, $\int \phi_\mu \; \phi_\nu \; d\tau$ where ϕ_μ and ϕ_ν are two molecular orbitals. Each molecular orbital forms a base in case of nondegeneracy or, in case of degeneracy, one function of the degenerate basis set of an IR of the point group of the molecule. Let us suppose that ϕ_μ and ϕ_ν respectively belong to two different IR's, Γ_μ and Γ_ν of the molecular point group. The product $\phi_\mu\phi_\nu$ is one of the basis set of product functions belonging to the direct product representation $\Gamma^{(\mu\times\nu)}$. Alternatively the product $\phi_\mu\phi_\nu$ may also be viewed as a general function in the direct product space spanned by a basis comprising the different linear combinations of the original binary product functions. Now remembering the condition that is satisfied during symmetry operation on functions (Sec. 3.5) it is easy to see that if R be a group element of the pointgroup of the molecule, then

$$\phi_\mu\phi_\nu \quad = \quad R(\phi_\mu\phi_\nu)$$

$$g\int \phi_\mu\phi_\nu d\tau = \int\sum_R R(\phi_\mu\phi_\nu)d\tau, \text{ where g is the order of the group.}$$

$$\text{i.e.,} \int \phi_\mu\phi_\nu d\tau = \frac{1}{g}\int\sum_R R(\phi_\mu\phi_\nu)d\tau \tag{6.10}$$

Incidentally it may be pointed out the operator $\frac{1}{g}\sum_R R$ is merely the projection operator for the totally symmetric IR of the molecular point group, i.e.,

$$\mathcal{P}A_1 = \frac{1}{g}\sum_R \chi_{A_1}(R)\widehat{R} = \frac{1}{g}\int\sum_R \widehat{R}$$

Hence one can write from Eq. (6.10)

$$\int \phi_\mu\phi_\nu d\tau = \int \mathcal{P}A_1(\phi_\mu\phi_\nu)d\tau \tag{6.11}$$

The Eq. (6.11) thus predicts that the integral will vanish if either the integrad itself or any of its component in some expanded form is not function that transforms as the totally symmetric IR, A_1, of the point group.

Now the answer as to whether or not it contains a component transforming as A_1 can be obtained by the reduction of the direct product representation in the basis set of the product functions. Thus applying Eq. (5.31) we have,

$$\begin{aligned} nA_1 &= \frac{1}{g}\sum_R \chi_{(\mu\times\gamma)}(R)\chi_{A_1^*}(R) \\ &= \frac{1}{g}\sum_R \chi_\mu(R).\ \chi_\nu(R) = \delta_{\mu\nu} \end{aligned} \tag{6.12}$$

wherein Theo. 5.9 and Theo. 5.5. have been employed.

It thus follows that a totally symmetric component (or the totally symmetric $\phi_\mu\phi_\nu$) will result subject to $\mu = \nu$, i.e., the IR $\Gamma_\mu = \Gamma_\nu$. It is only then that the integral will not vanish. This conclusion **teaches us a lesson, viz., integrals of the product of two functions belonging to different symmetries (two different IR's of the point group) must vanish.**

Frequently the integrand may consist of triple product of functions or two functions and an operator as mentioned in the beginning, viz., $\int \phi_\mu\phi_\nu\phi_\eta d\tau$ or $\int \phi_\mu F_\eta\phi_\nu d\tau$ where the subscripts μ, ν, η are indicative of the IR's of the point group. In the light of the previous discussion it is to be realised that the occurrence of a totally symmetric IR, A_1, in the direct product representation $\Gamma^{(\mu\times\nu\times\eta)}$ is essential for the survival of the integral. Equivalently also one may demand, in view of the lesson we have just learnt, that $\Gamma^{(\nu\times\mu)}$ should contain on reduction an IR, Γ_μ, i.e., the IR of the third function ϕ_μ for non-zero value of the integral.

The above statement is one of the conditions for the survival of the integral. Even the presence of an A_1 IR after resolution of the direct product may not eventually guarantee a non-zero integral value. The presence of A_1 IR should additionally be accompanied with the satisfaction of a second condition, viz., the factor functions ϕ_1 and ϕ_2 ought to have the same transformation properties as the first one. The imposition of the second condition is a consequence of the relations (5.7D - corollary). This also applies to the functions involving three functions ϕ_μ, F_η and ϕ_ν customarily, however, we only look out for the presence of A_1 IR and do not go in for the more severe test. The consideration of this section will be highly helpful in deducing the different section rules of chapter eight onwards.

Chapter 7

Qualitative Applications And Assignment Of Symmetry To Wave Functions

7.1 General

The present and the following chapters will encompass the applications of the symmetry principles and the group representation theory in different fields of chemical interest. It will be shown how qualitative ideas of symmetry can foretell some of the molecular properties. Furthermore quantitative use of the theories and principles marshalled in the previous chapters enables the formulation of relations and simple prediction of final results that can be arrived at after much computational effort by a quantum chemist. Group theory thus often aids a theoretical chemist to cut short the quantitative route of quantum calculations.

The approach to practical applications in each chapter will be preceded, wherever necessary, by a brief description of the relevant quantum mechanical aspects. It will be seen that in all applications of quantitative or semiquantitative nature the following recipe, either in its entirety or in fragments, will play a role:

(i) Given a basis set in the form of vectors or functions, to find the transformations under the symmetry operations of the group by employing conventional method or the method of inspection.
(ii) To find the characters of the reducible representation.
(iii) To find the IR's of the symmetry group carried by the basis set.
(iv) To find the combination of the bases which are symmetry adapted for the IR's of (iii). (Projection operator method or method of inspection).
(v) To make judicious use of direct product representation (Sec. 5.7) and the principle of vanishing of integrals (Sec. 6.4) where necessary.

Chapters 4, 5, and 6 abound in examples typifying one or the other of the processes of the above recipe. It will be seen how their coordinated use takes the shape of practical utilisation. To begin with we describe in some general terms a few qualitative applications.

7.2 Qualitative Applications

1. Dipole moment__ Dipole moment is a vectorial quantity. If a molecule possessing a permanent dipole moment has a certain symmetry, then the permissible symmetry operations upon the molecule will throw it in different equivalent configurations. But will these operations affect the dipole moment vector? Dipole moment vector must remain stationary or unmoved in direction during all such symmetry operations as otherwise its vectorial character will be impaired. This means that the dipole moment vector must lie on every possible symmetry element, viz., axis of symmetry, planes of symmetry. Consider the water molecule (C_{2v}) having the dipole moment vector lying along the bisector of the <HOH. Symmetry operations E, C_2, $\sigma_{v^{xz}}$, $\sigma_{v^{yz}}$ – all permit the dipole moment vector to remain unshifted. Similar is the case with methyl chloride (C_{3v}) where the dipole moment vector lies in the direction of $\overrightarrow{C - Cl}$ axis. It also lies on all the σ_v's of the molecule.

The invariance of the dipole moment vector under symmetry operations permits the following generalisations to be drawn:

(i) Molecules possessing inversion centre (i) cannot have permanent dipole moment since the length of a dipole moment vector cannot be thought of as squeezed to a point representing the symmetry element, i. Thus C_6H_6, $PtCl_4$ are all nonpolar.

(ii) Molecules belonging to D_n (or higher symmetry groups $D_{n\hbar}$, D_{nd}) are bereft of any permanent moment. A$C_{2'}$ operation will at once invert any vector not lying along the operational axis. Thus trans dichloro ethylene ($D_{2\hbar}$), $PCl_5(D_{3\hbar})$, staggered ethane (D_{3d}), carbon dioxide ($D_{\propto\hbar}$) are all without any permanent dipole moment.

(iii) Molecules belonging to symmetry groups possessing more than one principal axis of symmetry (T_d or $O_\hbar$) do not have permanent moment.

Thus examination of symmetry is of value in arriving at conclusions on the dipole character of the molecule.

2. Optical activity. The general common feature that is ascribed to a molecule capable of showing optical activity is that the mirror image of the molecule should not be superimposable on the former. This non-superimposability is guaranteed by an organic compound in which any carbon atom is connected to four different atoms or groups via single bonds. Often, however, it is asserted, though not strictly correctly, that a molecule should be completely asymmetric to show optical activity. Complete asymmetry means absence of symmetry elements. But trans −1, 2-dichlorocyclopropane (Fig. 7.1) which posse a C_2−axis is not quite asymmetric and yet it possesses optical activity.

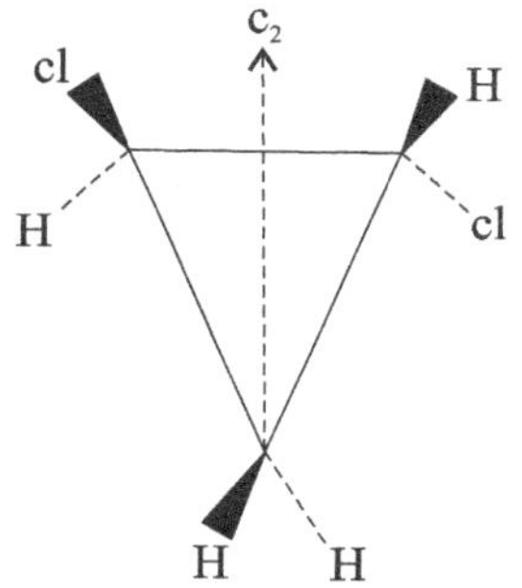

Fig. 7.1: Trans-1, 2-diclorocyclopropane possessing a C_2 axis.

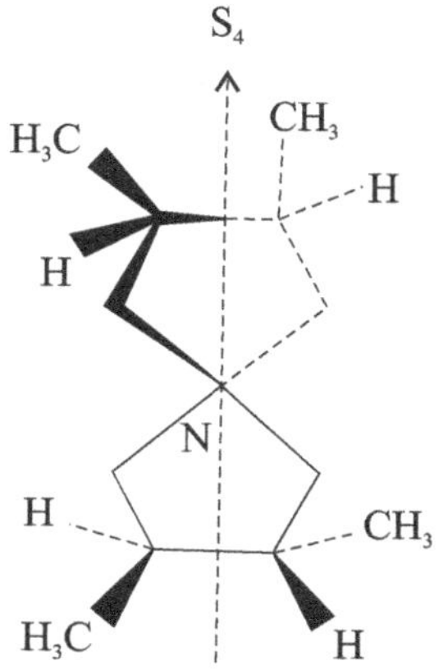

Spiro compound possessing S_4 axis.

The correct criterion of non-superimposability is linked to the absence of an improper axis of rotation, S_n. From symmetry viewpoint, if a molecule lacks an S_n axis, it should be optically active. Absence of S_1 or S_2 means absence of reflection plane, $\sigma_\hbar$ or inversion centre, i, respectively. So if an i or $\sigma_\hbar$ be detected in a molecule,

it turns out to be optically inactive. But many molecules, wherein neither i nor $\sigma_\hbar$ is present, are still optically inactive. One such example is the spiro compound shown in Fig. 7.1(a). Close scrutiny, however, reveals the presence of an S_4 axis which renders it optically inactive. It may thus be concluded that the absence or the presence of an S_n axis in a molecule is the fundamental criterion for the optical activity or inactivity respectively of the molecule.

7.3 Tagging Symmetry Labels To Wave Functions And Orbitals

1. Vibrational wave functions of diatomic molecules

(a) Quantum Mechanical Aspects-The vibrational motion of a diatomic molecule can be substituted by an equivalent one dimensional oscillation of a single particle of effective mass μ. The solution of the Schrodinger equation leads to a series of Hermite orthogonal functions, $e^{-\xi^2/2}H_n(\xi)$ as the wave functions with positive integral values of n(including zero) for the ground and excited states.

The normalised wave function is

$$\psi_n = \left(\frac{\sqrt{\beta/\pi}}{2^n n!}\right)^{1/2} e^{-\xi^2/2} H_n(\xi) = \left(\frac{\sqrt{\beta/\pi}}{2^n n!}\right)^{1/2} e^{-\beta x^2} H_n(\sqrt{\beta}z) \quad (7.1)$$

where $H_n(\xi) = (-1)^n e^{\xi^2} \frac{d^n}{d\xi^n}(e^{-\xi^2})$ is called Hermite polynomial of degree n in ξ. $\beta = \frac{2\pi}{h}\sqrt{\mu k} = \frac{4\pi^2\mu\nu}{h}$, k being the force constant and $\xi = \sqrt{\beta}x$. The energy eigenvalues, $E_n = \left(n + \frac{1}{2}\right)h\nu$, are all nondegenerate. The functional parts of the wavefunctions, excluding the normalising factors, of the first few quantum states characterised by n=0, 1, 2 and 3, are:

$$\begin{aligned} \psi'_0 &= e^{-\xi^2/2}H_0(\xi) = e^{-\xi^2/2};\ \psi'_1 = e^{-\xi^2/2}.2\xi \\ \psi'_2 &= e^{-\xi^2/2}(4\xi^2 - 2);\ \psi'_3 = e^{-\xi^2/2}(8\xi^{3'} - 12\xi) \end{aligned} \quad (7.2)$$

(b) Symmetry aspects :– Any diatomic molecule must belong either to $C_{\infty v}$, or $D_{\infty h}$ molecular point groups. Let us first consider the $C_{\infty v}$ case (heteronuclear diatomic molecule) having the symmetry elements E, $2C_\infty(\phi)$ and $\infty\ \sigma_v$'s. We choose the internuclear axis to be the z-axis so that the vibrational coordinate $\xi = \sqrt{\beta}z$ [cf. Eq. (7.1)]. Since the wave functions are all nondegenerate, each will form a base for one or the other of the one dimensional IR's of the point group $C_{\infty v}$. In the first place we consider the symmetry of the ground state wave function $\psi'_0 = e^{-\xi^2/2}$. Transformations of this base function under symmetry operations [Recipe (i) Sec. 7.1] all result in $1.e^{-\xi^2/2}$. This result is easily perceived in that the transformation of $e^{-\xi^2/2}$ depends on the nature of ξ^2 (i.e., upon z^2) and since z is invariant under all $C_{\infty v}$ operations, so is z^2 and hence also $e^{-\xi^2/2}$. The ground state wavefunction ψ'_0 thus belongs to the totally symmetric IR, Σ^+.

This result helps to find IR's of the other quantum states of vibration.

Example 7.1.

Find the IR's of the first and the second vibrationally excited state wavefunctions of the diatomic molecules belonging to $C_{\propto v}$ symmetry group.

Here we are concerned with $\psi_{1'} = e^{-\xi^2/2}.2\xi$ and $\psi_{2'} = e^{-\xi^2/2}(4\xi^2-2)$. Since $\psi_{1'}$ is the product of two functions, $e^{-\xi^2/2}$ and 2ξ, the representation $\Gamma_{\Psi_{1'}} = \Gamma_{\Psi_{0'}} \otimes \Gamma_{2\xi}$, i.e., a direct product representation with $\psi_{0'}$ and 2ξ as base functions. As the latter form bases for totally symmetric IR, Σ^+, it follows that $\psi_{1'}$, belongs to Σ^+. Similar conclusion is reached with respect to the second vibrationally excited state $\psi_{2'}$.

Determination of the IR's of the vibrational wavefunctions of homonuclear diatomic molecules (point group $D_{\propto h}$) is not at all a difficult job. Sticking to the choice of z-axis as aligned with the internuclear axis and midpoint as the origin, the symmetry operations to be performed on the basis set for this point group are E, $2C_{\propto}(\phi)$, $\propto \sigma_v$, i. $2S_{\propto}(\phi)$, $\propto C_{2'}$. Using z (and hence ξ) as the base function, transformation properties readily show this to belong to the IR, Σ_u^+. Use of $e^{-\xi^2/2}$ as the base function leads to its classification under the totally symmetric IR, Σ_g^+. The reason for this assignment is that the representation with ξ^2 as the base can be expressed as a direct product representation. $\Gamma_{\xi^2} = \Gamma_\xi \otimes \Gamma_\xi$ which is just $\Sigma_u^+ \otimes \Sigma_u^+ = \Sigma_g^+$

Example 7.2.

What are the symmetries of the vibrational wavefunctions of the N_2 molecule in the quantum states n=2 and 3 respectively.

N_2 molecule belongs to the point group $D_{\propto h}$ and the relevant vibrational wavefunctions excluding the normalising factors are

$$\begin{aligned} \psi_{2'} &= e^{-\xi^2/2}(4\xi^2 - 2) \quad \text{for n} = 2 \\ \text{and } \psi_{3'} &= e^{-\xi^2/2}(8\xi^3 - 12\xi) \quad \text{for n} = 3 \end{aligned}$$

The first function $= e^{-\xi^2/2} \times 4\xi^2 - 2e^{-\xi^2/2}$. Using our knowledge that both $e^{-\xi^2/2}$ and ξ^2 transform as Σ_g^+, the representation with the product function $\left(e^{-\xi^2/2} \cdot 4\xi^2\right)$ as base will be, according to the direct product rule, an IR, Σ_g^+ The second part $-2e^{-\xi^2/2}$ also transforms as Σ_g^+. Hence the IR of $\psi_{2'}$ is Σ_g^+.

Turning to $\psi_{3'}$ it is seen that while $e^{-\xi^2/2}$ transforms as Σ_g^+, $8\xi^3$ and 12ξ, imitating the behaviour of z^3 and z respectively, transform as $\Sigma_u{}^+$. The net set of resulting characters indicate the IR to be $\Sigma_u{}^+$.

It will be noted from the discussion and the examples that the vibrational wavefunctions of diatomic heteronuclear molecules (point group $C_{\infty v}$) are all of Σ^+ symmetry and those of homonuclear ones (point group $D_{\infty h}$) alternate between $\Sigma_g{}^+$ and $\Sigma_u{}^+$. The symmetries of the vibrational states of polyatomic molecules are discussed in Chapter 8.

2. Molecular orbitals

(a) Quantum Mechanical aspects: The treatment of molecular orbitals has been made at various levels of sophistication. These include Huckel $\pi-$MO's, SCFπMO's, SCF MO's of Hartree and Fock, SCFLCAO MO's of Roothaan with numerous parametric modifications, viz., CNDO, INDO, PNDO. NDDO and MINDO all of which involve the concept of one-electron orbitals. In the LCAO MO theory, each SCF MO is a linear combination of suitable atomic orbitals centred on the atoms constituting the molecules, i.e., $\psi_\mu = \sum_i C_{i\mu}\phi_i$, where ϕ_i's are the atomic orbitais.

The set of self consistent coefficients $C_{i\mu}$'s for any MO ψ_μ is written in the form of a column matrix, called eigenvector matrix, and the basis set of atomic orbitals, ϕ_i's in the form of a row matrix. Thus $\psi_\mu = \mathbf{\Phi C}_\mu$. In the simplified SCF LCAO MO methods, the atomic orbiials that are taken into consideration for linear combination are limited to the valence shells of the atoms and the basis is called valence shell basis.

(b) Symmetry aspects: Even with a complete lack of knowledge of the intricacies of the quantum mechanical methods for finding the SCF LCAO MO's, one can use the principle of symmetry to predict what IR's would characterise the MO's that emerge from the quantum mechanical treatment. The only information that one uses in doing so is the set of atomic orbitais that are to be linearly combined. The principle of determining the IR's depends upon the transformation properties and the subsequent follow-up [cf. recipe (i), (ii) and (iii)] or the use of the projection operator technique. Adequate illustrations for assignment of symmetry tags to the crystal field orbitals or group orbitals have already been given in many worked out examples (Examples 5.2, 5.4, 5.10, 5.11, 5.13, 5.14, 6.4). The principle of labelling a molecular orbital is not anything different. To spare ourselves the monotony, we do not intend to multiply instances here except treating one example selected from the CNDO SCF LCAO MO domain.

Example 7.3.

One of the twelve CNDO SCF LCAO molecular orbitals of the planar hydrazine molecule, built up with the valence shell atomic orbitals comprising the eight 2s, $2p_x$, $2p_y$, $2p_z$ orbitals on the two N-atoms and the four 1s orbitals of the four H-atoms, has the eigenvector matrix given by:

$$\begin{pmatrix} -0.4627 \\ -0.0000 \\ -0.0000 \\ 0.0743 \\ 0.4627 \\ 0.0000 \\ 0.0000 \\ 0.0743 \\ 0.3744 \\ 0.3744 \\ -0.3744 \\ -0.3744 \end{pmatrix}$$

The elements of the matrix are sequentially associated with the atomic orbitals designated serially by ϕ_1, ϕ_2, ϕ_3, ϕ_4, ϕ_5, ϕ_6, ϕ_7, ϕ_8, ϕ_9, ϕ_{10}, ϕ_{11} and ϕ_{12}. The last four represents the four H−1s orbitals and the first eight ϕ's are those of 2s, $2p_x$, $2p_y$ and $2p_z$ on the first N-atom and an identical set in identical sequence on the second N-atom. What is the IR of this molecular orbital?

We note first of all that N_2H_4 molecule belongs to D_{2h} point group. Let us set up the following coordinate axes for the molecular system where the atoms are numbered as mentioned in the problem (see Fig. 7.2).

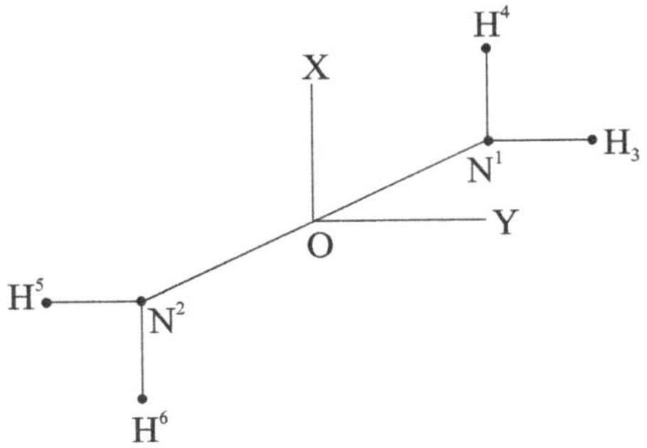

Fig. 7.2: Axial system in N_2H_4

Here N^2N^1 line forms the z-axis in the plane of the paper containing also the y-axis. The positive direction of the x-axis points vertically upwards.

We may lighten the burden of the problem by just satisfying ourselves with the knowledge of the way the four hydrogen orbitals combine amongst themselves under the different projection operators of the IR's. Of these the one which tallies with the combination as inferred from the eigenvector matrix, viz., ϕ_9, ϕ_{10} with coefficients opposite in sign to those preceding ϕ_{11} and ϕ_{12}, will at once tell us the IR of the whole *mo*. We may either use the four orbital basis set (ϕ_9 ϕ_{10} ϕ_{11} ϕ_{12}) or even a single base function, ϕ_9 since the latter will mix up with the other H-orbitals under the group operations of D_{2h}. The transformation properties with ϕ_9 as the base function are given below.

	$\widehat{E}$	$\widehat{C_2(z)}$	$\widehat{C_2(y)} = \sigma_h$	$\widehat{C_2(x)}$	i	$\widehat{\sigma^{xy}}$	$\widehat{\sigma^{xz}}$	$\widehat{\sigma^{yz}}$
ϕ_9	ϕ_9	ϕ_{10}	ϕ_{12}	ϕ_{11}	ϕ_{11}	ϕ_{12}	ϕ_{10}	ϕ_9

The projection operators for the different IR's of D_{2h} are $\wp A_g$, $\wp B_{1g}$, $\wp B_{2g}$, $\wp B_{3g}$, $\wp A_u$, $\wp B_{1u}$, $\wp B_{2u}$ and $\wp B_{3u}$. When these are constructed and applied on ϕ_9 separately, it is found that only $\wp B_{1u}$ gives the right combination for ϕ_9, ϕ_{10}, ϕ_{11} and ϕ_{12} in respect of their signs as indicated in the eigenvector matrix. The calculations with two projection operators $\wp B_{2g}$ and $\wp B_{1u}$ are shown below. The first annihilates the function and the second one projects the desired combination. The other projection operators may also be tried, but these lead to combinations different from the desired one.

$$\{\wp B_{2g}\}\,\phi_9 = \frac{1}{8}\sum_R \chi B_{2g}(R)\hat{R}\phi_9 = \frac{1}{8}\Big[1.\hat{E}\phi_9 - 1.\hat{C}_2(z)\phi_9 + 1.\hat{C}_2(y)\phi_9$$
$$-1.\hat{C}_2(x)\phi_9 + 1.1\phi_9 - 1.\hat{o}^{xy}\phi_9 + 1.\hat{\sigma}^{xz}\phi_9 - 1.\hat{\sigma}^{yz}\phi_9\Big]$$
$$= \frac{1}{8}\left[\phi_9 - \phi_{10} + \phi_{12} - \phi_{11} + \phi_{11} - \phi_{12} + \phi_{10} - \phi_9\right] = 0$$

$$\{\wp B_{lu}\} = \frac{1}{8}\Big[1.\hat{E}\phi_9 + 1.\hat{C}_2(z)\phi_9 - 1.\hat{C}_2(y)\phi_9 - 1.\hat{C}_2(x)\phi_9 - 1.i\phi_9$$
$$-1.\hat{\sigma}^{xy}\phi_9 + 1.\hat{\sigma}^{xz}\phi_9 + 1.\hat{\sigma}^{yz}\phi_9]$$
$$= \frac{1}{8}\left[\phi_9 + \phi_{10} - \phi_{12} - \phi_{11} - \phi_{11} - \phi_{12} + \phi_{10} + \phi_9\right]$$
$$= \frac{1}{4}\left[\phi_9 + \phi_{10} - \phi_{11} - \phi_{12}\right] \text{ which is a correct combination.}$$

The IR of the molecular orbital is thus found to be B_{lu}.

The fact that $\{\wp B_{2g}\}\,\phi_9$ is zero shows that if there be any molecular orbital of IR B_{2g} of hydrazine molecule it cannot have the hydrogen orbitals in the linear combination. It should comprise some of the orbitals centred on the two N-atoms. Further calculation shows that a B_{2g} molecular orbital consisting of two $2p_x$ atomic orbitals does exist for the molecule. So lightening the burden of the problem carries the risk of generating wrong notion about the occurrence of molecular orbitals of another symmetry, viz., B_{2g}! A correct method would have been to employ the projection operators on five base functions, say, ϕ_1, ϕ_2, ϕ_3, ϕ_4 and ϕ_9

Chapter 8

Molecular Vibrations, Normal Co-Ordinates, Selection Rules-Infra Red And Raman Spectra

8.1 General Remarks

The vibrations of polyatomic molecules are very complex and do not permit of any simple approach for understanding the nature of infra red spectra. Nevertheless induction of symmetry ideas goes a long way towards understanding and analysing the infra red and the vibrational Raman spectra. The prerequisites for any worthwhile treatment of vibrational wave functions and spectra are the concept, the principle of determination and the symmetry-based classification of the normal modes of vibrations of polyatomic molecules and their interconnection with the quantum mechanical vibrational wavefunctions.

But what are normal coordinates of vibrations? The concept of normal modes will gradually unfold itself as one describes the mathematical technique for determining these. While qualitative ideas of normal modes will be given in the initial stages, the formal definition will be deferred until the stage is ripe for it. Before delving into detail of the normal modes, we refer to a matrix theorem without proof. The implications of the theorem are relevant for the mathematical technique used to determine the normal modes.

Suppose we have two homogeneous quadratic functions that are expressible separately as matrix products $\widetilde{\mathbf{X}}\mathbf{AX}$ and $\widetilde{\mathbf{X}}\mathbf{BX}$ where $\mathbf{X}$ represents the column matrix of the variables (coordinates) and $\mathbf{A}$ and $\mathbf{B}$ are two real symmetric matrices. It is usually possible to effect a change of variables $\mathbf{X}=\mathbf{PY}$, such that the matrix $\mathbf{P}$ changes $\mathbf{A}$ and $\mathbf{B}$ to diagonal forms by an orthogonal transformation. As a result the original quadratic forms are transformed into a sum of squares involving the new variables $\mathbf{Y}$. [One may recall a some what parallel process of block diagonalisation of a reducible representation by a change of basis (Sec. 4.6; 5.6)]. A theorem proposes indirectly the way to determine the vital matrix $\mathbf{P}$.

Theorem 8.1.

*If **A** and **B** be two real symmetric matrices, then the roots λ (all distinct) of the equation* $|\mathbf{A}-\lambda\mathbf{B}|=0$ *are such that* $\widetilde{\mathbf{P}}\boldsymbol{AP}$ *and* $\widetilde{\mathbf{P}}\boldsymbol{BP}$ are both diagonal matrices. $\widetilde{\mathbf{P}}\boldsymbol{AP}=diag.\ (\lambda_1\ \lambda_2......\lambda_n)=\Lambda$ *and* $\widetilde{\mathbf{P}}\boldsymbol{BP}=\boldsymbol{1}_n$ a *unit matrix.*

To help grasp the contents of the foregoing paragraph we briefly indicate a suitable illustration. Consider two homogeneous quadratic forms $-y^2+2yz-2xy$ and $2x^2+2y^2+3z^2-4yz-4xz+2xy$.

$$\text{Here } \mathbf{X}=\begin{pmatrix} x \\ y \\ z \end{pmatrix} \mathbf{A}=\begin{pmatrix} 0 & -1 & 0 \\ -1 & -1 & 1 \\ 0 & 1 & 0 \end{pmatrix} \text{ and } \mathbf{B}=\begin{pmatrix} 2 & 1 & -2 \\ 1 & 2 & -2 \\ -2 & -2 & 3 \end{pmatrix}$$

as are evident from the numerical coefficients. It is thus possible to write

$$(-y^2+2yz-2xy)=(x\ y\ z)\begin{pmatrix} 0 & -1 & 0 \\ -1 & -1 & 0 \\ 0 & 1 & 0 \end{pmatrix}\begin{pmatrix} x \\ y \\ z \end{pmatrix} = \widetilde{\mathbf{X}}\mathbf{AX}$$

$$(2x^2+2y^2+3z^2-4yz-4zx+2xy)$$
$$=(x\ y\ z)\begin{pmatrix} 2 & 1 & -2 \\ 1 & 2 & -2 \\ -2 & -2 & 3 \end{pmatrix}\begin{pmatrix} x \\ y \\ z \end{pmatrix} = \widetilde{\mathbf{X}}\mathbf{BX}$$

To find the matrix **P**, we take a hint from Theo. 8.1 and try to find the eigenvalues and the normalised eigenvectors of the matrix $(\mathbf{A}-\lambda\mathbf{B})$ by setting the determinant $|\mathbf{A}-\lambda\mathbf{B}|=0$ (cf. Sec. 3.3). The eigenvalues in this case are=0, 1 and −1 and the corresponding

$$\text{eigenvectors } \mathbf{C} \text{ are } \begin{pmatrix} \frac{1}{\sqrt{2}} \\ 0 \\ \frac{1}{\sqrt{2}} \end{pmatrix}, \begin{pmatrix} 0 \\ \frac{1}{\sqrt{2}} \\ \frac{1}{\sqrt{2}} \end{pmatrix} \text{ and } \begin{pmatrix} \frac{1}{\sqrt{3}} \\ \frac{1}{\sqrt{3}} \\ \frac{1}{\sqrt{3}} \end{pmatrix} \text{ respectively.}$$

$$\text{The diagonalising matrix } \mathbf{P} \text{ is therefore, (cf. Sec. 3.3) } \begin{pmatrix} \frac{1}{\sqrt{2}} & 0 & \frac{1}{\sqrt{3}} \\ 0 & \frac{1}{\sqrt{2}} & \frac{1}{\sqrt{3}} \\ \frac{1}{\sqrt{2}} & \frac{1}{\sqrt{2}} & \frac{1}{\sqrt{3}} \end{pmatrix}$$

The transformation of variables is $\mathbf{X} = \begin{pmatrix} x \\ y \\ z \end{pmatrix} = \mathbf{PY}$

$$= \begin{pmatrix} \frac{1}{2} & 0 & \frac{1}{\sqrt{3}} \\ 0 & \frac{1}{\sqrt{2}} & \frac{1}{\sqrt{3}} \\ \frac{1}{\sqrt{2}} & \frac{1}{\sqrt{2}} & \frac{1}{\sqrt{3}} \end{pmatrix} \begin{pmatrix} x' \\ y' \\ z' \end{pmatrix}$$

Example 8.1.

Show that the above transformation really leads to a sum of squares of the two quadratic expressions. What are the new variables?

$$x = \left(\frac{1}{\sqrt{2}}x' + \frac{1}{\sqrt{3}}z'\right), \; y = \left(\frac{1}{\sqrt{2}}y' + \frac{1}{\sqrt{3}}z'\right) \text{ and}$$
$$z = \left(\frac{1}{\sqrt{2}}x' + \frac{1}{\sqrt{2}}y' + \frac{1}{\sqrt{3}}z'\right)$$

On substitution in the quadratic expressions, the first turns into

$$-\left[\frac{1}{\sqrt{2}}y' + \frac{1}{\sqrt{3}}z'\right]^2 + 2\left[\frac{1}{\sqrt{2}}y' + \frac{1}{\sqrt{3}}z'\right]\left[\frac{1}{\sqrt{2}}x' + \frac{1}{\sqrt{2}}y' + \frac{1}{\sqrt{2}}z'\right]$$
$$-2\left[\frac{1}{\sqrt{2}}x' + \frac{1}{\sqrt{3}}z'\right]\left[\frac{1}{\sqrt{2}}y' + \frac{1}{\sqrt{3}}z'\right]$$
$$= \frac{1}{2}y^{2'} - \frac{1}{3}z^{2'} = \text{a sum of squares.}$$

Similarly the second quadratic expression, on substitution of the equivalents for x, y and z in it, turns into

$$\frac{1}{2}x'^2 + \frac{1}{2}y^{2'} + \frac{1}{3}z^{2'}, \text{ a sum of squares.}$$

The new set of variables. **Y**, are obtained from $\mathbf{P}^{-1}\mathbf{X}$. This process yields

$$x' = \frac{1}{\sqrt{2}}(x+y), \; y' = \frac{1}{\sqrt{2}}(y+z) \text{ and } z' = \frac{1}{\sqrt{3}}(x+y+z)$$

Of the two quadratic functions, if one is a function of x, y and z but the other, a quadratic homogeneous function of $\dot{x}$, $\dot{y}$ and $\dot{z}$, the foregoing process of reduction to diagonal forms will still apply. It is this principle of which advantage is taken in determining the normal modes (Sec. 8.2)

8.2 Vibrations of Molecules. Normal Modes Of Vibrations.

1. Classical Mechanical Aspects (a) Qualitative description:

The motions of a molecule are translational, rotational and vibrational in nature. In vibrational motions the atoms are displaced from their mean equilibrium positions giving rise to potential and kinetic energies which are fluctuating with time. But due to intramolecular forces and coupling of motions, the vibrational motion of a molecule assumes a very complex nature. Thus were it possible to make a direct observation of the molecular vibratory motion, it would look as an apparently aperiodic one with no display of rythm. We may be tempted to compare this complex vibrational motion, a function of the configurational coordinates of the constituent atoms, with a complex function mathematically amenable to a splitting into Fourier series terms. Viewed in this sense, the overall vibratory motion of a molecule at any temperature may be looked upon as a superposition of a number of independent simple harmonic modes of vibrations of the molecule as a whole. Each such harmonic mode of vibration, having a characteristic frequency, is called a normal mode of vibration. What then are the qualitative features of such normal modes and how many of these are there for a molecule?

(i) For any such mode of vibration it is reasonable to assume that there would be at any time t a corresponding mathematical displacement by an amount Q of the molecule although the centre of gravity of the molecule remains stationary. This Q, called the normal coordinate or displacement, is related to the then displacements of some or all of the component atoms of the molecule.

(ii) During the execution of vibrations along any normal mode, there exists some harmony in the displaced positions of the participating atoms. All the participating atoms are in their extreme positions simultaneously or mean positions at the same time.

(iii) The total number of normal modes of vibrations for an N-atom nonlinear molecule is (3N-6). [3N-5 for a linear molecule]. Since 3N coordinates are necessary to specify completely the configuration of a molecule, the remaining six modes of motions of a nonlinear molecule are accounted for by three translational and three rotational motions.

(**b**) Quantitative Formulation: The configuration of an N-atom molecule can be fixed by choosing a set of 3N coordinates in different ways. Of these different possible ways, how are we to hit upon the way so that (3N-6) of the chosen coordinates correspond to vibratory normal coordinates? Certainly one cannot do it at the very start, but manages to arrive at the lot starting from some conveniently chosen set of reference axial systems. Consider N-number of axial systems (say cartesian) having the equilibrium positions of the atoms as the origins. The displacements of N-atoms at any time when resolved in the three component axial directions are, say, ξ_1, ξ_2, ξ_3, ...ξ_i....ξ_{3N} where ξ_1, ξ_2, ξ_3 are those associated with first atom, ξ_4, ξ_5 and ξ_6 those with the second and so on. To simplify calculations, a mass-weighted set of coordinates q_1, q_2, q_3....q_i.....q_{3N} are used where $\sqrt{m_i}\xi_i = q_i$. This set of $q_i's$, therefore, completely defines the configuration of the molecule.

It is now necessary to find general expressions for both the potential and the kinetic energies as these are the quantities which occur in the hamiltonian function and hence the Lagrangian needed to formulate the classical equations of motion. Since potential energy is a function of the positions of the atoms we can, assuming small displacements, express it in a Taylor series.

$$\begin{aligned} V(...\xi_i...\xi_j...) &= V_0(...\xi_i...\xi_j) + \sum_i \left(\frac{\delta V}{\delta \xi_i}\right)_0 \xi_i \\ &+ \frac{1}{2}\sum_i \sum_j \left(\frac{\delta^2 V}{\delta \xi_i\, \delta \xi_j}\right)_0 \xi_i \xi_j + \cdots\cdots \end{aligned} \tag{8.1}$$

Setting the zero of potential energy scale to coincide with the absolute value when the atoms are in their equilibrium positions (i.e., $V_0 = 0$), assuming complete harmonic motion $\left[\text{i.e., } \left(\frac{\delta V}{\delta \xi_i}\right)_0 = 0\right]$ and neglecting expansion terms beyond the third in (8.1), we have

$$V(...\xi_i...\xi_j...) = \frac{1}{2}\sum_i \sum_j \left(\frac{\delta^2 V}{\delta \xi_i\, \delta \xi_j}\right)_0 \xi_i \xi_j = \frac{1}{2}\sum_{i,\,j} f_{ij}\xi_i\xi_j \tag{8.2}$$

$$\text{or } V(...q_i...q_j...) = \frac{1}{2}\sum_i \sum_j \left(\frac{\delta^2 V}{\delta_{qi}\, \delta_{qj}}\right) q_i q_j = \frac{1}{2}\sum_{i,\,j} F_{ij} q_i q_j \tag{8.3}$$

where F_{ij} =(ij)th element of the force constant matrix

$$= \frac{1}{\sqrt{m_i m_j}} f_{ij}, \text{ since } q_i = \sqrt{m_i}\xi_i$$

Similarly the kinetic energy, T, when free from explicit time dependence can be written as

$$T = \frac{1}{2}\sum_{jk} a_{jk}\dot{\xi}_j\dot{\xi}_k \tag{8.4}$$

$$\text{where } a_{jk} = \sum_i m_i \left(\frac{\overrightarrow{\delta r_i}}{\delta\xi_j}\frac{\overrightarrow{\delta r_i}}{\delta\xi_k}\right)$$

$$\begin{aligned}
\therefore T &= \frac{1}{2}\sum_{jk}\left[\sum_i m_i \left(\frac{\overrightarrow{\delta r_i}}{\delta q_j}\frac{\overrightarrow{\delta r_i}}{\delta q_k}\right)\sqrt{m_j m_k}\frac{\dot{q}_j\dot{q}_k}{\sqrt{m_j m_k}}\right] \\
&= \frac{1}{2}\sum_{jk}\left[\sum_i m_i \left(\frac{\overrightarrow{\delta r_i}}{\delta q_j}\frac{\overrightarrow{\delta r_i}}{\delta q_k}\right)\right]\dot{q}_j\dot{q}_k \\
&= \frac{1}{2}\sum_{jk} M_{jk}\dot{q}_j\dot{q}_k
\end{aligned} \tag{8.5}$$

Here M_{jk} = an element of Inertia matrix= $\frac{a_{jk}}{\sqrt{m_j m_k}}$
It is thus seen from Eqs. (8.3 and 8.5) that both the potential and the kinetic energies are homogeneous quadratic functions of the coordinates q_i's and velocities $\dot{q}_i$'s respectively involving the force constant matrix **F** and the inertia matrix **M**. Theorem 8.1 (cf. con cluding portion of Sec. 8.1) enables simultaneous diagonalisation of **F** and **M** making it possible for both V and T to be expressed as summation of squared terms only in Q_i's and in $\dot{Q}_i$'s respectively. This process* essentially means a change of variables from q_i's to Q_i's, the new set of normal coordinates and Q_i's are merely suitable linear combinations of the q_j's (cf. Example 8.1). We have thus

$$V = \frac{1}{2}\sum_i \lambda_i Q_i^2 \text{ and } T = \frac{1}{2}\sum_i \dot{Q}_i^2 \tag{8.6}$$

In almost all cases of applications of symmetry ideas to vibration, one will not be required to find explicit forms of Q_i's in terms of q_j's. Nevertheless should one be aware of the underlying principle of the change–over to the normal coordinates. We shall not pursue the classical mechanics of vibrations any further and conclude with the formal definition of a normal coordinate.

"A normal coordinate of a vibrating molecule is a displacement Q which is some suitable linear combination of the corresponding the then mass weighted atomic displacements q_1,

$q_2 \dots q_i \dots$ such that the molecular motion along this mathematical coordinate Q is a simple one dimensional harmonic motion with potential energy $\frac{1}{2}\lambda Q^2$". Of the 3N Q's so found out for an N-atom nonlinear molecule, (3N-6) will represent normal coordinates and the remaining six, corresponding to eigenvalues $\lambda = 0$ each, represent translational and rotational displacements and are still (euphemistically) called normal coordinates.

The above definition of normal mode of a molecule has been coined from a more general definition valid for a solid. An extended solid, whose constituents are held together through electrical forces, is capable of vibrations at different definite frequencies with synchronous oscillations of its different parts. Each of such definite frequency vibrations is termed a normal mode. The most general motion of such a solid is a linear super position of some or all of the possible normal modes of vibrations. This motion of normal mode may be extended to the standing waves of an elastic stretched string or to an enclosed volume of air in an organ-pipe. Inclusion is also thus feasible for an electrically bound molecule or a complex ion.

* To have recourse to Theo. 8.1 for diagonalisation is an alternative equivalent to solving for Q_i's using Lagrange's equations of motions and hence transformation to new coordinates.

2. Wave Mechanical Aspects:

If $\psi(Q_1, Q_2, \dots Q_{3N-6})$ represents the vibrational wavefunction of a polyatomic molecule, where Q_i's are the normal coordinates, then the Schrodinger equation $H\psi = E\psi$, with E as the total vibrational energy, can be resolved into a number of component equations. Assuming $\psi = \prod_i^{(3N-6)} \phi_i(Q_i)$, a product of component wave-functions, $\phi_i(Q_i)$, each of which is just a function of one normal coordinate only and $E = \sum_i \in_i$, where $\in_i$ is the vibrational energy due to harmonic vibration along the normal mode Q_i, the wave equation, with the concurrent use of Eq. (8.6), takes the form

$$\sum_i \left(-\frac{\hbar^2}{2} \frac{\partial^2}{\partial Q_i^2} + \frac{1}{2}\lambda_i Q_i^2 \right) \prod_i \phi_i(Q_i) = \sum \in_i \prod_i \phi_i(Q_i) \qquad (8.7)$$

On dividing by $\prod_i \phi_i(Q_i)$ and remembering $\phi_i(Q_i)$ to be a function of Q_i only and independent of the other normal modes, Eq. (8.7) splits into

(3N-6) number of component equations, one for each normal coordinate, viz.,

$$\begin{aligned} \frac{1}{\phi_i}\left(-\frac{\hbar^2}{2}\frac{\partial^2}{\partial Q_i^2}+\frac{1}{2}\lambda_i Q_i^2\right)\phi_i &= \in_i \\ \text{i.e.,}\quad \frac{\partial^2\phi_i}{\partial Q_i^2}+\frac{2}{\hbar^2}\left(\in_1-\frac{1}{2}\lambda_i Q_i^2\right)\phi_i &= 0 \end{aligned} \tag{8.8}$$

The solutions of Eq. (8.8) are the Hermite orthogonal functions of the form

$$\phi_{i(n)} = \left(\frac{\sqrt{\beta/\pi}}{2^n n!}\right)^{\frac{1}{2}} e^{-\xi_i^{2/2}} H_n(\xi_i)$$

where the variable $\xi_i = \sqrt{\beta}Q_i$ with $\beta = \frac{2\pi}{\hbar}\nu = \frac{2\pi}{h}\sqrt{\lambda_i}$ and $H_n(\xi_i) = (-1)^n e^{\xi_i^2}\frac{\partial^n}{\partial \xi_i^n}e^{-\xi_i^2}$. This polynomial is the Hermite polynomial of degree n. The energy eigenvalue in the n-th quantum state due to vibration in the i-th mode is $\in_i(n) = \left(n+\frac{1}{2}\right)hv_i$.

The net wavefunction ψ can immediately be written down as a product. When n=0 for all modes, the state is called ground state, i.e., $\psi_0 = \phi_1(0).\ \phi_2(0).....\phi_i(0)....\phi_{3N-6(0)}$. The first excited state is one where vibrational excitation of just one normal mode, say Q_j, has taken place with all the other modes remaining in their respective zeroth quantum level. Such a state is called a fundamental state of the molecule and may be represented as $\psi_{f(j)} = \phi_1(0).\ \phi_2(0)...\phi_i(0)...\phi_j(1)...\phi_{3N-6(0)}$. It is evident there are other possible fundamental states depending on the mode which is excited.

Perturbation theory tells us that the electric dipole transition from one state to another takes plate if the transition moment integral

$$\int \psi_{m^*}\mu\psi_n d\tau \neq 0 \tag{8.9}$$

Here μ =dipolemoment= $e\overrightarrow{r} = e(\overrightarrow{x}+\overrightarrow{y}+\overrightarrow{z})$, ψ_{m^*} is the complex conjugate of the wavefunction of the final state and ψ_n, the initial one. A transition of a molecule from the ground vibrational state to one of the fundamental states, in which the j-th normal mode is excited, i.e., $\psi_{f(j)} \leftarrow \psi_0$, is termed a fundamental transition. We may thus have many fundamental transitions depending on what mode is excited. However, the number of fundamental transition bands observed in the vibrational spectra is not necessarily equal to the number of normal coordinates of vibrations. It may be less than this depending on the degeneracies of the modes. In fact, selection rule permitting it, the number is theoretically equal to the number of distinct λ (force constant) eigenvalues obtainable in the diagonalisation of the potential energy matrix (Sec. 8.2).

If one mode is doubly excited from the ground state, the transition is called the first overtone of the corresponding fundamental; if, however, the mode is triply excited, the state and the transition are called the second overtone.

Combination tones are the transitions in which two or more modes are simultaneously excited from the ground vibrational level to the fundamental levels, i.e., with both Q_i and Q_j modes being excited from n=0 to n=1.

8.3 Normal Modes of Vibrations. Symmetry Aspects

The process of working out quantitative expressions for the normal co-ordinates, Q_i's. involving the diagonalisation technique is a laborious one. However, even a lack of quantitative knowledge of Q_i's does not preclude us from obtaining useful information about vibrational spectra if we only know the symmetry species of the normal coordinates. To intimate the readers we may first state a result of prime importance: '*Each normal coordinate, Q_i, of a molecule forms a basis of an IR of the point group of the molecule. In case of degenerate vibrations, the associated normal coordinates. Q_j, Q_k..., will together transform as an IR of the molecular point group*'. The analogy to wavefunctions or molecular orbitals is immediately uppermost in one's mind. While a general proof proceeds along some tortuous path, this can be established more readily for specific cases where the expressions of Q's are known. We shall, however, accept the generalisation without offering any proof. The symmetry of the molecule sneaks into the discussion of its vibrations through the opening provided by the above principle and is fully exploited in the following applications.

(1) To find the distribution of the normal coordinates over the IR's i.e., how many of the normal coordinates belong to an IR and which of the IR's of the molecular point group are relevant to vibrations.
(2) To find the number of fundamentals (and hence degeneracies) and

also to predict the selection rules for infra-red fundamental and Raman fundamental transitions.

(3) To have qualitative ideas of the nature of deformations (stretching of bonds, deformations of angles) during vibrations along different normal modes belonging to different IR's.

(4) To explain facts connected with overtones and combination tones.

We shall illustrate these by a few examples first confining the use of symmetry to application no 1 and then gradually enlarging its scope to cover 2, 3 and 4. Relevant and necessary discussions will be made in between to help greater appreciation. The applications in the case of linear molecules will be taken up later (Sec. 8.4).

Example 8.2.

To which IR's do the normal coordinates of vibrations of BCl_3 molecule belong?

The BCl_3 molecule belongs to D_{3h} point group and is represented in Fig. 8.1 in which the atoms are numbered 1, 2, 3 and 4. Let d_1, d_2, d_3 and d_4 (Fig. 8.1) denote the respective displacements of the atoms from their equilibrium positions at some instant. Each of these displacements can be split into three cartesian components in the coordinate frames set up at each of the atoms. Let the X and Y axes of all the coordinate frames be in the plane of the molecule and all these are in parallel orientation. The positive directions of all the Z-axes point upwards. Let e_1, e_2 and e_3 represent the unit displacement vectors along the cartesian axes set up at atom no. 1, ξ_1, ξ_2, and ξ_3 the respective components of displacement d_1 and q_1, q_2, q_3 their mass-weighted magnitudes. Similarly e_4, e_5, e_6 and ξ_4, ξ_5, ξ_6 and q_4, q_5, q_6 are respectively the unit vectors, components of displacement d_2 and mass weighted coordinate values associated with atom 2. Corresponding sets of quantities exist for atoms 3 and 4 also (Fig. 8.1).

We now try to find the representation of the group D_{3h} in the displacement space spanned by e_1 e_2 e_3 e_4...$e_{11}e_{12}$. Since by now we have grown sufficiently adept in anticipating transformation effect, we skip writing detailed transformation result here. Instead we merely find the contributions to the overall character value by the individual basis vectors by noting their shifts (or nonshifts) under the symmetry group

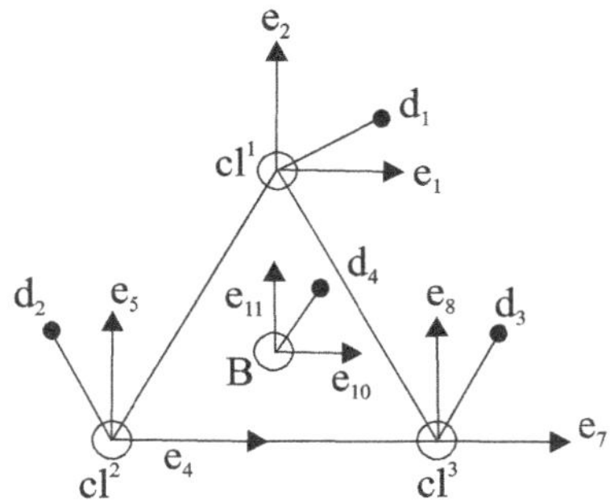

Fig. 8.1: Displacements of the atoms in BCl_3 molecule resolved along cartesian axes.

operations (Sec. 5.8) [Readers experiencing difficulty in following the data should first write out the full transformation results to confirm the entries in the table]. One must remember that while the symmetry operations are applied only the vectors are affected, the atoms (I to 4) stay put in their original positions as in Fig. 8.1.

Contribution to characters by the basis set (e_1, e_2...$e_{11}e_{12}$) under symmetry operations of D_{3h}.

	$\hat{E}$	$\hat{C}_3$	$\hat{C}_3^2$	$\hat{C}_{2'}$	$\hat{C}_{2''}$	$\hat{C}_{2'''}$	$\hat{\sigma}_h$	$\hat{S}_3$	$\hat{S}_3^5$	$\hat{\sigma}_{v'}$	$\hat{\sigma}_{v''}$	$\hat{\sigma}_{v'''}$
e_1	1	0	0	-1	0	0	1	0	0	-1	0	0
e_2	1	0	0	1	0	0	1	0	0	1	0	0
e_3	1	0	0	-1	0	0	-1	0	0	1	0	0
e_4	1	0	0	0	$\frac{1}{2}$	0	1	0	0	0	$\frac{1}{2}$	0
e_5	1	0	0	0	$-\frac{1}{2}$	0	1	0	0	0	$-\frac{1}{2}$	0
e_6	1	0	0	0	-1	0	-1	0	0	0	1	0
e_7	1	0	0	0	0	$\frac{1}{2}$	1	0	0	0	0	$\frac{1}{2}$
e_8	1	0	0	0	0	$-\frac{1}{2}$	1	0	0	0	0	$-\frac{1}{2}$
e_9	1	0	0	0	0	-1	-1	0	0	0	0	1
e_{10}	1	$-\frac{1}{2}$	$-\frac{1}{2}$	-1	$\frac{1}{2}$	$\frac{1}{2}$	1	$-\frac{1}{2}$	$-\frac{1}{2}$	-1	$\frac{1}{2}$	$\frac{1}{2}$
e_{11}	1	$-\frac{1}{2}$	$-\frac{1}{2}$	1	$-\frac{1}{2}$	$-\frac{1}{2}$	1	$-\frac{1}{2}$	$-\frac{1}{2}$	1	$-\frac{1}{2}$	$-\frac{1}{2}$
e_{12}	1	1	1	-1	-1	-1	-1	-1	-1	1	1	1
χ^{red}	12	0	0	-2	-2	-2	4	-2	-2	2	2	2

These character values in the basis set ($e_1\ e_2 \ldots e_{11}\ e_{12}$) are also the same as will occur in the transformations of coordinates ($\xi_1,\ \xi_2 \ldots \xi_{11} \xi_{12}$) and hence of ($q_1\ q_2 \ldots q_{11} q_{12}$). Since the normal coordinates $Q_1,\ Q_2 \ldots Q_{11},\ Q_{12}$ are connected, via q_i's, with the displacement $\xi_1,\ \xi_2 \ldots \xi_{11},\ \xi_{12}$ in the form of linear relations, the matrix representations of $D_{3\hbar}$ in the ξ basis and in the **Q** basis are related through a similarity transformation (Sec. 4.3). The trace of a matrix being invariant under a similarity transformation (Theo. 4.2), it is evident that the trace values of the reducible representation compiled above are also those in the space of the normal coordinates, i.e., the basis set **Q**. Now employing the relation $n_i = \frac{1}{g} \sum\limits_R \chi_{i^*}(R) \chi^{red}(R)$ of Theorem 5.8 and the $\chi^{red}(R)$ data, we find the occurrence of the following IR's.

$$
\begin{aligned}
n_{A1}{}' &= \frac{1}{12}[1.12 + 3.1.(-2) + 1.4 + 2.(-2) + 3.1.2] = 1 \\
n_{A2}{}' &= \frac{1}{12}[1.12 + 3.(-1).(-2) + 1.4 + 2.1.(-2) + 3.(-1).2] = 1 \\
n_{A1}{}'' &= \frac{1}{12}[1.12 + 3.1.(-2) - 1.4 + 2.(-1).(-2) + 3.(-1).2] = 0 \\
n_{A2}{}'' &= \frac{1}{12}[1.12 + 3.(-1).(-2) - 1.4 + 2.(-1).(-2) + 3.1.2] = 2 \\
n_{E}{}' &= \frac{1}{12}[2.12 + 2.4 + 2.(-1).(-2)] = 3 \\
n_{E}{}'' &= \frac{1}{12}[2.12 - 2.4 + 2.1.(-2)] = 1
\end{aligned}
$$

It is thus found $\Gamma^{red} = A_1 \oplus A_2 \oplus 2A_2{}'' \oplus 3E' \oplus E''$. Of these IR's, those representing translational displacements in the three directions can either be found out independently or can be ascertained from the character table. These are E' and $A_2{}''$. Similarly using the rotational vectors R_x, R_y and R_z, as the basis, the IR's spanned by these are detected to be E'' and $A_2{}'$. Thus deleting one E', one $A_2{}''$, E'' and $A_2{}'$ from the set of IR's obtained before, we are left with the IR's, viz., $A_1{}'$, $A_2{}''$, $2E'$ which represent the six normal modes of vibrations of the BCl_3 molecule. Since two E' representations occur, there are two distinct pairs of degenerate vibrations associated with four of the six normal coordinates. The other two, $A_1{}'$ and $A_2{}''$, are non-degenerate vibrations characterized by two different λ values.

In finding the character values of the reducible representation of the foregoing example, we could have used one typical group element of each class instead of considering all the elements of $D_{3\hbar}$ since the characters of all the elements of a particular class are the same. In the examples to follow we shall utilize just one typical element for ascertaining the character.

Example 8.3.

Find the symmetry species (i.e., IR's) of the normal coordinates of vibrations of $[CoF_6]^{3-}$ ion.

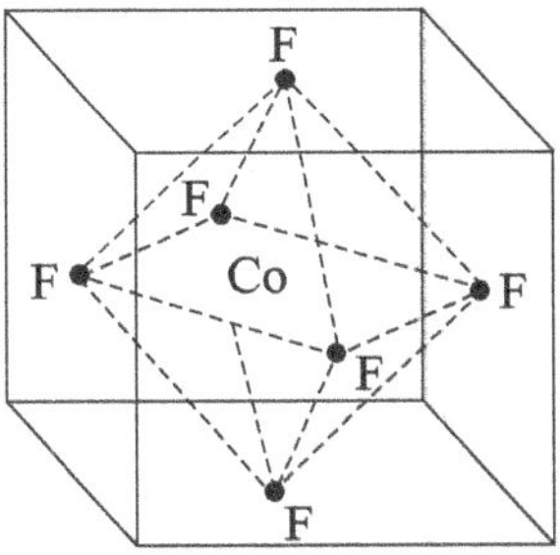

Fig. 8.2: Octahedral $[CoF_6]^{3-}$ ion inscribed in a cube.

This ion is of $O_\hbar$ point group and possesses 21 normal modes including translations and rotations. Setting up the cartesian reference frames at each of the atoms and numbering these 1 to 21 serially, we can find the effect of the classes of symmetry operations upon the twentyone unit vectors along the cartesian axes. The operations are E, $8C_3$, $6C_4$, $3C_2$, $6C_2'$, i, $3\sigma_\hbar$, $8S_6$, $6S_4$ and $6\sigma_d$. Consider the complex ion inscribed in a cube (Fig. 8.2) where the ligand atoms occupy the mipoints of the faces of the cube.

Transformations	Effect	Character
E	All vectors unshifted	21
$8C_3$	Shift of all vectors. Effective shifts of X, Y and Z vectors at Co through $\pi/2$ each.	0
$6C_4$	One suitably chosen C_4 leaves the Z-vectors of the apical atoms and at Co atom unchanged. Contribution due to $\pi/2$ shifts of X and Y vectors is zero.	3

Transformations	Effect	Character
$3C_2$	It leaves Z-vectors at the apical atoms and at Co atom unchanged, but inverts X and Y vectors.	−3
$6C_2'$	All the vectors are shifted but the Z-vectors at the Co atom is inverted. Contribution from shifts of X and Y vectors at Co is zero.	−1
$3\sigma_\hbar$	One suitable $\sigma_\hbar$ inverts the Z-vectors, but leaves X and Y vectors unchanged at the Co atom and at the four square planar atoms.	5
i	All the vectors are shifted, but those at Co atom are only inverted.	−3
$8S_6$	All vectors are shifted	0
$8S_4$	Only the Z-vector at the Co atoms is inverted.	−1
$6\sigma_d$	A suitable σ_d keeps the Y-vectors at Co atom and at two square base F atoms unchanged, all the others are affected.	3

The above results are summarized below:

1	E	$8C_3$	$6C_4$	$3C_2$	$6C_2{}'$	$3_{\sigma\hbar}$	i	$8S_6$	$6S_4$	$6\sigma_d$
χ^{red}	21	0	3	−3	−1	5	−3	0	−1	3

The rest part of the job is a mechanical one of feeding the character values of the reducible representation along with those of the IR's of O_h point group into the relation $n_i = \frac{1}{g}\sum_R \chi^*i(R)\chi^{red}(R)$. One is thus led to final results which show that the normal coordinates are spread over the following IR's.

$$\Gamma^{red} = A_{1g} \oplus E_g \oplus 3T_{1u} \oplus T_{1g} \oplus T_{2g} \oplus T_{2u}$$

deleting the IR's T_{1u} (one number) and T_{1g} for the translation and rotational motions, the normal coordinates of vibrations have symmetry species A_{1g}, E_g, $2T_{1u}$, T_{2g} and T_{2u} associated with six distinct values of λ and hence of distinct frequencies. There are four triply degenerate sets of vibrations of which two distinct sets (frequencies different) belong to T_{1u} symmetry.

Internal Coordinates As Basis Sets:

What we have accomplished in the foregoing examples is to find the characters of the representation in the displacement space carried by e_1, e_2, e_3, e_4.... set and then to argue about the nonchangeability of the characters when the representation is made in the normal coordinate space, this being the same displacement space with a different basis set. To simplify matters one can use internal coordinates, viz., changes in bond lengths and deformations of interbond angles instead of the displacements of the atoms as the basis set. True one would need 3N number of internal coordinates to cover all the degrees of freedom. But we are generally interested in vibrations and can work with (3N-6) number of suitably chosen internal coordinates spanning the vibrational subspace of the total configuration space. The process of choosing such infernal coordinates is to use bond length changes of all the classically bonded atoms and then to select additionally as many independent angle-deformations as will make a total count of (3N-6) including the bond length changes. For example in the NH_3 molecule of point group C_{3v}, there will be six normal modes of vibrations. We may try to represent this point group C_{3v} in the vibrational space spanned by the internal coordinates. These are the three bondlength changes $\vec{S}_1$, $\vec{S}_2$ and $\vec{S}_3$ of the bonds between the N-atom and the hydrogen atoms 1, 2 and 3 respectively. The remaining three internal coordinates are the deformations α_1, α_2 and α_3 of the equilibrium interbond angles, viz., $<H^1NH^2$, $<H^2NH^3$ and $<H^3NH^1$. The vectors $\vec{S}_1$, $\vec{S}_2$ and $\vec{S}_3$. will be collinear with the respective bonds and their transformation properties under symmetry operations are the same as those of the bonds. Similarly the transformation properties of α_1 α_2 and α_3 can he ascertained from the behaviour of the interbond angles under symmetry operations. But never will there be a mix-up of the bond length changes with the deformations of the angles. In other words the matrix representation $\mathbf{D}(R)=\mathbf{D}_s(R)\oplus\mathbf{D}_\alpha(R)$, a direct sum of representations in the spaces spanned by S_1, S_2, S_3 and by α_1, α_2 and α_3.

We now deal with the symmetry species of the normal coordinates of ammonia molecule using the above set of internal coordinates.

Example 8.4.

How are the vibrational modes of NH_3 molecule distributed over the IR's?

The following table may be drawn up for the characters of the reducible representation of the point group C_{3v} in the basis set of the foregoing internal coordinates of NH_3 molecule.

	E	$2C_3$	$3_{\sigma v}$
S_1	1	0	1
S_2	1	0	0
S_3	1	0	0
α_1	1	0	1
α_2	1	0	0
α_3	1	0	0
χ^{red}	6	0	2

On resolution into the IR's of C_{3v}, it is found

$$\begin{aligned} n_{A1} &= \frac{1}{6}[1.6 + 3.1.2.] = 2 \\ n_{A2} &= \frac{1}{6}[1.6 + 3(-1).2] = 0 \\ n_E &= \frac{1}{6}[2.6] = 2 \end{aligned}$$

Hence two normal modes are of species A_1 and there are two distinct degenerate pairs of vibrations of symmetry E associated with the four other normal coordinates.

While the segmentary use of the internal coordinates provides means for applications (see later), it may not be always easy to find the character set of the reducible representations using as the basis all the (3N-6) internal coordinates. As regards simple molecules of C_{2v} and C_{3v} the entire basis set of internal coordinates permits of ready reckoning of contributions to character values. For many others of higher symmetries (e.g., CH_4) the task is not easy. However, in all cases, the use of smaller basis set of internal coordinates (e.g., only the changes in bond lengths) leads to a large amount of qualitative information as to the nature of vibrations of the molecule.

Infra Red And Raman Fundamental Transitions

Selection Rules

We now turn to our second set of applications of symmetry principles in vibrations. Fundamental transitions (sec. 8.2) constitute the most important changes in infra red and Raman spectroscopy. To start with we can predict the symmetry of the ground state vibrational wavefunction. We know $\psi_0 = \prod_i \phi_i(o)(Q_i)$ and $\phi_{i'o}(Q_i)$ is of the type $e^{-\xi_i^{2/2}} H_o(\xi_i)$ where the function is written without the normalization factor and $\xi_i = \sqrt{\beta} Q_i$ (Sec. 8.2). Since $H_o(\xi_i) = 1$ and $e^{-\xi_i^{2/2}}$ transforms as a totally symmetric representation A_1 it turns out that the ground state vibrational wavefunction belongs to the IR, A_1. Now for a fundamental infra red transition involving the normal coordinate Q_j to occur, it is necessary [Eq. (8.9)].

$$\int \psi_{f(j)^*} \mu \psi_0 d\tau \text{ be non zero.}$$

The dipole moment, μ at any time during vibration can be resolved in the directions of $(e_1 e_2 e_3)$ of the coordinate axes set up with the origin at some stationary point under all symmetry operations.

$$\begin{aligned} \text{Now } \mu = \sum_{i=1}^{3} \vec{e_i} \mu_i \text{ where } \mu_1 &= \sum_k Z_k x_k \\ \mu_2 &= \sum_k Z_k y_k \\ \text{and } \mu_3 &= \sum_k Z_k z_k \end{aligned}$$

where x_k, y_k and z_k are the displacements of the atom K of the molecule in the directions of the cartesian axes set up at the equilibrium position of the kth atom and incidentally parallel to e_1, e_2 and e_3 respectively of the molecular axial system, Here ϵ represents the absolute magnitude of electronic change and Z_k, a numerical magnitude carrying its own sign, positive or negative.

$$\begin{aligned} \text{Again } \mu_1 &= \in \sum_k Z_k x_k = \in Z_1 x_1 + \in Z_2 x_2 + ...+ \in Z_k x_k + ... \\ &= \in [x_1^{(1)} + x_1^{(2)} + ... + x_1^{(k)} + ...] \\ &= \in x \end{aligned}$$

where x, the net sum, is some value of the coordinate along the molecular axial direction e_1. The foregoing way of expressing $z_k, x_k = x_1{}^{(k)}$ etc. means the magnitude of $Z_k x_k$(*+ve or −ve*) will have a similar projection of the same magnitude and sign, viz., x_1^k on the parallel e_1 direction of the molecular axis, Like μ_1, similar expressions can be written down for $\mu_2 = \epsilon y$ and $\mu_3 = \epsilon z$.

The transition moment integral should then be

$$\int \left[\phi^*_{j(1)} \Pi_{i \neq j} \phi^*_{i(0)} \in (x + y + z) \Pi_i \phi_{i(o)}\right] d\tau \neq 0$$

for the specific fundamental transition to be feasible. The $\phi's$ in the integrand, it is to be understood, are in their normalised forms.

Since $\phi_i(0)$'s are all orthonormal and fully symmetric, the condition that the fundamental transition be possible demands the survival of the integral viz.,

$$\int \phi^*_{j(1)}(x + y + z)\phi_j(0) d\tau \neq 0.$$

A reference to the condition for the survival of quantum integrals (Sec. 6.4) requires the product $\phi^*_{j(1)} x$ $\left(\text{or } \phi^*_{j(1)} y \text{ or } \phi^*_{j(1)z}\right)$ belong to an IR which is the same as that of the third function $\phi_{j(0)}$, i.e., an IR, A_1. Furthermore, $\phi^*_{j(1)}$, being $e^{-\xi_j^{2/2}} H_1(\xi_j)$, transforms as ξ_j and hence Q_j [cf. Chap. 7]. The conclusion, therefore, emerges that if the normal mode Q_j transforms in the same way as x, y or z in the relevant molecular point group, only then will the A_1 symmetry of the product $\phi^*_{j(1)}$ x[or $\phi^*_{j(1)}$ y or $\phi^*_{j(1)}$z] be ensured permitting a nonzero value of the transition moment integral. The selection rule for infra red fundamental transition may thus be stated in terms of symmetry principle: ***For the occurrence of an infra red fundamental band involving the normal coordinate Q_j, the latter should transform in the same way (i.e., should have the same IR) as anyone, some or all of the vectors x, y, z in the molecular point group.***

The fundamental vibrational Raman transition can take place if the average polarisability of the molecule during transition is different from zero. Since polarisability is proportional to squared terms or binary products of vectors x^2, y^2, z^2, xy, yz, zx it is necessary that the integral $\int \phi^*_j(1) p \phi_j(0) d\tau$, which determines the fundamental Raman transition, survive. Here p is the polarisability operator involving one or some of the terms x^2, y^2, z^2, xy, yz and zx. Arguing in the manner as has been done for infra red transition, **the selection rule for fundamental Raman transition implies that the normal coordinate Q_j should transform in the same way (i.e., should have the same IR) as any or some of the quadratic functions x^2, y^2, z^2, xy, yz, and zx.**

The normal mode, designated by its IR, which undergoes ir or Raman excitation to the fundamental level, is called ir active fundamental or Raman active fundamental respectively.

Quite often both these selection rules are stated differently. The direct product representation of $\left\{\phi^*_{j(1)} r \phi_{j(0)}\right\}$ should contain a totally symmetric component A_1 for infra red activity. Similarly for Raman active fundamental involving the normal coordinate Q_j, the direct product representation of $\left\{\phi^*_{j(1)} r^2 \phi_{j(0)}\right\}$ should on resolution yield an A_1 component. That these two statements are quite equivalent to the previous two will be evident from a follow-up of the contents of Sec. 6.4 dealing with nonvanishing quantum integrals. We are now poised to tackle a few concrete problems of ir and Raman fundamental transitions vis a vis the selection rules.

Example 8.5.

Of the different normal modes of vibrations of H_2O molecule which are ir active and which ones Raman active? How many fundamentals of each type are possible?

In the first place we have to find the IR's of the normal modes [cf. Example 8.2-8.4] and to examine their ir and Raman activity. The characters of the reducible representation of H_2O molecule (C_{2v}) in the basis set of internal coordinates [bond length changes s_1, s_2 of the OH links and α the deformation of the angle HOH] spanning the vibrational subspace of the configuration space are

	E	C_2	$\sigma_v^{(xz)}$	$\sigma_v^{(yz)}$
χ^{red}	3	1	1	3

and its resolution leads to $\Gamma^{red} = 2A_1 \oplus B_2$. Of the three normal coordinates of vibrations two are totally symmetric and one is of B_2 symmetry.

Infra red activity: Since the vectors z and y transform as A_1 and B_2, the latter are ir active fundamentals.

Raman activity: Character table shows x^2, y^2 and z^2 transform as A_1 and yz as B_2. Hence the Raman active fundamentals are $2A_1$ and B_2.

There will thus be two ir active fundamentals and three Raman active fundamentals. The ir active fundamentals will be y and z-directions polarized.

Example 8.6.

Which vibrations of BCl_3 molecule and $[C_0F_6]^{3-}$ ion are ir active and Raman active?

BCl_3 molecule: It is found (Example 8.2) that the normal vibrational coordinates belong to A_1', A_2'' and 2E′. While (x, y) belong to E′ in $D_{3\hbar}$, z transforms as A_2''. Hence the ir active fundamentals are A_2'' and E′.

The Raman active fundamentals are A_1 and E′ as can be inferred from the character table.

$[CoF_6]^{3-}$ ion: The normal coordinates are found to belong to A_{1g}, E_g, $2T_{1u}$, T_{1g}, and T_{2u} (see Example 8.3). The triply degenerate T_{1u} is carried by the basis set (x, y, z). The IR's A_{1g}, E_g and T_{1g} reflect the transformation properties of some or others of the quadratic terms involved in the polarisability factor. Therefore, we can conclude

ir active fundamentals: $2T_{1u}$ (two different triply degenerate vibrational mode)
Raman active fundamentals: A_{1g}, E_g, T_{1g}
Inactive mode: T_{2u}
Here the Raman active and the ir active fundamentals are mutually exclusive.

Centrosymmetric Systems And Laporte's Selection Rule.

Example 8.6 demonstrates that the normal modes which are ir active are not Raman active and vice versa. This exclusive result is consistent with a general selection rule, called the Laporte's rule valid for centrosymmetric systems, i.e., molecules or ions possessing an inversion centre, i. In fact, so far as vibrational fundamental transitions are concerned, Laporte's rule is just a fallout of the selection rules already stated earlier in terms of symmetry principle. The rule states that "**electric dipole transitions can occur only between symmetric (gerade) and antisymmetric (ungerade) states and not between states of the same symmetry in a centrosymmetric molecule or ion**". Here the words symmetric and antisymmetric are used in the restricted sense of behaviour with respective to inversion operation only. For an ir active fundamental the integrand $\left\{\phi^*_{j(1)}(\vec{x} + \vec{y} + \vec{z})\phi_{j(0)}\right\}$ should be symmetric in order that the integral may not vanish. Since x, y and z are all antisymmetric in a centrosymmetric system, one of the two functions $\phi^*_{j(1)}$ and $\phi_{j(0)}$ should be antisymmetric (the other being symmetric) to render the integrand assume symmetric nature.

For Raman activity the integrand to be considered is $\left\{\phi^*_{j(1)}(x^2 \text{ etc.})\phi_{j(0)}\right\}$. Since the quadratics x^2, y^2, z^2, xy, yz and zx are all symmetric under inversion $\phi^*_{j(1)}$ and $\phi_{j(0)}$ should be of the same type (both gerade or both ungerade) for the non zero value of the transition moment integral. Therefore, the Raman active IR's having the same symmetry of the quadratics (x^2 etc.) have to be symmetric in contrast with the ir active IR's. The two sets of IR's are thus mutually exclusive for molecules or ions with the inversion centre.

Normal Coordinates And Nature Of Deformations

We now turn to the third category of applications of symmetry to glean qualitative information on bond stretching and angular deformations in normal modes of vibrations and also in in-plane and out-of-plane vibrations of planar molecules. We have referred to the use of the internal coordinates spanning the vibrational space part of the total configuration space. One may even use a vibrational subspace carried by the smaller basis sets of bond length changes or the basis set comprising angular deformation only. (The latter basis set is not easy to use, however, in all cases). This sort of segmentary use of internal coordinates is typified in the following examples.

Example 8.7.

Identify the in-plane normal modes of vibrations of BCl_3 molecule.

To find the IR's of the normal modes, we use the internal coordinates $\vec{S_1}$, $\vec{S_3}$ and $\vec{S_3}$, representing the bond length changes as the basis set. The characters of the reducible representation are shown:

	E	$2C_3$	$3C_2'$	$\sigma_\hbar$	$2S_3$	$3\sigma_0$
S_1	1	0	1	1	0	1
S_2	1	0	0	1	0	0
S_3	1	0	0	1	0	0
χ^{red}	3	0	1	3	0	1

Γ^{red} on resolution is found to comprise A_1' and E'. Hence the in-plane stretching (contracting) modes of vibrations are of A_1' and E' symmetries.

For the in-plane bending modes, we use α_1, α_2 and α_3, the deformations of the angles Cl^2BCl^3, Cl^3BCl^1 and Cl^1BCl^2 respectively (Fig. 8.1), as the base vectors clearly remembering that these three increments or decrements in angles in one plane cannot all be independent. The characters of the Γ^{red} are once again detected to be 3, 0, 1, 3, 0, 1 comprising the IR's A_1' and E' of which the completely symmetric A_1' is a spurious result. This is because of our neglect of the condition of interdependence of the three deformation angles while drawing up the character table of the reducible representation. One finds that the in-plane bending (angle deforming) vibrations in BCl_3 are of E' symmetry.

Since E' vibrations are seen to occur in both the basis sets, E' vibrations are really of mixed type involving both in-plane bond stretching (contracting) and angle bending. Thus in final analysis the in-plane vibrations of BCl_3 molecule are

A_1'– purely bond stretching (contracting) type
Two E'– both bond stretching and angle bending type (degenerate pair).

Example 8.8.

What is the out-of-plane vibration in the BCl_3 molecule?

Example 8.7 accounts for five vibrations (including the degenerate pair) out of the permissible six using internal set of coordinates as basis vectors for in-plane motions. Use of displacements as vectors (Example 8.2) reveals that the normal coordinates are spread over the IR's A_1', A_2'' and $2E'$. Striking out the in-plane vibrations, A_1' and two E's of

Example 8.7, we find that the out-of-plane vibration belongs to A_2''.

That A_2'' is an out-of-plane vibration can also be argued in the following way. The $\chi(\sigma_h)$ of the IR A_2'' has a value -1. This means that an A_2'' vibration must be antisymmetric with respect to reflection in the horizontal plane and since this antisymmetry cannot be shown by any in-plane vibration, A_2'' must represent an out-of-plane vibration of the BCl_3 molecule.

Example 8.9.

Label the bond stretching and angle bending modes of vibrations in NH_3 molecule.

NH_3 belongs to C_{3v}. We use the vibrational subspace spanned by the bond length changes $\vec{S}_1$, $\vec{S}_2$ and $\vec{S}_3$ (cf. Example 8.4). The IR's involved, as already shown before, are A_1 and E. A second use of the basis set α_1, α_2 and α_3 also yields the IR's A_1 and E. Since the IR's are of common symmetries, it is to be understood that both the A_1 vibrations involve bond stretchings (contractions) and angle deformations. So also do the two degenerate pairs of E vibrations.

Combination Tones And Overtones

We explain here briefly and qualitatively the usefulness of symmetry principles in explaining the occurrence of absorption bands of combination tones and overtones. If the vibrations pertaining to one or more normal modes are excited simultaneously to their respective fundamental levels, the resulting state is a combination level, also called a combination tone. With the simple harmonic approximation for potential energy of vibration, the probability of attainment of a combination tone is rather low permitting the appearance, if any, of weak absorption bands only. But absorption bands of reasonable intensities do arise due to excitation to combination tone state. The wave function in this latter state may be written as $\left\{\phi_{j(1)}\phi_{k(1)}\Pi_{i\neq j,\ k}\phi_{i(0)}\right\}$ and should have energy $h(\nu_j + \nu_k)$ above the ground state value. The symmetry of the state will be given by the IR's appearing in the direct product representation $\Gamma_j \otimes \Gamma_k$ in the basis of product function $\phi_{j(1)}\phi_{k(1)}$. Strictly speaking, however, the vibrations of the molecular system involve a little anharmonicity which contributes to a perturbation hamiltonian H'.

This perturbation (cf. Sec. 6.3) removes the degeneracy and gives rise to a number of distinct but closely spaced levels about the unperturbed energy level. If any or more of the IR's representing the closely spaced but distinct levels have any of the translational vectors as basis, then corresponding ir absorption consequent upon transition to combination tone will result. The definition of overtone has already been given. The principle of formation of overtone bands is similar to that of combination tone, but the analysis is not that easy.

8.4 Symmetry In Vibrations Of Linear Molecules

Linear molecules belong either to $C_{\propto v}$ or $D_{\propto \hbar}$ point groups. We shall first describe the IR's, ir active and Raman active fundamentals of the unsymmetrical linear molecules. Since the principle is the same as that of the nonlinear molecules we directly treat the specific case of linear N_2O molecule.

Example 8.10.

Find the symmetry species of the normal coordinates, ir and Raman active fundamentals of N_2O molecule. Also qualitatively describe the nature of vibrations (bond stretching and angular deformations).

N_2O molecule (structure N≡N→O) is linear and belongs to $C_{\propto v}$ having the group elements E, $2C_{\propto}(\phi)$, $\propto \sigma_v$. We set up coordinate axes with unit vectors $e_1(^1)$ $e_2(^1)$, $e_3(^1)$ at the equilibrium position of the oxygen atom, $e_1(^2)$ $e_2(^2)$ $e_3(^2)$ at the equilibrium site of the middle N-atom and $e_1(^3)$, $e_2(^3)$, $e_3(^3)$ at the equilibrium location of the terminal N-atom. We also set up molecular axis with unit vectors e_1, e_2, e_3 having its origin at any point on the line joining the nuclei and e_3 collinear with the nuclear axis, which is the z-axis. All these axial systems centred at different points are in parallel orientations with $e_3(^1)$, $e_3(^2)$, $e_3(^3)$ all collinear with e_3 along the nuclear axis. The displacement vectors of each atom at any time during vibrations are resolved along $e_1(^p)$ $e_2(^p)$ and $e_3(^p)$ where the p's are all 1's all 2's and all 3's.

To find the IR's of the normal coordinates we resort to transformation properties of the coordinate sets or of the basis sets under $C_{\propto v}$. We choose two sets $(e_3(^1),\ e_3(^2),\ e_3(^3))$ set describing vibrations in a direction parallel to the axis of the molecule and the set $(e_1(^1),\ e_2(^1),\ e_1(^2),\ e_2(^2),\ e_1(^3),\ e_2(^3))$ to describe motion in a direction normal to the axis.

$$\text{That is } \Gamma = \Gamma_{||} + \Gamma_{\perp} \tag{8.10}$$

The characters for the two sets are (cf. Sec. 5.5)

	E	$2C_{\propto(\phi)}$	$\propto \sigma_v$
$(e_1(^3)e_2(^3)e_3(^3))$	3	3	3

$$\Gamma_{||}^{\text{red}} = 3\ \Sigma^+ \text{ from character table} \tag{8.11}$$

Again

	E	$2C_{\propto}(\phi)$	$\propto \sigma_v$
$(e_1(^1)e_2(^1)e_1(^2)e_2(^2)e_1(^3)e_2(^3))$	6	$3(1+2\cos\phi)$	0

$$\text{where } \Gamma_{\perp}^{\text{red}} = 3\Pi \text{ (from character table)} \tag{8.12}$$

The molecule being linear has (3N-5), i.e., 4 normal coordinates of vibrations. Of the $3\sum^+$ IR's in (8. 11) one is due to z-directional translational motion (e_3 direction) leaving $\Gamma_{ll}^{red} = 2\sum^+$ as denoting the normal coordinates. Again subtracting the degenerate IR, $\prod$, representing translations (e_1 and e_2 direction) and another degenerate IR, Π, for the two rotational modes one is left with a single IR, Π, denoting a pair of degenerate vibration normal to the nuclear axis. That is relation (8.12) shorn of translational and rotational components stands as

$$\Gamma_{\perp} = \Pi$$

Infra red and Raman activity: The transformation properties of the translation vectors and of the polarisability components as shown in the character table of $C_{\propto v}$, lead to the following:

	IR's of the fundamentals	
ir active	$\sum^+$ (two distinct bands)	$\prod$
Raman active	$\sum^+$ (two distinct bands)	$\prod$

Nature of Vibrations: The two parallel modes of vibrations, viz., $2\sum^{+}$ are only of bond stretching contracting types in as much as these are vibrations along the e_3, i.e., z-direction of the molecule. The degenerate pair of vibrations $\prod$ represent motions causing departure from the equilibrium linear state and hence involve bending modes of vibrations.

We can qualitatively draw the vibrational modes from the conclusions reached above (Fig. 8.3).

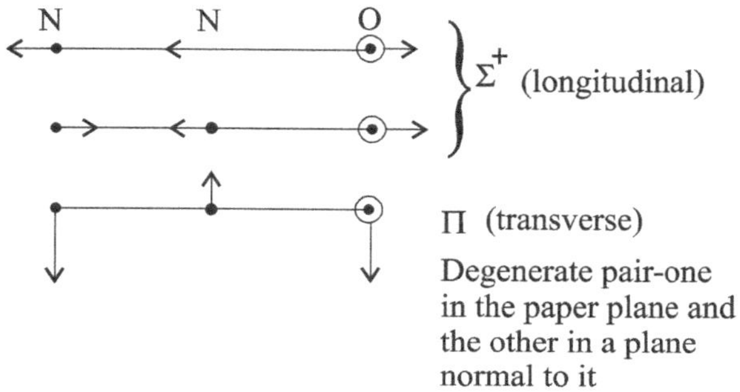

Fig. 8.3: Nature of vibrations in N_2O molecule

It turns out that for an n-atom linear molecule the reducible representation for vibrations (including translations and rotations) of the molecule breaks up as $\Gamma = n\sum^{+} + n\prod$. There will thus be (n-1) $\sum^{+}$ longitudinal vibrations where one $\sum^{+}$ is dropped to account for e_3-directional translation. Similarly counting out the other two translational and two rotational motions, the transverse type vibrations will be (2n-4) in number divided into (n-2) pairs of degenerate modes.

We now illustrate the vibrational behaviour of a symmetrical linear molecule, say, carbon disulphide CS_2. The procedure is the same as that of the unsymmetrical molecule and depends upon the selection of two basis sets one comprising vectors in the direction of the nuclear axis and the other consisting of vectors perpendicular to the nuclear axis.

Example 8.11

Find the ir and the Raman active fundamentals in the vibrations of carbon disulphide molecule.

CS_2 $\left(\text{structure } \begin{matrix} 3 & & 2 & & 1 \\ S & = & C & = & S \end{matrix}\right)$ is linear, symmetrical and belongs to the $D_{\propto\hbar}$ group having the elements E, $2C_\propto(\phi) \propto \sigma_v$, i, $2S_\propto(\phi)$, and $\propto C_2'$. Cartesian coordinate axes with unit vectors $e_1(^1)$, $e_2(^1)$ $e_3(^1)$ and $e_1(^2)$, $e_2(^2)$, $e_3(^2)$ and $e_1(^3)$, $e_2(^3)$, $e_3(^3)$ are set up at the equilibrium locations of the atoms 1, 2 and 3 in parallel orientations. The $e_1(^3)$, $e_2(^3)$ and $e_3(^3)$ axes all lie along the internuclear axis representing the z-direction of the molecule. The axial system e_1, e_2, e_3 of the molecule as a whole is set up with the origin at the midpoint of the internuclear axis and in the same parallel orientations as the other axial systems. In this particular axample, the set $(e_1e_2e_3)$ coincides with the set $e_1(^2)$ $e_2(^2)$ $e_3(^2)$. The displacements of the atoms at any time during vibration are resolved into mass-weighted components along the axial directions associated with each atom and their transformation properties will give rise to characters same as those obtained from conjoint transformations of $e_1(^p)$ $e_2(^p)$ and $e_3(^p)$. Choosing two sets of base vectors (i) $e_3(^1)$, $e_3(^2)$ $e_3(^3)$ and (ii) $e_1(^1)$ $e_2(^1)$ $e_1(^2)$ $e_2(^2)$, $e_1(^3)$, $e_2(^3)$ we can obtain characters by using the shift rule (Sec. 5.8). But we can still cut it short by employing the following argument. Since $C_{\propto v}$ is a subgroup of $D_{\propto\hbar}$ and generates for an n-atom linear molecule $n\sum^+$ and $n\prod$ IR's with the types of basis sets indicated here, under $D_{\propto\hbar}$ the same types of basis sets will generate g-and u-natured IR's in a $D_{\propto\hbar}$ type linear molecule.

$$\text{(i) with n = even, the IR's are } \Gamma_{\parallel} = \frac{n}{2}\Sigma_g{}^+ \oplus \frac{n}{2}\Sigma_u{}^+$$
$$\text{and } \Gamma_{\perp} = \frac{n}{2}\Pi_g \oplus \frac{n}{2}\Pi_u \quad (8.13)$$

$$\text{(ii) when n = odd; } \Gamma_{\parallel} = \frac{n-1}{2}\Sigma_g{}^+ + \frac{n+1}{2}\Sigma_u{}^+ \quad (8.14)$$
$$\text{and } \Gamma_{\perp} = \frac{n-1}{2}\Pi_g \oplus \frac{n+1}{2}\Pi_u$$

Since CS_2 consists of three atoms, the relation (8.14) will apply

$$\Gamma_{\parallel} = \Sigma_g{}^+ \oplus 2\Sigma_u{}^+ \quad \text{(for longitudinal motions)}$$
$$\Gamma_{\perp} = \Pi_g \oplus 2\Pi_u \quad \text{(for transverse vibrations)}$$

Subtracting the IR's for three translational and the two rotational modes, the normal coordinates of vibrations have the following IR's

$$\Gamma_{\parallel} = \Sigma_g{}^+ \oplus \Sigma^+{}_u \text{ and } \Gamma_{\perp} = \Pi_u$$

Infra red and Raman activities of the normal modes can be tabulated after noting the character values of $D_{\propto\hbar}$ and observing how x, y, z or the quadratics x^2 etc. are distributed as bases of the IR's.

	Fundamentals		
ir activity	Σ_{u+}		Π_u
Raman activity		Σ^+g	

We thus notice a confirmation of the general finding that in centrosymmetric systems, ir active normal modes are Raman inactive and vice versa. The principle of exclusion holds good here as expected.

Chapter 9

Hybrid Orbitals, Symmetry Orbitals And Molecular Orbitals

9.1 Introduction :

The description of the items of this chapter presupposes our familiarity with the concept of "atomic orbitals" which are the components of the various orbital terms appearing in the caption. In the first phase of the present chapter our discussion will be confined to a symmetry treatment of the hybrid orbitals. The symmetry orbitals and the molecular orbitals will appear in its wake.

Hybrid orbitals are just an artificial concept aimed at helping our qualitative understanding of the stereodisposition and the directional nature of the chemical bonds in a molecule. In quantitative terms, these are mere mathematical constructs being some suitable linear combination of atomic orbitals designed to match a preplanned symmetry of their distribution in space. For the CH_4 molecule, the preplanned symmetry is a tetrahedral one and the hybrid orbitals of the carbon atom are so formed from its atomic orbitals that the functions have graphical behaviour in keeping with this symmetry. In ethylene molecule $D_{2\hbar}$, where all the angles made by the σ-bonds at the two carbon atoms are $\frac{2\pi}{3}$ each, the symmetry of each σ-bond set is $D_{3\hbar}$ corresponding to that of a planar trigonal one. This serves as the preplanned symmetry in conformity with which mathematical construction of the hybrid orbitals is effected. In many instances the preplanned symmetry may be one of trigonal bipyramid distribution or of an octahedral distribution. The hybrid orbitals of the central atom, prior to bond formation, may be deemed to be such overall SALC's as to satisfy collectively the overall symmetries of trigonal bipyramid or of an octahedron and are not just the mere basis set for one or the other IR's of the point group (Chap. 5). Fig. 9.1 depicts the symmetry of some hybrids of a central atom. To distinguish from the conventional sense of the word SALC's, the hybrid set will be called "point group SALC's".

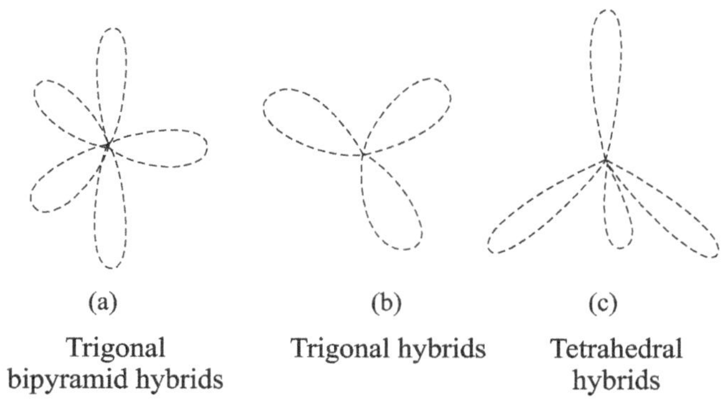

Fig. 9.1: Trigonal bipyramid hybrids, Trigonal hybrids, Tetrahedral hybrids

The utility of the construction of hybrids is in the formation of valence bond orbitals or of localised molecular orbitals through overlaps with other atomic orbitals of the constituent atoms of the molecule. Though constituted of a number of atomic orbitals, the nature of the roles of hybrids is similar to that of the atomic orbitals.

9.2 Principle Of Constructing Hybrid Orbitals.

The ideas of symmetry step in when we remember that compositionally the hybrids are merely point group SALC's. The stages in the construction of the hybrids are:

(i) to know the transformation properties of the basis set of hybrids under symmetry operations of the point group.
(ii) to find the IR's spanned by the basis hybrids.
(iii) to find, with the help of character table, the atomic orbitals of the concerned atom which transform as the IR's found under (ii).
(iv) to compose the point group SALC's with desirable atomic orbitals as indicated under (iii). The formation of these SALC's is generally effected through matrix inversion.

Let us now consider the feasibilities of realising these stages. In stage (i) we are confronted with an apparent (but not serious) difficulty of not knowing initially the functional forms of the hybrids upon which symmetry operations are to be applied. This difficulty can easily be circumvented when we find (Fig. 9.1) that the prominent feature of the hybrids is their directional character. This particular feature may be displayed by considering a set of vectors shooting out from a central atom in the directions of the individual hybrid orbitals. The transformation properties of these set of vectors (not necessarily all linearly independent) will reflect those of the hybrids. Hence these vectors may serve as proxies for the hybrids. Thus for tetrahedral hybridization, as in the formation of CH_4, we choose four vectors ($\vec{e}_1$ $\vec{e}_2$ $\vec{e}_3$ $\vec{e}_4$) at a mutual inclination of $109°28'$ and directed in space from some central point-the location of the carbon atom. For hybrids suiting a trigonal bipyramid symmetry, the choice is for five vectors ($\vec{e}_1$ $\vec{e}_2$ $\vec{e}_3$ $\vec{e}_4$ $\vec{e}_5$) of which $\vec{e}_1$, $\vec{e}_2$, $\vec{e}_3$ are directed to the vertices of an equilateral triangle in a plane and the $\vec{e}_4$, $\vec{e}_5$ pair project perpendicularly up and down from the centroid of the equilateral triangle. Being in the same three-dimensional physical space naturally all these vectors are not linearly independent. The transformation properties or, more importantly, the contribution to the character values of the transformation matrices can be obtained by noting the shifts or nonshift of the individual vectors under the group operations.

Stage (ii) is a mere formality of applying Theo. 5.8 to know the IR's and stage (iii) following it needs no comment. The last stage is one of forming the point group SALC's, i.e., the overall point group symmetry adapted linear combinations.

As hybrids are just an artificial measure preparatory to bond formation in molecules and as chemists are primarily (though not exclusively) interested in sigma (σ) and pi (π) bonds, the hybrids are thus classified into σ-hybrids and π-hybrids. The appropriateness of the qualifying symbols σ and π depends on their subsequent use as components for σ-bond formation or π-bond formation respectively.

The molecules (or ions) in which hybrid orbitals are conceived to be formed before bond formation are numerous. Some molecules or ions such as CH_4, SO_4^{-2}, NO_3^-, PCl_5, $[CoF_6]^{3-}$, $CHCl_3$ contain one focal (central) atom surrounded by a number of other atoms. The surrounding atoms are not mutually connected in the classical sense of bonding. In such molecules hybrid orbitals are constructed for the central atom. The vectors (e_1 e_2.....e_n), which act as the basis set and are substitutes for the hybrids, all shoot out from the central atom. In many cases, however, more than one focal atom may be present, such as, $H_2C{=}CH_2$ with two focal atoms or C_6H_6 with six focal atoms. The hybrids formed in such cases are suited to form σ-bonds.

In some molecules or ions again, viz., PCl_3, BCl_3, $[Co(CN)_6]^{3-}$ there exist additional opportunities of forming hybrid orbitals from linear combination of suitable atomic orbitals on the central atoms. These hybrid orbitals, unlike the σ-hybrids, are so juxtaposed in space as can easily effect lateral overlaps with suitable orbitals of the ligand atoms forming π-bond. Such hybrids are termed $\pi-$hybrids. We shall discuss the two sets of hybrids in different sections.

9.3 Hybrids For $\sigma-$ Bond Formation.

This section is just a practical application of the principles enunciated in the previous one. Since concrete examples will be more helpful than a generalized one, we shall illustrate by selecting a few molecules involving different hybridization schemes.

1. **Trigonal Hybridisation.**

Example 9.1.

Find the qualitative and the quantitative forms of the $\sigma-$ hybrids of the boron atom in BCl_3 molecule.

The adjoining figure (Fig. 9.2) shows the axial set-up of BCl_3 mole-

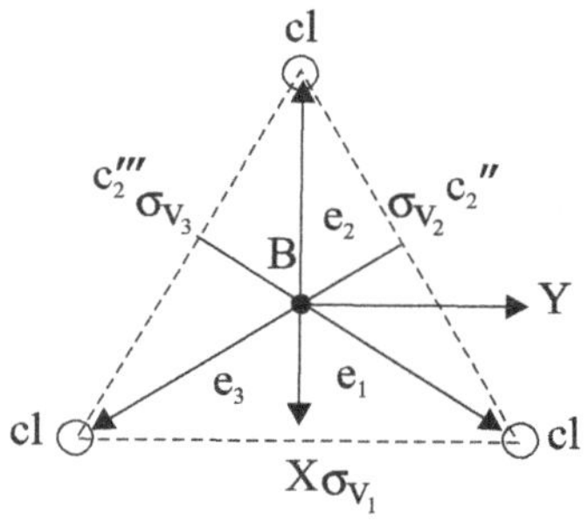

Fig. 9.2: BCl_3 with symmetry elements and the hybrid vectors.

cule of the point group D_{3h}. Let $\vec{e}_1$, $\vec{e}_2$, $\vec{e}_3$ represent the vectors pointing along the three major positive lobes of the hybridised orbitals.

Stage (i). using (e_1 e_2 e_3) as the basis, the transformation properties under $D_{3\hbar}$ are shown below.

	$\hat{E}$	$\hat{C}_3$	$\hat{C}_3^2$	$\hat{C}_2'$	$\hat{C}_2''$	$\hat{C}_2'''$	$\hat{\sigma}_\hbar$	$\hat{S}_3$	$\hat{S}_3^5$	$\hat{\sigma}_{v1}$	$\hat{\sigma}_{v2}$	$\hat{\sigma}_{v3}$
e_1	e_1	e_2	e_3	e_3	e_2	e_1	e_1	e_2	e_3	e_3	e_2	e_1
e_2	e_2	e_3	e_1	e_2	e_1	e_3	e_2	e_3	e_1	e_2	e_1	e_3
e_3	e_3	e_1	e_2	e_1	e_3	e_2	e_3	e_1	e_2	e_1	e_3	e_2
$\chi_{\hbar yb}$	3	0	0	1	1	1	3	0	0	1	1	1

Stage (ii). Use of the relation $n_i = \frac{1}{g_i}\sum_R \chi_i^*(R)\chi^{red}(R)$ and the above set of character values obtained in the reducible representation, $\Gamma^{\hbar yb}$, yields for the $D_{3\hbar}$ point group

$$\Gamma^{\hbar yb} = A_1' + E'$$

Stage (iii) Looking up the character table one finds that under $D_{3\hbar}$ the following, atomic orbitals form bases for the IR's A_1' and E'.

$$\begin{aligned} A_1' &\rightarrow s,\ d_{z2} \\ E' &\rightarrow (p_x,\ p_y);\ (d_{xy},\ d_{x^2-y^2}) \end{aligned}$$

Since in the formation of three hybridised orbitals a minimum of three basis orbitals are needed, we conclude that these trigonally directed hybrids must consist of sp^2 mixing, sd^2 mixing, dp^2 mixing or d^3 mixing.

We now allow our chemical intuition to decide between these possibilities. Since the boron atom has the electronic configuration $ls^2 2s^2 p^1$ in the ground state, any sd^2 type of hybridisation would require a promotion of one s and one p electrons to 3d levels. This will involve an input of large energy (since the principal quantum number changes from 2 to 3) unlikely to be compensated for by lowering of energy by subsequent bond formation through overlap with the chlorine orbitals. Our conclusion, therefore, rests on $sp^2[s,\ p_x,\ p_y]$ hybridisation as this would just mean a prepromotion of a 2s electron to 2p state for hybridisation to be effective. In BCl_3, hybrids involving d-orbitals are not permissible from the viewpoint of energy.

Stage (iv). Formation of point group SALC's—the quantitative forms of the hybrid orbitals.

We do not attempt a direct formation of these SALC's from s, p_x, p_y orbitals of boron atom but instead follow an indirect procedure. Let us denote the orthogonal hybrids (composition yet unknown) by ψ_1, ψ_2 and ψ_3 and the atomic orbitals s, p_x and p_y by ϕ_1, ϕ_2 and ϕ_3 respectively.

We construct projection operators $\wp A_1'$ and $\wp E'$ and apply these on the hybrids to find their linear combinations of A_1' and E' symmetries. These linear combinations may then be equated with $\phi_1(s)$ and some combinations of ϕ_2 and ϕ_3 (p_x and p_y) respectively.

$$\left.\begin{aligned} \wp^{A_1'} &= \tfrac{1}{12}\Big[\hat{E} + \hat{C}_3 + \hat{C}_3^2 + \hat{C}_2' + \hat{C}_2'' + \hat{C}_2''' + \hat{\sigma}_\hbar + \hat{S}_3 + \hat{S}_3^5 + \hat{\sigma}_{v1} \\ &\qquad + \hat{\sigma}_{v2} + \hat{\sigma}_{v3}\Big] \\ \wp^{E'} &= \tfrac{2}{12}\Big[2\hat{E}' - \hat{C}_3 - \hat{C}_3^2 + 2\hat{\sigma}_\hbar - \hat{S}_3 - \hat{S}_3^5\Big] \end{aligned}\right\} \begin{matrix} a \\ b \end{matrix} \tag{9.1}$$

The transformations of ψ_1, ψ_2 and ψ_3 under $D_{3\hbar}$ being the same as those of $\vec{e}_1$ $\vec{e}_2$ and $\vec{e}_3$, we have

$$\left.\begin{aligned} \wp A_1'\psi_1 &= \tfrac{1}{12}[\psi_1 + \psi_2 + \psi_3 + \psi_3 + \\ &\psi_2 + \psi_1 + \psi_1 + \psi_2 + \psi_3 + \psi_3 + \psi_2 + \psi_1] \\ &= \tfrac{1}{3}(\psi_1 + \psi_2 + \psi_3) \\ \text{Similarly,} & \\ \wp E'\psi_1 &= \tfrac{1}{3}[2\psi_1 - \psi_2 - \psi_3] \\ \text{and } \wp E'\psi_2 &= \tfrac{1}{3}[2\psi_2 - \psi_3 - \psi_1] \end{aligned}\right\} \tag{9.2}$$

Writing (9.2) in normalized form, we equate the first one with ϕ_1 and the latter two as some linear combinations (normalized) of ϕ_2 and ϕ_3. Thus.

$$\left.\begin{aligned} \phi_1 &= \tfrac{1}{\sqrt{3}}(\psi_1 + \psi_2 + \psi_3) \\ \tfrac{1}{\sqrt{a^2+b^2}}(a\phi_2 + b\phi_3) &= \tfrac{1}{\sqrt{6}}(2\psi_1 - \psi_2 - \psi_3) \\ \text{and } \tfrac{1}{\sqrt{c^2+d^2}}(c\phi_2 + d\phi_3) &= \tfrac{1}{\sqrt{6}}(2\psi_2 - \psi_1 - \psi_3) \end{aligned}\right\} \tag{9.3}$$

The factors $\frac{1}{\sqrt{a^2+b^2}}$ and $\frac{1}{\sqrt{c^2+d^2}}$ are the two normalising factors. We are to pick the values of a, b, c and d by examining the effect of some suitably chosen symmetry operations. Using the Fig. 9.2 and the transformation table, we find

$$\begin{aligned} \sigma_{v1}\left[\frac{1}{\sqrt{c^2+d^2}}(c\phi_2 + d\phi_3)\right] &= \sigma_{v1}\left\{\frac{1}{\sqrt{6}}(2\psi_2 - \psi_1 - \psi_3)\right\} \\ = \frac{1}{\sqrt{6}}(2\psi_2 - \psi_1 - \psi_3) &= \frac{1}{\sqrt{c^2+d^2}}(c\phi_2 + d\phi_3) \\ \text{i.e., } \frac{1}{\sqrt{c^2+d^2}}(c\phi_2 - d\phi_3) &= \frac{1}{\sqrt{c^2+d^2}}(c\phi_2 + d\phi_3), \end{aligned}$$

($\phi_3(\equiv p_y)$ changes sign under σ_{v1}).
whence d=0

$$\pm\,\phi_2 = \frac{1}{\sqrt{6}}(2\psi_2 - \psi_1 - \psi_3) \tag{9.4}$$

Since all the constituents of the RHS of Eq. (9.4) have only negative components of ϕ_2, it immediately follows

$$\phi_2 = \frac{1}{\sqrt{6}}(\psi_1 + \psi_3 - 2\psi_2) \tag{9.5}$$

In a like manner

$$\sigma_{v3}\left[\frac{1}{\sqrt{a^2+b^2}}(a\phi_2 + b\phi_3)\right] = \sigma_{v3}\left\{\frac{1}{\sqrt{6}}(2\psi_1 - \psi_2 - \psi_3)\right\}$$
$$= \frac{1}{\sqrt{6}}(2\psi_1 - \psi_2 - \psi_3) = \frac{1}{\sqrt{a^2+b^2}}(a\phi_2 + b\phi_3) \tag{9.6}$$

Since from Fig. 9.2,

$$\left.\begin{aligned}\sigma_{v3}\phi_2 &= -\cos 60^\circ\,\phi_2 + \cos 30^\circ\,\phi_3 = -\tfrac{1}{2}\phi_2 + \tfrac{\sqrt{3}}{2}\phi_3\\ \text{and }\sigma_{v3}\phi_3 &= \tfrac{\sqrt{3}}{2}\phi_2 + \tfrac{1}{2}\phi_3\end{aligned}\right\} \tag{9.7}$$

We may write, using (9.7) and (9.6),

$$\frac{1}{2}(-a + \sqrt{3}b)\phi_2 + \frac{1}{2}(\sqrt{3}a + b)\phi_3 = a\phi_2 + b\phi_3$$

Equating the coefficients of ϕ_2 and ϕ_3, one finds a$= \frac{1}{\sqrt{3}}$b and its substitution in Eq. (9.6) leads to

$$\pm\,\frac{1}{2}(\phi_2 + \sqrt{3}\phi_3) = \frac{1}{\sqrt{6}}(2\psi_1 - \psi_2 - \psi_3)$$

Noting that ψ_1 and $-\psi_3$ have positive components along ϕ_3, it is possible to write

$$\frac{1}{2}(\phi_2 + \sqrt{3}\phi_3) = \frac{1}{\sqrt{6}}(2\psi_1 - \psi_2 - \psi_3) \tag{9.8}$$

Use of Eq. (9.5) in relation (9.8) enables one to write

$$\phi_3 = \frac{1}{\sqrt{2}}(\psi_1 - \psi_3) \tag{9.9}$$

Collecting the derived relations (9.3), (9.5) and (9.9), it is seen that the atomic orbitals are given by

$$\left.\begin{aligned}\phi_1 &= \tfrac{1}{\sqrt{3}}\psi_1 + \tfrac{1}{\sqrt{3}}\psi_2 + \tfrac{1}{\sqrt{3}}\psi_3\\ \phi_2 &= \tfrac{1}{\sqrt{6}}\psi_1 - \tfrac{2}{\sqrt{6}}\psi_2 + \tfrac{1}{\sqrt{6}}\psi_3\\ \phi_3 &= \tfrac{1}{\sqrt{2}}\psi_1 - \tfrac{1}{\sqrt{2}}\psi_3\end{aligned}\right\} \tag{9.10}$$

or, written compactly in matrix form

$$\Phi = \Psi\mathbf{T}, \text{ where } \Phi = (\phi_1\phi_2\phi_3), \ \Psi = (\psi_1\psi_2\psi_3)$$

and

$$\mathbf{T} = \begin{pmatrix} \frac{1}{\sqrt{3}} & \frac{1}{\sqrt{6}} & \frac{1}{\sqrt{2}} \\ \frac{1}{\sqrt{3}} & -\frac{2}{\sqrt{6}} & 0 \\ \frac{1}{\sqrt{3}} & \frac{1}{\sqrt{6}} & -\frac{1}{\sqrt{2}} \end{pmatrix}$$

$\psi = \Phi\mathbf{T}^{-1}$. Now since the transformation from ψ set to Φ set is one involving two orthogonal basis sets, the $\mathbf{T}$ matrix should be orthogonal, i.e., $\mathbf{T}^{-1} = \widetilde{\mathbf{T}}$

$$\begin{aligned}(\psi_1\ \psi_2\ \psi_3) &= (\phi_1\ \phi_2\ \phi_3)\widetilde{\mathbf{T}} \\ &= (\mathrm{s}\ \mathrm{p_x}\ \mathrm{p_y}) \begin{pmatrix} \frac{1}{\sqrt{3}} & \frac{1}{\sqrt{3}} & \frac{1}{\sqrt{3}} \\ \frac{1}{\sqrt{6}} & -\frac{2}{\sqrt{6}} & \frac{1}{\sqrt{6}} \\ \frac{1}{\sqrt{2}} & 0 & -\frac{1}{\sqrt{2}} \end{pmatrix}\end{aligned} \tag{9.11}$$

which are the required trigonal hybrids.

Before we turn to a second example, the following observations may be noted. Since the hybrids are normalized and also orthonormal, being the sum of orthonormal atomic orbitals centred on the focal atom, a hybrid $\psi_\mathrm{i} = \mathrm{a_i}\phi_1 + \mathrm{b_i}\phi_2 + \mathrm{c_i}\phi_3 + \mathrm{d_i}\phi_4 + ...$ allows writing

$$\sum_\mathrm{i} \mathrm{m_i}^2 = 1 \text{ for m} = \mathrm{a,\ b,\ c...} \tag{9.12}$$

Mutual orthogonality of hybrids ψ_i, ψ_j demands

$$\mathrm{a_i a_j + b_i b_j + c_i c_j + d_i d_j} + ... = 0 \tag{9.13}$$

These two conditions along with the piecemeal uses of selected symmetry operations are often employed in finding the coefficients to be associated with the atomic orbitals in their linear combinations to give the quantitative expressions for the hybrid orbitals.

Finally the trigonal hybrids given by (9.11) are not unique. Their expressions depend on the orientation of the hybrids (i.e., of the proxy vectors $\mathrm{e_1\ e_2\ e_3}$) relative to the initial choice of the axial directions x, y and z. If instead of the choice indicated in Fig. 9.2.. we choose the set of axes and hybrids as in Fig. 9.3, the hybrids will be given by

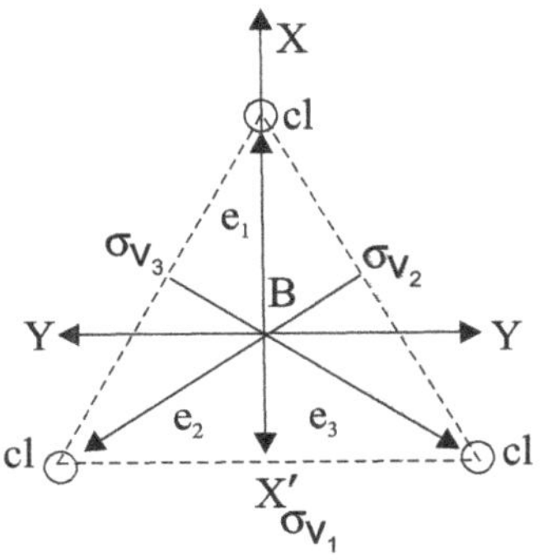

Fig. 9.3: New choice of axial directions in BCl_3 molecule.

$$(\psi_1\ \psi_2\ \psi_3) = (s\ p_x\ p_y)\begin{pmatrix} \frac{1}{\sqrt{3}} & \frac{1}{\sqrt{3}} & \frac{1}{\sqrt{3}} \\ \frac{2}{\sqrt{6}} & -\frac{1}{\sqrt{6}} & -\frac{1}{\sqrt{6}} \\ 0 & \frac{1}{\sqrt{2}} & -\frac{1}{\sqrt{2}} \end{pmatrix} \tag{9.14}$$

2. Tetrahedral hybridisation :

Example 9.2

Obtain the nature and the quantitative forms of equivalent tetrahedral hybrids.

We may assume that the transformation properties of ψ_1 ψ_2 ψ_3 and ψ_4 hybrids are the same as of the proxy vectors $\vec{e}_1$, $\vec{e}_2$ $\vec{e}_3$ and $\vec{e}_4$ under T_d symmetry (Fig. 9.4).

The mechanical process of transformation (stage i) and reduction of reducible representation (stage ii) lead to $\Gamma^{hyb} = A_1 \oplus T_2$. Character table of T_d shows the following distribution of the atomic orbitals as base vectors under A_1 and T_2.

$$A_1 \rightarrow s$$
$$T_2 \rightarrow (p_x,\ p_y,\ p_z) \text{ and } (d_{xy},\ d_{xz},\ d_{yz})$$

Symmetry considerations, therefore, suggest the hybrids to be of sp^3 or sd^3 types or a mixture of the two. Stage (iii) is thus completed. Let us try to express the sp^3 hybrids in quantitative forms for which we have to turn to stage (iv). This stage of forming the point group SALC's is, however, on many occasions replaced by the method of inspection and inference (cf. Examples Chapter 5). This is quite a permissible

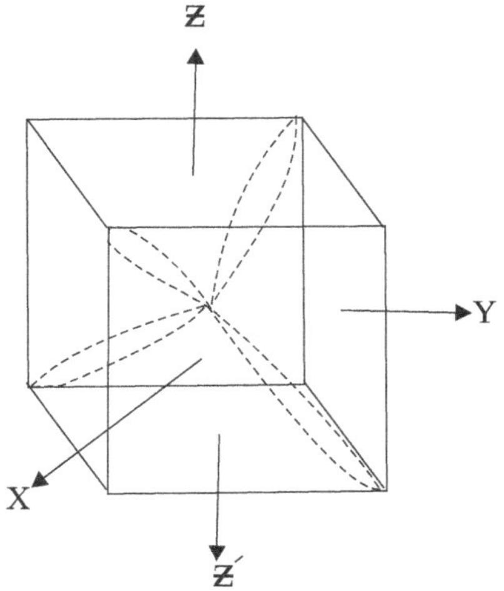

Fig. 9.4: Tetrahedral hybrids with axial set-up

technique in group theory and will be adopted here instead of the formal procedure.

Looking at Fig. 9.4, it is found that the locations of the main lobes of the hybrids ψ_1, ψ_2, ψ_3 and ψ_4 are in the first, third, eighth and the sixth octants respectively of a sphere circumscribed about the cube. Remembering the positive and the negative directions of x, y and z axes and hence of the distribution of the p_x, py and p_z orbitals of the central atom in these regions, one can write the composition of the hybrids as

$$\left.\begin{aligned} \psi_1 &= a_1 s + b_1 p_x + c_1 p_y + d_1 p_z \\ \psi_2 &= a_2 s - b_2 p_x - c_2 p_y + d_2 p_z \\ \psi_3 &= a_3 s + b_3 p_x - c_3 p_y - d_3 p_z \\ \psi_4 &= a_4 s - b_4 p_x + c_4 p_y - d_4 p_z \end{aligned}\right\} \tag{9.15}$$

Since each hybrid is poised symmetrically with respect to x, y and z axes and hence with respect to p_x, p_y and p_z orbitals [the signs having been taken care of in relations (9.15)] the coefficients $b_1=b_2=b_3=b_4=c_1=c_2=c_3=c_4=d_1=d_2=d_3=d_4=b$(say). The s-orbital being spherically symmetric, it is also evident that $a_1=a_2=a_3=a_4=a$ (say). Relations (9.15) then reduce to

$$\left.\begin{aligned} \psi_1 &= a\,s + b(p_x + p_y + p_z) \\ \psi_2 &= a\,s + b(-p_x - p_y + p_z) \\ \psi_3 &= a\,s + b(p_x - p_y - p_z) \\ \psi_4 &= a\,s + b(-p_x + p_y - p_z) \end{aligned}\right\} \tag{9.16}$$

Utilising the normalisation property of ψ_1 and the mutual orthogonality of ψ_1 and ψ_2, we have

$$\left.\begin{aligned} a^2 + 3b^2 &= 1 \\ \text{and } a^2 - b^2 &= 0 \end{aligned}\right\} \tag{9.17}$$

$$\text{whence } a = b = \pm\frac{1}{2}$$

Therefore the normalized set of sp^3 hybrids are

$$\left.\begin{aligned} \psi_1 &= \tfrac{1}{2}(s + p_x + p_y + p_z) \\ \psi_2 &= \tfrac{1}{2}(s - p_x - p_y + p_z) \\ \psi_3 &= \tfrac{1}{2}(s + p_x - p_y - p_z) \\ \psi_4 &= \tfrac{1}{2}(s - p_x + p_y - p_z) \end{aligned}\right\} \tag{9.18}$$

(9.19)

As in the above example, it is also possible to avoid the use of projection operators in the case of trigonal hybridisation of Example 9.1. But, unlike the sp^3 set, the sp^2 set will involve some use of trigonometry in that the hybrids are not symmetrically oriented with respect to x and y axes.

The octahedral hybridisation, which is frequently assumed in coordination complexes, will constitute our next worked out example.

3. Octahedral Hybridisation

Example 9.3

Find the nature and the quantitative forms of the hybrids of the cobalt atom necessary for the formation of the complex ion $[Co(CN)_6]^{3-}$

This ion belongs to the point symmetry group $O_\hbar$. The hybrid orbitals of cobalt meant for σ−bond formation should, therefore, be six in number and be so distributed in space as to match $O_\hbar$, symmetry. Let the hybrids be denoted by the matrix $\Psi = (\psi_1\psi_2\psi_3\psi_4\psi_5\psi_6)$. The octahedral distribution of their major positive lobes is shown in Fig. 9.5.

A drill along the stages (i), (ii) and (iii), consisting of transformations of the basis hybrids, characters of the reducible representation and inference from the character table, shows that

$$\Gamma^{hyb} = A_{1g} \oplus T_{1u} \oplus E_g$$

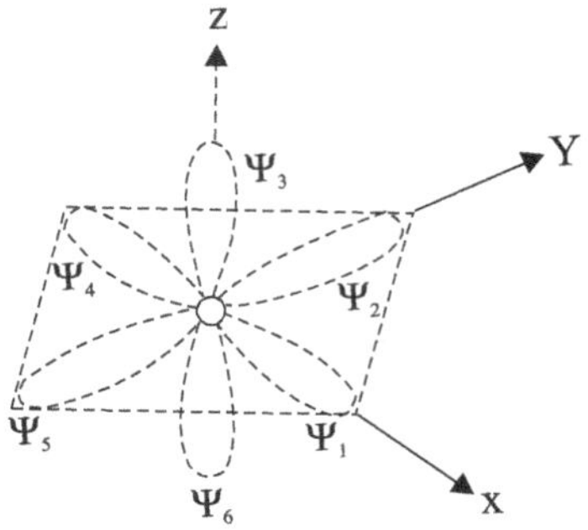

Fig. 9.5: Octahedral hybrids.

with the s-orbital transforming as A_{1g}, $(p_x\ p_y\ p_z)$ orbitals as T_{1u} and $(d_{(x^2-y^2)}, d_z^2)$ like E_g. The possible hybridisation is, therefore, d^2sp^3 type.

Constructing the projection operators $\wp A_{1g}$, $\wp T_{1u}$ and $\wp E_g$ and applying these on the hybrids it is possible to form SALC's of the hybrids, to correlate these with the atomic orbitals s, p_x, p_y, p_z, d_z^2 $d_{(x^2-y^2)}$ and finally to resort to matrix inversion to arrive at the quantitative expressions of the hybrids.

As a simplifying alternative[10] we may also adopt the method of inspection. Thus if $\Phi = (\phi_1\phi_2\phi_3\phi_4\phi_5\phi_6) = (s\ p_x\ p_y\ p_z, d_z^2 d_{x^2-y^2})$ represents the matrix of the atomic orbitals, then remembering the shapes of the orbitals and their distribution along the axial directions, we can immediately write

$$\begin{aligned}
s &= \phi_1 = \frac{1}{\sqrt{6}}(\psi_1+\psi_2+\psi_3+\psi_4+\psi_5+\psi_6)\\
p_x &= \phi_2 = \frac{1}{\sqrt{2}}(\psi_1-\psi_4)\\
p_y &= \phi_3 = \frac{1}{\sqrt{2}}(\psi_2-\psi_5)\\
p_z &= \phi_4 = \frac{1}{\sqrt{2}}(\psi_3-\psi_6)\\
d_z^2 &= \phi_5 = \frac{1}{\sqrt{12}}(2\psi_3+2\psi_6-\psi_1-\psi_2-\psi_4-\psi_5)\\
d_{x^2-y^2} &= \phi_6 = \frac{1}{2}(\psi_1+\psi_4-\psi_2-\psi_5)
\end{aligned}$$

In setting this up mentally, the only difficulty one experiences is in the writing of the combination for d_z^2 orbital. But recalling the fact that there exist two positive lobes at the two ends of the z-axis and a concentric negative shell including a central positive zone of low intensity in the xy plane, it is not hard to imagine a combination of ψ's approximating such a description. Thus from (9.19) the basis sets Φ and ψ are related as.

$$\begin{array}{r} \Phi = \psi\mathbf{T} \text{ where } \mathbf{T} = \\ \text{The hybrid matrix} \\ \psi = \Phi\mathbf{T}^{-1} \text{ and since} \\ \mathbf{T} \text{ is orthogonal, } \mathbf{T}^{-1} = \widetilde{\mathbf{T}}. \end{array} \begin{bmatrix} \frac{1}{\sqrt{6}} & \frac{1}{\sqrt{2}} & 0 & 0 & -\frac{1}{\sqrt{12}} & \frac{1}{2} \\ \frac{1}{\sqrt{6}} & 0 & \frac{1}{\sqrt{2}} & 0 & -\frac{1}{\sqrt{12}} & -\frac{1}{2} \\ \frac{1}{\sqrt{6}} & 0 & 0 & \frac{1}{\sqrt{2}} & \frac{2}{\sqrt{12}} & 0 \\ \frac{1}{\sqrt{6}} & -\frac{1}{\sqrt{2}} & 0 & 0 & -\frac{1}{\sqrt{12}} & \frac{1}{2} \\ \frac{1}{\sqrt{6}} & 0 & -\frac{1}{\sqrt{2}} & 0 & -\frac{1}{\sqrt{12}} & -\frac{1}{2} \\ \frac{1}{\sqrt{6}} & 0 & 0 & -\frac{1}{\sqrt{2}} & \frac{2}{\sqrt{12}} & 0 \end{bmatrix}$$

Hence

$$\begin{aligned} \psi_1 &= \frac{1}{\sqrt{6}}\mathrm{s} + \frac{1}{\sqrt{2}}\mathrm{p_x} - \frac{1}{\sqrt{12}}\mathrm{d_z^2} + \frac{1}{2}\mathrm{d_{(x^2-y^2)}} \\ \psi_2 &= \frac{1}{\sqrt{6}}\mathrm{s} + \frac{1}{\sqrt{2}}\mathrm{p_y} - \frac{1}{\sqrt{12}}\mathrm{d_z^2} - \frac{1}{2}\mathrm{d_{(x^2-y^2)}} \\ \psi_3 &= \frac{1}{\sqrt{6}}\mathrm{s} + \frac{1}{\sqrt{2}}\mathrm{p_z} + \frac{1}{\sqrt{3}}\mathrm{d_z^2} \\ \psi_4 &= \frac{1}{\sqrt{6}}\mathrm{s} - \frac{1}{\sqrt{2}}\mathrm{p_x} - \frac{1}{\sqrt{12}}\mathrm{d_z^2} + \frac{1}{2}\mathrm{d_{(x^2-y^2)}} \\ \psi_5 &= \frac{1}{\sqrt{6}}\mathrm{s} - \frac{1}{\sqrt{2}}\mathrm{p_y} - \frac{1}{\sqrt{12}}\mathrm{d_z^2} - \frac{1}{2}\mathrm{d_{(x^2-y^2)}} \\ \psi_6 &= \frac{1}{\sqrt{6}}\mathrm{s} - \frac{1}{\sqrt{2}}\mathrm{p_z} + \frac{1}{\sqrt{3}}\mathrm{d_z^2} \end{aligned} \tag{9.20}$$

The following represent some results of hybridisation schemes that are applied in the bond formation of molecules. Chemical sense plays its role in selecting the proper type from amongst the possible few. Symmetry only gives the general results.

Geometry of bonds	Point group of the hybrid set	Nature of hybridization
Linear	$C_{\propto v}$	sp, dp
Planar trigonal	$D_{3\hbar}$	sp^2, sd^2, dp^2, d^3
Tetrahedral	T_d	sp^3, sd^3
Tetragonal plane	$D_{4\hbar}$	dsp^2, d^2p^2
Trigonal bipyramid	$D_{3\hbar}$	d^3sp, dsp^3
Octahedral	$O_\hbar$	d^2sp^3

9.4 Hybrids For $\pi-$Bond Formation.

It has already been mentioned that in molecules like PCl_3 there exists the possibility of $\pi-$bonds due to lateral overlaps of the suitably positioned ligand orbitals with the hybrid orbitals of the central atom (B or P). Such hybrids, if formed, are termed $\pi-$hybrids. We shall illustrate the nature of $\pi-$hybrids of the phosphorus atom in PCl_3.

The molecule is assumed planar and belongs to D_{3h} point group. Hence the $\pi-$hybrids, if formable, should conform to this symmetry. Fig. 9.6 demonstrates the set-up of the coordinate axes at the P atom

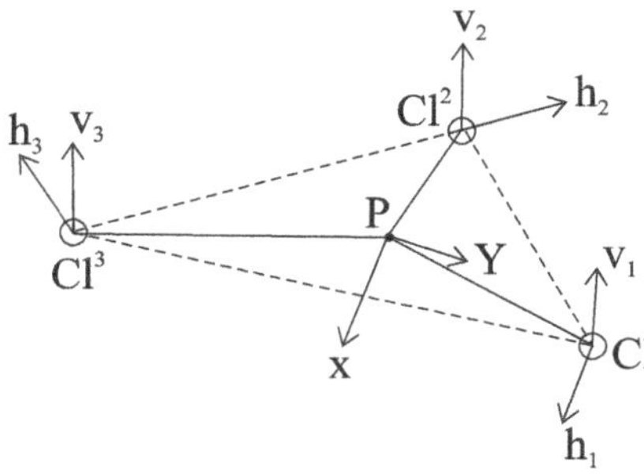

Fig. 9.6: Axial set-up and hybrid directions in PCl_3.

similar to that in BCl_3 (Fig. 9.2).

It is known that the sigma bonds denoted by the lines PCl^1, PCl^2 and PCl^3 are formed by overlaps of suitable orbitals of Cl^1, Cl^2 and Cl^3 with the sp^2 $-\sigma$hybrids of the P atom. These hybrids have already used s, p_x and p_y orbitals of the P atom. Now if in addition the P atom is also to be $\pi-$bonded to each *Cl* atom separately through the former's own $\pi-$hybrids, there can be a maximum of six such imaginable $\pi-$hybrids on the P atom, one pair of hybrids being earmarked for a maximum of two $\pi-$bonding with each Cl atom. Let the vectors $\vec{v}_1$, perpendicular to the molecular plane and $\vec{h}_1$ in the molecular plane and normal to the first σ bond direction PCl^1 represent the directions of the first pair of π hybrids of the phosphorus atom. Two points need be noted in this connection.

(i) To avoid clumsiness in the figure the hybrid vectors, instead of being shown at the P atom, are shown in parallel orientation at the atom Cl^1.

(ii) Since the $\pi-$bonds are always accompanied by nodal surfaces, the two nodal surfaces in the $\pi-$bonding with Cl^1 will be the molecular plane (to which $\vec{v_1}$ is normal) and the vertical plane passing through PCl^1 line to which $\vec{h_1}$ is normal.

The other two pairs of such $\pi-$hybrid vectors of P, needed for possible $\pi-$bondings separately with Cl^2 and Cl^3 and shown for convenience at the two Cl atoms, are $\vec{v_2}$, $\vec{h_2}$ and $\vec{v_3}$, $\vec{h_3}$ respectively. All these six hybrid vectors in reality shoot out from the P atom. Of these six, $\vec{v_1}$, $\vec{v_2}$, $\vec{v_3}$ constitute a vertical set and $\vec{h_1}$, $\vec{h_2}$, $\vec{h_3}$ make up an horizontal basis set.

These six hybrid vectors, as the basis set, are subjected to transformations of the point group $D_{3\hbar}$, to give the reducible representations. Since the vertical set and the horizontal set do not mix, the total hybrid representation is already in block diagonal form, i.e., we can write

$$\Gamma^{hyb} = \underset{\text{(vertical)}}{\Gamma^{hyb}} \oplus \underset{\text{(horizontal)}}{\Gamma^{hyb}}$$

The χ values for the two component representations can be separately obtained by noting the shift of vectors under the group operations.

$D_{3\hbar}$	$\hat{E}$	$2\hat{C}_3$	$3\hat{C}'_2$	$\hat{\sigma}_\hbar$	$2\hat{S}_3$	$3\hat{\sigma}_v$
$\chi_{vertical}$	3	0	−1	−3	0	1

This decomposes into IR″s A''_2 and E''. The basis atomic orbitals are p_z for A''_2 and (d_{xz}, d_{yz}) for E''. Hence the vertical $\pi-$hybrids of the P atom are of pd^2 type and are three in number capable of forming $\pi-$bonds with the three Cl atoms and having the molecular plane as the nodal surface. For the Γ^{hyb} (horizontal)

$D_{3\hbar}$	E	$2\hat{C}_3$	$3\hat{C}'_2$	$\hat{\sigma}_\hbar$	$2\hat{S}_3$	$3\hat{\sigma}_v$
$\chi_{horizontal}$	3	0	−1	3	0	−1

The decomposition is

$$\underset{\text{(horizontal)}}{\Gamma^{hyb}} = A'_2 \oplus E'$$

with basis atomic orbitals $(d_{(x^2-y^2)}d_{xy})$ for E′ and none for A'_2. This means there cannot be formed three equivalent $\pi-$hybrids of the horizontal set. Only two hybrids can form in the plane giving rise to two $\pi-$bonds shared equally by three chlorine atoms.

9.5 Symmetry Orbitals, Molecular Orbitals : Introduction.

It was already explained in Chapter 6 that the molecular orbitals form bases for the IR's of the molecular point group. How is this interconnection between symmetry demand and the forms of mo's going to be of any value in practical applications? Let us go a little more deeply into the quantum mechanical equations whereby we may reach some vantage point to make us feel the utility of such interconnection. The readers who are not conversant with the SCF methods may skip this and the next sections to join us later in Sec. 9.7.

Schrodinger equation for a molecular system (2n electron) is

$$H\psi(1,\ 2,\ .....p.\ q) = E\psi(1,\ 2,\ ....pq...) \tag{9.21}$$

where, in atomic units, the total electronic hamiltonian

$$\begin{aligned} H = H_1 + H_2 &= \sum_p H^{core}(p) + \sum_{p<q} \frac{1}{r_{pq}} \\ &= \left\{ \sum_p -\frac{1}{2}\nabla p^2 - \sum_p \sum_\alpha \frac{z_\alpha}{r_{\alpha p}} \right\} + \sum_{p<q} \frac{1}{r_{pq}} \end{aligned} \tag{9.22}$$

The total wavefunction ψ is a function of the coordinates of all the electrons, each figure or letter in the parenthesis e.g., 1, 2,... p, q, etc. represents a set of three coordinates such as $(x_1\ y_1\ z_1)$, $(x_2y_2z_2)...(x_p\ y_p\ z_p)...$

Eq. (9.21) in its pristine form is insoluble. In the SCF method, Hartree and Fock made two assumptions, one regarding the total wavefunction ψ and the other regarding a virtual substitution of H by an H^{eff} valid for an average smoothed-out charge distribution. It is assumed that ψ can be represented (for a closed-shell molecule) by a determinant of spin orbitals which are themselves the binary products of single-electron molecular orbitals and spin wavefunctions α or β.

$$\Psi = \frac{1}{[(2n)!]^{1/2}} \sum_{p} (-1)^{p} P \{\psi_1(1)\alpha(1)\psi_1(2)\beta(2)\psi_2(3)\alpha(3)\psi_2(4)\beta(4)...\} \tag{9.23}$$

Here $\sum_{p}(-1)p\ P$ is the permutation operator for exchange of electron coordinates and ψ's are the molecular orbitals assumed to be mutually orthogonal but yet unknown in forms. The suffixes 1, 2, 3...i, j, ...p, q.... represent tags denoting different molecular orbitals. The molecular orbitals are to be obtained theoretically from a variation equation obtainable from (9.22) and (9.23) along with the Lagrange's undetermined multipliers. The equation is

$$\delta G = \delta[E - 2\sum_{ij} \in_{ij} S_{ij}]\delta \left\{ \sum_{i} 2H_{ii} + \sum_{i}\sum_{j}(2J_{ij} - K_{ij}) - 2\sum_{i}\sum_{j} \in_{ij} S_{ij} \right\} = 0 \tag{9.24}$$

$$\begin{aligned} \text{where } E &= \langle \Psi \mid H \mid \Psi \rangle = \sum_{i} 2 \int \psi_i^*(p) H^{core}(p) \psi_i(p) d\tau \\ &+ \sum_{i}\sum_{j} 2 \int \psi_i^*(1)\psi_j^*(2) \frac{1}{r_{12}} \psi_i(1)\psi_j(2) d\tau \\ &- \sum_{i}\sum_{j} \int \psi_i^*(1)\psi_j^*(2) \frac{1}{r_{12}} \psi_j(1)\psi_i(2) d\tau \end{aligned}$$

The expressions for H_{ii}, J_{ij} and K_{ij}, are thus evident. S_{ij}'s are the overlap integrals and $2\in_{ij}$'s, the Lagrange's multipliers. Uptil now the original hamiltonian is left untarnished. But Eq. (9.24) is unmanageable without a redefinition of the hamiltonian. It is shown that H can be approximated as

$$H = \sum_{p} H^{core}(p) + H_2 \approx H^{eff} = \sum_{p} \left[H^{core}(p) + \sum_{j} \{2j_j(p) - K_j(p)\} \right] \tag{9.25}$$

=sum of effective one-electron hamiltonians

$$\left.\begin{aligned} \text{Here } J_j(1)\psi_i(1) &= \left[\int \psi_j^*(2) \tfrac{1}{r_{12}} \psi_j(2) d\tau_2 \right] \psi_i.(1) \\ \text{and } K_j(1)\psi_i(1) &= \left[\int \psi_j^*(2) \tfrac{1}{r_{12}} \psi_i(2) d\tau_2 \right] \psi_j(1) \end{aligned}\right\} \tag{9.26}$$

Insertion of Eq. (9.25) in (9.24) and subsequent handling lead to a series of Hartree Fock equations

$$F\psi_i = \left[H^{core}(p) + \sum_j \{2J_j(p) - K_j(p)\}\right] \psi_i = \in_i \psi_i \qquad (9.27)$$

F is called the Fock operator. These integro-differentia equations, if at all solved, will yield both the energies and the mo's which are the exact Hartree Fock orbitals. We pause here for a moment to think of symmetry. The true hamiltonian H is totally symmetric. It is assumed that H^{eff} and F are also totally symmetric. Hence the exact Hartree Fock molecular orbitals will have the symmetries of the IR's of the molecular point group.

Eq. (9.27) in most instances proves to be unwieldy once again and is in need of further simplification. Roothaan[11] made an assumption that the mo's, ψ_i's, can be expressed as LCAO's, i.e., $\psi_i = \sum_\mu \phi_\mu C_{\mu i}$ where the greek suffixes μ, ν are the atomic orbital tags. This assumption reduces the Fock equations (9.27) to what are called Roothaan equations.
O for all μ's =

$$\sum_\nu (F_{\mu\nu} - \in_i S_{\mu v})C_{vi} = 0 \text{ for all } \mu_s \qquad (9.28)$$

Here $F_{\mu v}$ is the $(\mu\nu)$th matrix element of the matrix operator $\mathbf{F}$ in the basis set of the atomic orbitals and is given by

$$\begin{aligned} F_{\mu\nu} &= \int \phi^*_\mu(1)F(1)\phi_\nu(1)d\tau_1 \\ &= \int \phi^*_\mu(1)\left[H^{core}(1) + \sum_j \{2J_j(1) - K_j(1)\}\right]\phi_\nu(1)d\tau_1 \end{aligned}$$

The solutions of (9.28) depend upon setting

$$\det|\mathbf{F} - \in_i \mathbf{S}| = 0 \qquad (9.29)$$

whence the orbital energies, eigenvectors and finally Roothaan's approximate molecular orbitals are obtained. These approximate mo's in LCAO forms also constitute bases of the IR's of the molecular point group. Not only do the Roothaan solutions but also the solutions of the various parametrised semitheoretical SCF methods fundamentally depend on this determinantal equation.

It is here that a quantum chemist seizes the opportunity offered by the symmetry principles to reduce the labour of solving Eq. (9.29) for energies. Symmetry helps us here in two ways:

1. Since it is known that the LCAO's are going to form the solutions, one can in advance formulate the suitable SALC's of atomic orbitals knowing the IR's of the molecular point group. These are symmetry orbitals $\phi'_s = \sum_{\mu} \phi_{\mu} \ C_{\mu s}$ Such symmetry orbitals (SALC's) belong to the same atomic orbitals function space. Combination of SALC's gives rise to the mo's.

2. By using the symmetry orbitals as the basis set for the matrix representation of the Fock operator **F** and with a proper sequential numbering of the symmetry orbitals, many of the Fock elements will vanish due to the rules of vanishing integrals (Chapter 6). This will blockdiagonalise the **F** matrix. Hence one will have to solve a number of small-dimensional determinantal equations for energy evaluation, although the iterative process cannot be avoided because of the occurrence of $J_j(p)$ and $K_j(p)$ operators in the **F** matrix.

9.6 $\pi-$Molecular Orbitals and Htickel Approximations: Introduction.

Many properties of various compounds, especially of the conjugated organic molecules, are essentially due to $\pi-$electrons and the the delocalised $\pi-$bonds. Sigma bonds and electrons in the sigma bonds constitute just a framework without any participation in such properties. The view[12] is held that the $\pi-$part can be quantitatively treated independently of the sigma part. This quantitative process starts from the bifurcation of the total hamiltonian and the factoring of the total wavefunction into component wavefunctions related to the sigma and π parts of the molecule.

Consider a molecule, containing n_σ electrons and n_π electrons, obeying Schrodinger equation $H\psi=E\psi$. Assuming $\psi = \psi^\sigma \psi^\pi$ and $H=H^\sigma+H^\pi$ we can split the original equation into two component equations.

$$\begin{aligned}
H^{\sigma}\psi^{\sigma} &= E_{\sigma}\psi^{\sigma} \text{ and } H^{\pi}\psi^{\pi} = E_{\pi}\psi^{\pi}, \text{ where} \\
H^{\sigma} &= \sum_{i=1}^{n_\sigma} -\frac{1}{2}\nabla_i^2 - \sum_i \sum_\alpha \frac{z_\alpha}{r_{\alpha i}} + \sum_{i<j} \frac{1}{r_{ij}} \\
\text{and } H^{\pi} &= \sum_{k}^{n_\pi} -\frac{1}{2}\nabla_k^2 - \sum_k \sum_\alpha \frac{z_\alpha}{r_{\alpha k}} + \sum_{j}^{n_\sigma} \sum_{k}^{n_\pi} \frac{1}{r_{jk}} + \sum_{k<l}^{n_\pi} \frac{1}{r_{kl}} \\
&= \sum_{k}^{n_\pi} H^{\pi\text{core}}(k) + \sum_{k<l}^{n_\pi} \frac{1}{r_{kl}} \qquad (9.30)
\end{aligned}$$

In the above expressions for the hamiltonians, the suffixes i, j refer to sigma electrons and k, l to $\pi-$electrons.

Expressing $\psi^{\pi} = \frac{1}{(n_\pi!)^{\frac{1}{2}}} \sum_p (-1)p\, P\, \{\psi_1^{\pi}(1)\alpha(1)\psi_1^{\pi}(2)\beta(2)\psi_2^{\pi}(3)\alpha(3)\psi_2^{\pi}(4)\beta(4)\cdots\}$

where ψ^{π}'s are the $\pi-$molecular orbitals and redefining an effective π hamiltonian, viz.,

$$H^{\pi} \approx H^{\pi\text{eff}} = \sum_p \left[H^{\pi\text{core}}(p) + \sum_l \{2J_l(p) - K_l(p)\} \right] \qquad (9.31)$$

it is possible to arrive, as in the previous section, at the series of Fock equations for the π system, viz.,

$$F^{\pi}\psi_i^{\pi} = \epsilon_i^{\pi}\, \psi_i^{\pi}. \qquad (9.32)$$

$$\text{where } \; F^{\pi}(p) = \left[H^{\pi\text{core}}(p) + \sum_l \{2J_l(p) - K_l(p)\} \right]$$

To obtain the equivalent Roothaan form of equation we have to suppose that the $-\pi$mo's are the linear combinations of the p_z, atomic orbitals contributed one apiece by the suitable carbon or the hetero atoms, i.e.,

$\psi_i^{\pi} = \sum_\mu \phi_\mu^{\pi} C_{\mu i}$. The final equation is then

$$\sum_\nu \left(F_{\mu\nu}^{\pi} - \epsilon_i^{\pi}\, S_{\mu\nu}^{\pi}\right) C_{\nu i} = 0 \qquad (9.33)$$

$$\text{i.e., } \; |\mathbf{F}^{\pi} - \epsilon^{\pi}\, \mathbf{S}^{\pi}| = 0 \qquad (9.34)$$

From the symmetry viewpoint we have the same two advantages as in Sec. 9.5, i.e., forming $\pi-$symmetry orbitals and $\pi-$molecular orbitals and secondly, effecting simplification of the Fock matrix by using the symmetry orbitals as the basis set.

The simple Huckel's parametrised $\pi-$mo method stems basically from (9.33). The parameters are

$$\begin{aligned} F^{\pi}_{\mu\mu} &= \alpha \text{ for all } \mu\text{'s.} \\ F^{\pi}_{\mu\nu} &= \beta \text{ when } \phi_\mu \text{ and } \phi_\nu \text{ are on adjacent carbon atoms and zero} \\ &\quad \text{otherwise.} \\ S^{\pi}_{\mu\text{v}} &= \delta_{\mu\nu} \end{aligned}$$

This represents the simplest version of $\pi-$mo theory and has been used widely by the organic chemists. Other semitheoretical SCF methods, such as PPP (Pariser Parr and Pople) method, also have their origins in Eq. (9.33).

9.7 Symmetry Orbitals, Group Orbitals And Molecular Orbitals.

Restricting our attention to symmetry only, we shall in this section deal with illustrations of building up the different orbitals as mentioned in the caption. Any SALC orbital is a symmetry orbital and may or may not be a molecular orbital. If corresponding to a given IR, there exist more than one linearly independent symmetry orbitals, a general linear combination of these symmetry orbitals will yield molecular orbitals. Group orbitals are also symmetry orbitals and are termed sometimes, more expressively, partial molecular orbitals. Linear combinations of group orbitals with suitable atomic orbitals lead to molecular orbitals.

Two types of mo's, the $\pi-$mo's and the $\sigma-$mo's, are of interest to the chemists. We shahl dwell upon the $\pi-$mo's of ring and open systems, mo's of sandwich compounds and a few mo's of simple systems. Construction of energy diagrams will be made only in a few interesting cases. Molecular orbitals of transition metal complexes will be dealt in Chapter 10.

A. π–MO's Of Ring Systems

Carbocyclic systems having a planar structure and of molecular formula C_nH_n (such as benzene, cyclopropene) form delocalised π–mo's. Assuming the molecular plane to be the xy plane, each carbon atom lets its p_z–atomic orbital combine with other p_z-orbitals to form π–mo's. The general method of forming the mo's involves the following stages: Stage

(i) Transformation of basis atomic orbitals to find the reducible representation.
(ii) Finding IR's in the reducible representation.
(iii) Use of projection operators to find the symmetry orbitals. (separate process also exists for cyclic molecules).
(iv) Combination of the symmetry orbitals into mo's.

For the cyclic C_nH_n systems, simpler method exists which will be illustrated by taking the important case of benzene first.

Example 9.4. Obtain the π–mo's of benzene.

Benzene molecule belongs to the point group $D_{6\hbar}$ and has a planar framework constituted by the carbon atoms, each of which is σ–bonded to a pair of carbon atoms and all jointly giving rise to a ring. The p_z orbitals of all the carbons are vertically disposed with respect to the ring plane and with the positive ends turned upwards. Let us number the carbon atoms 1 to 6 (Fig. 9.7) and the p_z atomic orbitals as ϕ_1 ϕ_2 ϕ_3 ϕ_4 ϕ_5 and ϕ_6.

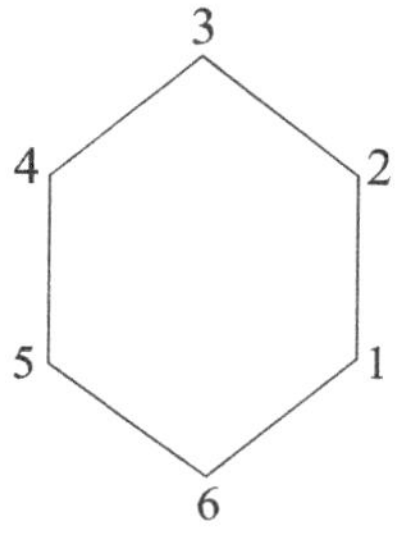

Fig. 9.7: Numbering the C-atoms of 3_7H_7.

The essentialities of the π–mo's of C_6H_6 of symmetry group D_{6h} are obtainable by considering the cyclic subgroup C_6 only of which the various symmetry operations are C_n^m, n=6 and m ranges from 0 to 5, C_6^0 being C_6^6=E for benzene.

The transformation properties of stage

(i) for finding the reducible representation can easily be obtained for benzene. A general expression for $C_n^m \phi_\mu$ in any planar carbocycle (for an anticlock rotation of ϕ_μ's) is

$$\left.\begin{aligned} C_n^m \phi_\mu &= \phi_{m+\mu} \text{ when } m+\mu \leq n \quad \text{and} \\ &= \phi_{m+\mu-n} \text{ when } m+\mu > n \end{aligned}\right\} \tag{9.35}$$

Using the basis set Φ and Eq. (9.35) we have for C_6H_6 the final set of characters as shown here:

	$\widehat{E}$	$\widehat{C_6}$	$\widehat{C_3}$	$\widehat{C_2}$	$\widehat{C_3^2}$	$\widehat{C_6^5}$
χ^{rea}	6	0	0	0	0	0

(ii) Application of Theo. 5.8 shows that the reducible representation Γ^{red}=A+B+E_1+E_2

The E_1 representation, which is doubly degenerate, is known to consist of two one-dimensional IR's with complex characters. So is E_2. Hence the number of one-dimensional IR's spanned by the basis set is 6, the same as the number of p_z orbitals of benzene molecule and is also the same, as will be shown afterwards, as the number of π−mo's of benzene. The last statement is consequent upon the fact that each mo forms a base of an IR

It is at stage (iii) that we deviate from the general prescribed method and avoid the use of projection operators. It is noted that since all the IR's are one-dimensional, the respective characters themselves represent the transformation coefficients.

Let ψ_i be a π−mo of IR, Γ_i, of C_6H_6. Then

$$\psi_i = \sum_{\mu=1}^{6} \phi_\mu a_{\mu i} \tag{9.36}$$

For an operator C_n^m of the point group C_6,

$$C_n^m \psi_i \quad = \quad \sum_{\mu=1}^{6} C_n^m \phi_\mu a_{\mu i}$$

$$\text{i.e., } \chi_i(C_n^m)\psi_i = \sum_{\mu=1}^{6} C_n^m \phi_\mu a_{\mu i} \tag{9.37}$$

Utilizing Eqs. (9.35 and 9.36) in (9.37), we may write

$$\left.\begin{aligned} \sum_{\mu=1}^{6} \chi_i(C_n^m)a_{\mu i}\phi_\mu &= \sum_{\mu=1}^{6} \phi_{\mu+m}a_{\mu i} \text{ (when } \mu + m \leq n) \\ &= \sum_{\mu=1}^{6} \phi_\mu a_{(\mu-m)i} \text{ with } \mu > m \end{aligned}\right\} \tag{9.38a}$$

$$\begin{aligned} \text{If } (\mu + m) > n,\ \text{LHS} &= \sum_{\mu=1}^{6} \phi_{\mu+(m-n)}a_{\mu i} \\ &= \sum_{\mu=1}^{6} \phi_\mu a_{(\mu-\overline{m-n})i} \\ &= \sum_{\mu=1}^{6} \phi_\mu a_{(\mu+n-m)i} \end{aligned} \tag{9.38b}$$

When, however, $\mu > m$, $a_{\mu+n-m} \equiv a_{\mu-m}$, since $a_{\mu+n} \equiv a_\mu$. Hence, under such conditions, Eq. (9.38b) can be written as

$$\sum_{\mu=1}^{6} \chi_i(C_n^m)a_{\mu i}\phi_\mu = \sum_{\mu=1}^{6} \phi_\mu a_{(\mu-m)i} \tag{9.38}$$

which is identical with Eq. (9.38a) in form and condition.

Since the atomic orbitals, ϕ_μ's (viz., p_z's), are all linearly independent, one can equate the coefficients of ϕ_6 from Eqs. (9.38a or 9.38) for different possible values of m(0. to 5). This yields

$$\begin{aligned} \chi_i(C_n^0)a_{6_i} &= a_{6i} & \chi_i(C_n^3)a_{6i} &= a_{3i} \\ \chi_i(C_n^1)a_{6_i} &= a_{5i} & \chi_i(C_n^4)a_{6i} &= a_{2i} \\ \chi_i(C_n^2)a_{6_i} &= a_{4i} & \chi_i(C_n^5)a_{6i} &= a_{1i} \end{aligned} \tag{9.39}$$

Remembering n=6 for benzene, Eq. (9.36) can be recast with the help of (9.39) as

$$\begin{aligned} \psi_i = \sum_\mu a_{\mu i}\phi_\mu &= a_{6i}\left[\chi_i(C_6^5)\phi_1 + \chi_i(C_6^4)\phi_2 + \chi_i(C_6^3)\phi_3\right. \\ &\qquad \left.\chi_i(C_6^2)\phi_4 + \chi_i(C_6^1)\phi_5 + \chi_i(C_6^0)\phi_6\right] \end{aligned} \tag{9.40}$$

Putting a_{6i}=1 and looking up the character table of C_6, we may write the π−mo's (unnormalised) from (9.40) as

$$\begin{aligned}
\psi(\mathrm{a}) &= \phi_1+\phi_2+\phi_3+\phi_4+\phi_5+\phi_6 \\
\psi(\mathrm{b}) &= -\phi_1+\phi_2-\phi_3+\phi_4-\phi_5+\phi_6 \\
\psi(\mathrm{e}_1^1) &= \epsilon^*\phi_1-\epsilon\phi_2-\phi_3-\epsilon^*\phi_4+\epsilon\phi_5+\phi_6 \\
\psi(\mathrm{e}_1^2) &= \epsilon\phi_1-\epsilon^*\phi_2-\phi_3-\epsilon\phi_4+\epsilon^*\phi_5+\phi_6 \\
\psi(\mathrm{e}_2^1) &= -\epsilon\phi_1-\epsilon^*\phi_2+\phi_3-\epsilon\phi_4-\epsilon^*\phi_5+\phi_6 \\
\psi(\mathrm{e}_2^2) &= -\epsilon^*\phi_1-\epsilon\phi_2+\phi_3-\epsilon^*\phi_4-\epsilon\phi_5+\phi_6 \qquad (9.41)
\end{aligned}$$

The degenerate $\psi(\mathrm{e}_1)$ pair and the degenerate $\psi(\mathrm{e}_2)$ pair of (9.41) can be thrown into real forms, e.g.,

$$\begin{aligned}
\psi(\mathrm{e}_1^1)+\psi(\mathrm{e}_1^2) &= (\epsilon^*+\epsilon)\phi_1-(\epsilon+\epsilon^*)\phi_2-2\phi_3-(\epsilon^*+\epsilon)\phi_4+(\epsilon+\epsilon^*)\phi_5+2\phi_6 \\
&= \phi_1-\phi_2-2\phi_3-\phi_4+\phi_5+2\phi_6 \qquad (9.42)
\end{aligned}$$

$$\text{since } (\epsilon^*+\epsilon)=\left(\cos\frac{2\pi}{6}-\mathrm{i}\sin\frac{2\pi}{6}+\cos\frac{2\pi}{6}+\mathrm{i}\sin\frac{2\pi}{6}\right)=1.$$

Again $(\epsilon-\epsilon^*)=2\mathrm{i}\sin\frac{2\pi}{6}=2\mathrm{i}\frac{\sqrt{3}}{2}=\mathrm{i}\sqrt{3}$ and remembering this we have

$$\begin{aligned}
\left[\psi(\mathrm{e}_1^2)-\psi(\mathrm{e}_1^1)\right] &= (\epsilon-\epsilon^*)\phi_1+(\epsilon-\epsilon^*)\phi_2-(\epsilon-\epsilon^*)\phi_4-(\epsilon-\epsilon^*)\phi_5 \\
&= \mathrm{i}\sqrt{3}(\phi_1+\phi_2-\phi_4-\phi_5) \qquad (9.43)
\end{aligned}$$

As the symmetry of an orbital is not affected by multiplication or division by a numerical quantity, we can write (9.42) and (9.43) in normalised forms

$$\left.\begin{aligned}
\psi(\mathrm{e}_1\mathrm{a}) &= \tfrac{1}{\sqrt{12}}(\phi_1-\phi_2-2\phi_3-\phi_4+\phi_5+2\phi_6) \\
\psi(\mathrm{e}_1\mathrm{b}) &= \tfrac{1}{2}(\phi_1+\phi_2-\phi_4-\phi_5)
\end{aligned}\right\} \qquad (9.44)$$

The degenerate pair, $\psi(\mathrm{e}_2)$ can similarly be rendered into real forms.

Thus the final set of π−mo's in normalised forms are identifiable with the symmetry orbitals which, if used as basis set, completely. diagonalise the Huckel matrix. The π−mo's are

$$\left.\begin{aligned}
\psi(\mathrm{a}) &= \tfrac{1}{\sqrt{6}}(\phi_1+\phi_2+\phi_3+\phi_4+\phi_5+\phi_6) \\
\psi(\mathrm{b}) &= \tfrac{1}{\sqrt{6}}(-\phi_1+\phi_2-\phi_3+\phi_4-\phi_5+\phi_6) \\
\psi(\mathrm{e}_1\mathrm{a}) &= \tfrac{1}{\sqrt{12}}(\phi_1-\phi_2-2\phi_3-\phi_4+\phi_5+2\phi_6) \\
\psi(\mathrm{e}_1\mathrm{b}) &= \tfrac{1}{2}(\phi_1+\phi_2-\phi_4-\phi_5) \\
\psi(\mathrm{e}_2\mathrm{a}) &= \tfrac{1}{\sqrt{12}}(-\phi_1-\phi_2+2\phi_3-\phi_4-\phi_5+2\phi_6) \\
\psi(\mathrm{e}_2\mathrm{b}) &= \tfrac{1}{2}(\phi_1-\phi_2+\phi_4-\phi_5)
\end{aligned}\right\} \qquad (9.45)$$

Let us pause for a moment to reflect on what we have gained or missed in the foregoing treatment before proceeding to further illustrations.

Since C_6H_6 belongs to $D_{6\hbar}$, the π–orbitals of relations (9.45) do not reveal the full symmetries when represented simply as ψ(a), ψ(b) etc. A careful reader ought to be able to write down the full symmetries by a meticulous check of the character table of $D_{6\hbar}$ [mainly $\chi(C_2')$, $\chi(C_2'')$ and $\chi(i)$] and of the structures of the π–orbitals. These are $\psi(a_{2u})$, $\psi(b_{2g})$, $\psi(e_{1g}a)$, $\psi(e_{1g}b)$, $\psi(e_{2g}a)$ and $\psi(e_{2g}b)$. Our task in this respect would have been easier had we undertaken the initial extra labour of dealing with D_6 subgroup instead of the C_6 subgroup.

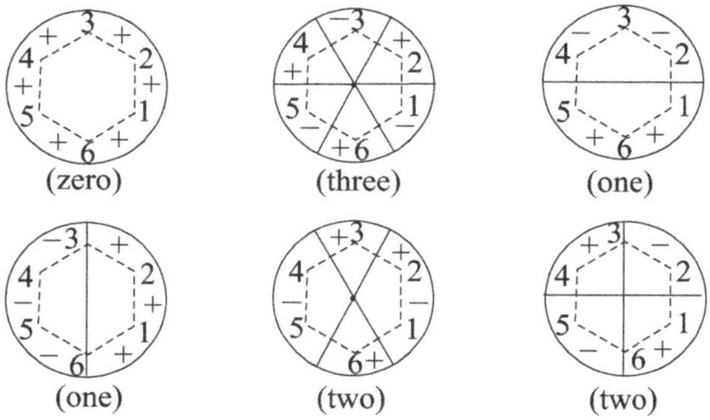

Fig. 9.8: π– nodal planes in benzene excluding the one in the molecular plane.

The quantitative forms of the orbitals (9.45) enable us to have qualitative ideas of the π–nodal planes in benzene molecule. These are apparent from the Fig. 9.8 which can be tentatively drawn having relations (9.45) in view.

The state ψ(b) with the highest number of nodal planes will have the highest π–energy. The next in order are the degenerate pair $\psi(e_2a)$ and $\psi(e_2b)$ with two nodal planes each.

The calculations of π energies and resonance energies from the mo's are merely a little formality. But since, in the present instance, these do not involve any further use of symmetry principles, these are omitted here.

One general principle has emerged from symmetry treatment of the $\pi-$mo's in all cyclic molecules:

Cyclic molecules having n number of p_z atomic orbitals will yield n number of $\pi-$mo's and each of the n IR's of the C_n point group will represent the symmetry of the $\pi-$mo's, one apiece.

This general principle, if taken advantage of, at once dispenses with the necessity of going through stages (i) and (ii) in finding the $\pi-$mo's. One immediately starts writing the $\pi-$mo's corresponding to the one-dimensional n IR's of the C_n point group (see Example 9.5).

There are also suitable instances of molecules which are not cyclic in the real sense but may be *deemed* to consist of a number of cyclic molecules (!) only from symmetry viewpoint. Even suitable openchain molecules or molecules with fused rings may come under the 'deemed cyclic' category and the advantages of cyclic molecule can be exploited in such instances (see Example 9.6). In these cases only $\pi-$symmetry orbitals or group orbitals are first generated, which can then be linearly combined to yield the $\pi-$mo's.

Example 9.5.

Find the $\pi-$molecular orbitals of cyclopentadienyl radical.

The radical $(C_5H_5)^{\cdot}$ has a (hypothetically symmetrical) planar structure with five C-atoms contributing their p_z-orbitals for $\pi-$bond formation. The radical belongs to the point group $D_{5\hbar}$ and is cyclic Hence, taking advantage of the general principle stated before, its symmetry orbitals (mo's in this case) can be formed directly by consulting the character table of C_5. The $\pi-$mo's (unnormalised) are

$$\left.\begin{aligned}
&\psi(a) = \phi_1 + \phi_2 + \phi_3 + \phi_4 + \phi_5 \\
&\psi(e_1^1) = a_{5e1'}\left[\chi_{e1}(C_5^4)\phi_1 + \chi_{e1}(C_5^3)\phi_2 + \chi_{e1}(C_5^2)\phi_3\right. \\
&\left.+\chi_{e1}(C_5^1)\phi_4 + \chi_{e_1}(e_5^0)\phi_5\right] \\
&= a_{5e_1'}(\epsilon^*\phi_1 + \epsilon^{2*}\phi_2 + \epsilon^2\phi_3 + \epsilon\phi_4 + \phi_5) \\
&\psi(e_1^2) = a_{5e1}^2(\epsilon\phi_1 + \epsilon^2\phi_2 + \epsilon^{2*}\phi_3 + \epsilon^*\phi_4 + \phi_5) \\
&\psi(e_2^1) = a_{5e2}{}^1(\epsilon^{2*}\phi_1 + \epsilon\phi_2 + \epsilon^*\phi_3 + \epsilon^2\phi_4 + \phi_5) \\
&\psi(e_2^2) = a_{5e2}^2(\epsilon^2\phi_1 + \epsilon^*\phi_2 + \epsilon\phi_3 + \epsilon^{2*}\phi_4 + \phi_5) \\
&\text{where } \epsilon = \left[\cos\tfrac{2\pi}{5} + \mathrm{i}\sin\tfrac{2\pi}{5}\right]
\end{aligned}\right\} \quad (9.46)$$

Taking the mutual sums and differences of the pairs of complex orbitals and dropping the numerical common coefficients (since this does not affect the symmetries of the orbitals) the real forms are obtained from the complex ones. These are with $w = 2\pi/5$

$$\left.\begin{array}{l}\psi(\mathrm{e_1a}) = [\phi_1 \cos \mathrm{w} + \phi_2 \cos 2\mathrm{w} + \phi_3 \cos 2\mathrm{w} + \phi_4 \cos \mathrm{w} + \phi_5] \\ \psi(\mathrm{e_1b}) = [\phi_1 \sin \mathrm{w} + \phi_2 \sin 2\mathrm{w} - \phi_3 \sin 2\mathrm{w} - \phi_4 \sin 2\mathrm{w}]\end{array}\right\} \quad (9.47)$$

$$\left.\begin{array}{l}\psi(\mathrm{e_2a}) = [\phi_1 \cos 2\mathrm{w} + \phi_2 \cos \mathrm{w} + \phi_3 \cos \mathrm{w} + \phi_4 \cos 2\mathrm{w} + \phi_5] \\ \psi(\mathrm{e_2b}) = [\phi_1 \sin 2\mathrm{w} - \phi_2 \sin \mathrm{w} + \phi_3 \sin \mathrm{w} - \phi_4 \sin 2\mathrm{w}]\end{array}\right\} \quad (9.48)$$

Additionally, we have from (9.46)

$$\psi(\mathrm{a}) = [\phi_1 + \phi_2 + \phi_3 + \phi_4 + \phi_5] \quad (9.49)$$

To find the normalising factors the values of cosw, cos2w, sinw, sin2w in Eqs. (9.47-9.49) are used. The set of $\pi-$orbitals can be written in normalised forms, viz.,

$$\begin{aligned}
\psi^\pi(\mathrm{a}) &= \frac{1}{\sqrt{5}}(\phi_1 + \phi_2 + \phi_3 + \phi_4 + \phi_5) \\
\psi^\pi(\mathrm{e_1a}) &= \sqrt{\frac{2}{5}}\,[(\cos \mathrm{w})\phi_1 + (\cos 2\mathrm{w})\phi_2 + (\cos 2\mathrm{w})\phi_3 + (\cos \mathrm{w})\phi_4 + \phi_5] \\
\psi^\pi(\mathrm{e_1b}) &= \sqrt{\frac{2}{5}}\,[(\sin \mathrm{w})\phi_1 + (\sin 2\mathrm{w})\phi_2 - (\sin 2\mathrm{w})\phi_3 - (\sin \mathrm{w})\phi_4] \\
\psi^\pi(\mathrm{e_2a}) &= \sqrt{\frac{2}{5}}\,[(\cos 2\mathrm{w})\phi_1 + (\cos \mathrm{w})\phi_2 + (\cos \mathrm{w})\phi_3 + (\cos 2\mathrm{w})\phi_4 + \phi_5] \\
\psi^\pi(\mathrm{e_2b}) &= \sqrt{\frac{2}{5}}\,[(\sin 2\mathrm{w})\phi_1 - (\sin \mathrm{w})\phi_2 + (\sin \mathrm{w})\phi_3 - (\sin 2\mathrm{w})\phi_4] \quad (9.50)
\end{aligned}$$

B. $\pi-$MO's Of Open Chain Systems

We now turn to such open chain π systems which, contrarily to all canons of chemistry, can be 'deemed cyclic' only to achieve easily the group orbitals from symmetry considerations. This "deemed cyclic" category permits reaping the benefits of the general principle laid down earlier concerning cyclic molecules. Trivinylmethyl radical, with the skeleton as shown with numbers in Fig. 9.9, provides one such instance. Other systems, such as trimethylene cyclopropane, tetramethylene cyclobutane, can be regarded as consisting of two 'cyclic systems' each from symmetry viewpoint. Their group orbitals and molecular orbitals can be found

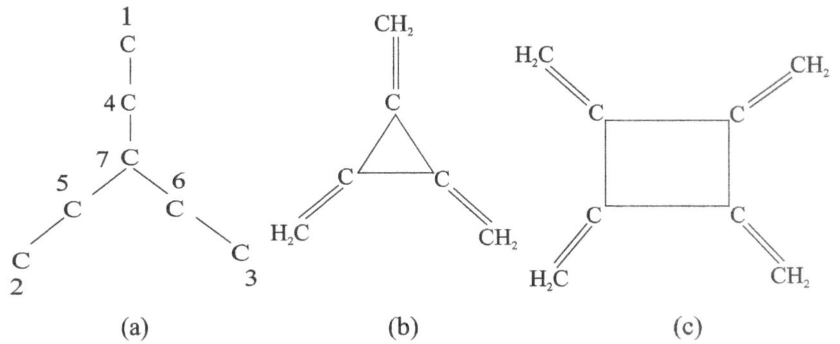

Fig. 9.9–(a) Skeleton of Trivinylmethyl radical (b) Trimethylene cyclopropane (c) Tetramethylene cyclobutane).

out in the same way as in trivinyl methyl radical worked out below.

Example 9.6

Find the $\pi-$mo's and $\pi-$energies of trivinylmethyl radical.

The skeleton of the radical is shown in Fig. 9.9(a) with the tag number of the C-atoms. It has the symmetry $D_{3\hbar}$. The group elements of the point group C_3 (a subgroup of $D_{3\hbar}$) are adequate for finding the present set of orbitals. We shall thus work with the point group C_3. Let ϕ_1 ϕ_2....ϕ_7 be the p_z atomic orbitals of these carbon atoms. It will be readily realised that under the group operations of C_3, ϕ_1 ϕ_2 and ϕ_3 exchange places, so also do the orbitals ϕ_4, ϕ_5 and ϕ_6. There is no mixing of the set (ϕ_1 ϕ_2 ϕ_3) with the set (ϕ_4 ϕ_5 ϕ_6). The orbital ϕ_7 remains fixed in its position. The behaviour of ($\phi_1\phi_2\phi_3$) set under C_3 is the same as would be displayed by the p_z orbitals of the carbon atoms 1, 2 and 3, if these were chemically bonded in a cyclic form. Hence from symmetry viewpoint we may regard the C-atoms 1, 2, 3 forming a 'hypothetical cyclic' compound. Exactly similar remarks would apply to the carbon atoms 4, 5 and 6.

Taking advantage of the general principles and noting that the character table of C_3 is

	E	C_3	C_3^2
A	1	1	1
E	1	ϵ^*	ϵ
	1	ϵ	ϵ^*

where $\epsilon = e^{i\frac{2\pi}{3}}$, we may at once write down the group orbitals for the two 'deemed cyclic' sets ($\phi_1\ \phi_2\ \phi_3$) and ($\phi_4\phi_5\phi_6$).

$$\left.\begin{array}{ll} g^{(1)} = \psi(a) = \phi_1 + \phi_2 + \phi_3 & g^{(4)} = \psi(a) = \phi_4 + \phi_5 + \phi_6 \\ g^{2} = \psi(e_1^1) = \epsilon\phi_1 + \epsilon^*\phi_2 + \phi_3 & g^{(5)} = \psi(e_1^1) = \epsilon\phi_4 + \epsilon^*\phi_5 + \phi_6 \\ g^{(3)} = \psi(e_1^2) = \epsilon^*\phi_1 + \epsilon\phi_2 + \phi_3 & g^{(6)} = \psi(e_1^2) = \epsilon^*\phi_4 + \epsilon\phi_5 + \phi_6 \end{array}\right\} \tag{9.51}$$

Writing the group orbitals in real forms with normalising factors, we have,

$$\left.\begin{array}{ll} g_1(a) = \frac{1}{\sqrt{3}}(\phi_1 + \phi_2 + \phi_3) & g_4(a) = \frac{1}{\sqrt{3}}(\phi_4 + \phi_5 + \phi_6) \\ g_2(e_1a) = \frac{1}{\sqrt{6}}(\phi_1 + \phi_2 + 2\phi_3) & g_5(e_1a) = \frac{1}{\sqrt{6}}(\phi_4 + \phi_5 + 2\phi_6) \\ g_3(e_1b) = \frac{1}{\sqrt{2}}(\phi_1 - \phi_2) & g_6(e_1b) = \frac{1}{\sqrt{2}}(\phi_4 - \phi_5) \end{array}\right\} \tag{9.52}$$

We have still one atomic orbital left for consideration, viz., ϕ_7 of the C-atom No. 7. It may be recalled (cf. Sec. 5.8) that the atomic orbitals of an atom, defined with respect to an origin at a point within the molecule which is the meeting point of all the symmetry elements of the molecules, transform as bases of the IR's of the molecular point group. Viewed in this light, it is found from the character table of C_3 point group that ϕ_7 (ap_z atomic orbital) belongs to the IR, A, of C_3. Hence $g_7(a)=\phi_7$.

To find the $\pi-$orbital energies, we have to solve (cf. Sec. 9.6) the equation

$$\sum_{v} \left(F^{\pi}_{\mu v} - \in^{\pi}_{i} S^{\pi}_{\mu v}\right) C_{vi} = 0 \tag{9.53}$$

In other words to find the $\pi-$energies we have to diagonalise the Fock $\pi-$matrix $\mathbf{F}^{\pi}$, i.e., to solve the determinantal equation

$$|\mathbf{F}^{\pi} - \in^{\pi} \mathbf{S}^{\pi}| = 0$$

the $\mathbf{F}^{\pi}$ is the matrix form of the Fock $\pi-$hamiltonian in the basis set of atomic orbital (viz. p_z). We can, instead, use the basis set of $\pi-$group orbitals (g_1, g_2,...g_7) since these also span the same p_z-atomic orbitals function space. Only we just remember the group orbitals keeping the orbitals of the same symmetry in sequential order. This ordering leads to block diagonalisation of the Fock matrix. With this end in view, let us serialise the orbits as

$$\{g_1 = G_1,\ g_4 = G_2,\ g_7 = G_3\}, \{g_2 = G_4,\ g_5 = G_5\}, \{g_3 = G_6,\ g_6 = G_7\}$$

The determinant then turns out to be

$$\begin{vmatrix}
f_{11}^{\pi} - \in & f_{12}^{\pi} & f_{13}^{\pi} & & & & \\
f_{21}^{\pi} & f_{22}^{\pi} - \in & f_{23}^{\pi} & & & & \\
f_{31}^{\pi} & f_{32}^{\pi} & f_{33}^{\pi} - \in & & & & \\
 & & & f_{44}^{\pi} - \in & f_{45}^{\pi} & & \\
 & & & f_{54}^{\pi} & f_{55}^{\pi} - \in & & \\
 & & & & & f_{65}^{\pi} - \in & f_{67}^{\pi} \\
 & & & & & f_{76}^{\pi} & f_{77}^{\pi} - \in
\end{vmatrix} = 0 \tag{9.54}$$

The vacant places have Zeros as the elements. Here the element

$f_{ij}^{\pi} =< G_i|F^{\pi}|G_j >$ and is to be distinguished from $F_{\mu\nu}^{\pi}$ where $F_{\mu\nu}^{\pi} =< \phi_\mu|F^{\pi}|\phi_\nu >$ involves the atomic p_z orbitals. Since the G_is are related, through g's, to the corresponding ϕ_μ's (i.e p_z's), each of the f_{ii}^{π}'s, f_{ij}^{π}'s can be expressed in terms of $F_{\mu\mu}^{\pi}$'s and $F_{\mu\nu}^{\pi}$'s.

We just confine ourselves to working out the details in respect of f_{11}^{π} and f_{12}^{π} only.

$$\begin{aligned}
f_{11}^{\pi} = \langle G_1|F^{\pi}|G_1\rangle &= \frac{1}{3}\langle(\phi_1+\phi_2+\phi_3)|F^{\pi}|(\phi_1+\phi_2+\phi_3)\rangle \\
&= \frac{1}{3}F_{11}^{\pi} + \frac{1}{3}F_{22}^{\pi} + \frac{1}{3}F_{33}^{\pi} + \frac{1}{3}F_{12}^{\pi} + \frac{1}{3}F_{13}^{\pi} + \frac{1}{3}F_{21}^{\pi} \\
&\quad + \frac{1}{3}F_{23}^{\pi} + \frac{1}{3}F_{31}^{\pi} + \frac{1}{3}F_{33}^{\pi}
\end{aligned}$$

If we incorporate Huckel's $\pi-$ orbital approximation viz.,$\mathrm{F}_{\mu\mu}^{\pi} = \alpha$, $\mathrm{F}_{\mu\nu}^{\pi} = 0$ or β (non adjacent and adjacent approximations for all μ's and ν's, we can write

$$\begin{aligned}
\mathrm{f}_{11}^{\pi} &= \frac{1}{3}(\mathrm{F}_{11}^{\pi} + \mathrm{F}_{22}^{\pi} + \mathrm{F}_{33}^{\pi}) \\
&= \alpha
\end{aligned}$$

Similarly, it can be shown

$$\begin{aligned}
f_{12}^{\pi} =< \mathrm{G}_1|\mathrm{F}^{\pi}|\mathrm{G}_2 > &= \frac{1}{3} < (\phi_1+\phi_2+\phi_3)|\mathrm{F}^{\pi}| < \phi_4+\phi_5+\phi_6 > \\
&= \frac{1}{3}(\beta+\beta+\beta) = \beta \text{ (adjacentcy approximation)}
\end{aligned}$$

It is then obvious that the (7×) determinant (9.54) is factored into one (3 × 3) and two (2 × 2) determinants only because symmetry orbitals (group orbitals), iustead of the pure p orbitals, have been used as the basis set. This is an example how symmetry principle simplifies the secular determinant. (cf. conclusion of sec. 9.5). Turning our attention to the (3 × 3) determinant first and the approximation and the nature of results shown and putting $\frac{\alpha-\epsilon}{\beta} = x$, the first two determinants reduce to

$$\begin{vmatrix} x & 1 & 0 \\ 1 & x & \sqrt{3} \\ 0 & \sqrt{3} & x \end{vmatrix} = 0 \tag{9.55}$$

and

$$\begin{vmatrix} x & 1 \\ 1 & x \end{vmatrix} = 0 \tag{9.56}$$

The third is not written, but let us a trivial exercise for the students. The solutions of (9.55) are $x = 0, \pm 2$ and of (9.56) $x = \pm 1$. The $\pi-$ energy values are thus

$$\begin{aligned} \epsilon_1\,(a) &= \alpha; \quad \epsilon_2\,(a) = \alpha - 2\beta; \quad \epsilon_3\,(a) = \alpha + 2\beta \\ \epsilon_4\,(e_1) &= \alpha - \beta; \quad \epsilon_5\,(e) = \alpha + \beta \end{aligned}$$

The remaining part of the problem consists of finding the linear combinations of (G_1 G_2...G_7) and hence of the atomic orbitals (ϕ_1, ϕ_2....ϕ_7) identifiable with the $\pi-$mo's. These steps do not involve any further application of symmetry principle, but depend only on the algebraic process of finding the coefficients C_{vi} of Eq. (9.53) for each specific value of ϵ_i. coupled with the normalisation condition $\sum_v C^2_{vi} = 1$. These algebraic steps are skipped here.

The final set of $\pi-$mo's are

$$\psi_1\pi(a) = \frac{1}{2}(\phi_1 + \phi_2 + \phi_3 - \phi_7); \; \epsilon = \alpha$$

$$\psi_2\pi(a) = \frac{1}{\sqrt{24}}(\phi_1 + \phi_2 + \phi_3 - 2\phi_4 - 2\phi_5 - 2\phi_6 + 3\phi_7); \; \epsilon = \alpha - 2\beta$$

$$\psi_3\pi(a) = \frac{1}{\sqrt{24}}(\phi_1 + \phi_2 + \phi_3 + 2\phi_4 + 2\phi_5 + 2\phi_6 + 3\phi_7); \; \epsilon = \alpha + 2\beta$$

$$\left.\begin{aligned} \psi_4\pi(e_1) &= \tfrac{1}{\sqrt{12}}(2\phi_1 - \phi_2 - \phi_3 - 2\phi_4 + \phi_5 + \phi_6) \\ \psi_5\pi(e_1) &= \tfrac{1}{2}(\phi_2 - \phi_3 - \phi_5 + \phi_6) \end{aligned}\right\} \epsilon = \alpha - \beta.$$

The sixth and the seventh $\pi-$orbitals may be obtained from the third determinant.

Employment of projection operators in the construction of symmetry orbitals and molecular orbitals may now be taken up. Of course in many such cases where projection operators are used, the simpler methods may turn out to be the methods of inspection and inference or the methods of carbocylic systems and 'deemed cyclic' systems treated earlier. As a very simple example of projection operator method, we shall consider the trans form of butadiene.

Example 9.7

Find the $\pi-$mo's and $\pi-$orbital energies of trans form of butadiene.

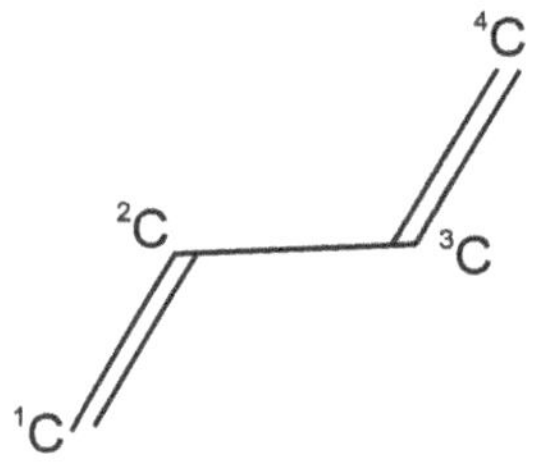

Let the p_z-orbitals of the four C-atoms as numbered in the structural formula be ϕ_1, ϕ_2, ϕ_3 and ϕ_4. The molecule belongs to $C_{2\hbar}$ point group. The full transformation properties are (the positive lobes of ϕ's point upwards).

	E	C_2	$\sigma_\hbar$	i
ϕ_1	ϕ_1	ϕ_4	$-\phi_1$	$-\phi_4$
ϕ_2	ϕ_2	ϕ_3	$-\phi_2$	$-\phi_3$
ϕ_3	ϕ_3	ϕ_2	$-\phi_3$	$-\phi_2$
ϕ_4	ϕ_4	ϕ_1	$-\phi_4$	$-\phi_1$
χ_{red}	4	0	-4	0

On reducing the representation we have

$$\Gamma^{red} = 2A_u + 2B_g$$

The projection operators

$$\begin{aligned}\wp \mathrm{A_u} &= \frac{1}{4}\left[1\hat{\mathrm{E}} + 1.\hat{\mathrm{C}}_2 - 1\hat{\sigma}_\hbar - 1.\hat{\mathrm{i}}\right]\\ \text{and } \wp \mathrm{B_g} &= \frac{1}{4}\left[\hat{\mathrm{E}} - \hat{\mathrm{C}}_2 - \hat{\sigma}_\hbar + \hat{\mathrm{i}}\right]\end{aligned}$$

With the help of these projection operators we now build up the symmetry orbitals choosing combinations $(\phi_1 + \phi_2)$ and $(\phi_1 - \phi_2)$ as two arbitrary functions in the function space. The reason for choosing ϕ_1 and ϕ_2 is that they are nonequivalent and do not mix under symmetry operations. Symbolising the SALC's as $\mathrm{S_1}$ $\mathrm{S_2}$ $\mathrm{S_3}$ and $\mathrm{S_4}$ we have the following set of expressions.

$$\begin{aligned}
\mathrm{S_1} = \wp \mathrm{A_u}(\phi_1 + \phi_2) &= \frac{1}{4}(\phi_1 + \phi_4 + \phi_1 + \phi_4) + \frac{1}{4}(\phi_2 + \phi_3 + \phi_2 + \phi_3)\\
&= \frac{1}{2}(\phi_1 + \phi_2 + \phi_3 + \phi_4)\\
\mathrm{S_2} = \wp \mathrm{A_u}(\phi_1 - \phi_2) &= \frac{1}{2}(\phi_1 + \phi_4) - \frac{1}{2}(\phi_2 + \phi_3)\\
&= \frac{1}{2}(\phi_1 - \phi_2 - \phi_3 + \phi_4)\\
\mathrm{S_3} = \wp \mathrm{B_g}(\phi_1 + \phi_2) &= \frac{1}{2}(\phi_1 - \phi_4) + \frac{1}{2}(\phi_2 - \phi_3)\\
&= \frac{1}{2}(\phi_1 + \phi_2 - \phi_3 - \phi_4)\\
\mathrm{S_4} = \wp \mathrm{B_g}(\phi_1 - \phi_2) &= \frac{1}{2}(\phi_1 - \phi_4) - \frac{1}{2}(\phi_2 - \phi_3)\\
&= \frac{1}{2}(\phi_1 - \phi_2 + \phi_3 - \phi_4)
\end{aligned}$$

On expressing the **F** matrix in the basis set of these symmetry orbitals, the secular determinant factors into two smaller determinantal equations, viz.,

$$\begin{vmatrix} \mathrm{F}_{11}^{\pi} - \epsilon & \mathrm{F}_{12}^{\pi} \\ \mathrm{F}_{21}{}^{\pi} & \mathrm{F}_{22}^{\pi} - \varepsilon \end{vmatrix} = 0 \text{ and } \begin{vmatrix} \mathrm{F}_{33}^{\pi} - \epsilon & \mathrm{F}_{34}^{\pi} \\ \mathrm{F}_{43}{}^{\pi} & \mathrm{F}_{44}^{\pi} - \varepsilon \end{vmatrix} = 0$$

Assuming Huckel parametrisation, putting α=0, and remembering that a matrix element $\mathrm{F_{ij}}^\pi = <\mathrm{S_i}\,|\mathrm{F}^\pi|\,\mathrm{S_j}>$ the solutions are found to be

$$\begin{aligned}\epsilon^\pi(\mathrm{a_u}) &= 1.62\beta & \epsilon^\pi(\mathrm{b_g}) = -1.62\beta\\ \epsilon^\pi(\mathrm{a_u}) &= -0.62\beta & \epsilon^\pi(\mathrm{b_g}) = 0.62\beta\end{aligned}$$

$$\text{Since} \begin{vmatrix} F^{\pi}_{11}-\epsilon & F^{\pi}_{12} \\ F_{21}{}^{\pi} & F^{\pi}_{22}-\epsilon \end{vmatrix} = \begin{vmatrix} \frac{3}{2}\beta-\epsilon & -\frac{1}{2}\beta \\ -\frac{1}{2}\beta & -\frac{1}{2}\beta-\epsilon \end{vmatrix} = 0$$

it is seen that $\mathbf{F}^{\pi}$ is not completely diagonalised indicating that the basis symmetry orbitals are not the $\pi-$mo's.

$$\psi^{\pi}_{1}(a_u) = c_1S_1 + c_2S_2 \text{ with } \epsilon^{\pi} = 1.62\beta.$$

The set of simultaneous equations to be solved are (in matrix form)

$$\begin{pmatrix} \frac{3}{2}\beta-\epsilon & -\frac{1}{2}\beta \\ -\frac{1}{2}\beta & -\frac{1}{2}\beta-\epsilon \end{pmatrix}\begin{pmatrix} c_1 \\ c_2 \end{pmatrix} = 0, \text{ where } \epsilon = \epsilon\pi = 1.62\beta$$

These equations, along with the normalisation condition $c_1^2 + c_2^2 = 1$, lead to $c_1 = 0.971$ and $c_2 = -0.228$.

Thus $\psi_1\pi(a_u) = 0.37\phi_1 + 0.599\phi_2 + 0.599\phi_3 + 0.37\phi_4$ for $\epsilon\pi = 1.68\beta$. Likewise it can be shown

$$\begin{aligned} \psi_2\pi(b_g) &= 0.599\phi_1 + 0.37\phi_2 - 0.37\phi_3 - 0.599\phi_4 \; (\epsilon\pi = 0.62\beta) \\ \psi_3\pi(a_u) &= 0.599\phi_1 - 0.37\phi_2 - 0.37\phi_3 + 0.599\phi_4 \; (\epsilon\pi = -0.62\beta) \\ \psi_4\pi(b_g) &= 0.37\phi_1 - 0.599\phi_2 + 0.599\phi_3 - 0.37\phi_4 \; (\epsilon\pi = -1.62\beta) \end{aligned}$$

C. MO's Of Metal Sandwich Compounds.

Many compounds are known where a metal atom is 'sandwiched' between two carbocyclic rings. Examples are ferrocene $(C_5H_5)_2Fe$, ruthanocene $(C_5H_5)_2Ru$, dibenzene chromium $(C_6H_6)_2Cr$ and the like. In these compounds, the metal atom is located on the principal symmetry axis of the respective point groups. The mo's of such sandwich compounds are essentially some suitable linear combinations of the $\pi-$group orbitals of the two carbocyclic rings and metal atomic orbitals of matching symmetries.

The extension of the terminology 'sandwich compounds' is also made to cover compounds in which the metal atom exists in between two different carbocylic rings as in (C_5H_5) (C_6H_6) Mn or even compounds where only one carbocyclic ring is present, e.g., $Cr(CO)_3(C_6H_6)$.

The technique of building up these mo's is not different from what we have already considered. In fact symmetry principles alone lead to

π-group orbitals (partial $\pi-$molecular orbitals) of the carbocyclic rings and classify the metal atomic orbitals. One then realises which metal orbitals have the potentialities of combining linearly with the $\pi-$group orbitals to form mo's. The actual quantitative compositions of the mo's, however, depend on evaluations of energies and determination of the eigenvectors by quantum mechanical methods.

Example 9.8: mo's of ferrocene molecule.

The molecule ferrocene, being an oft quoted example of metal sandwich compound, will serve as our illustrative species. We shall divide

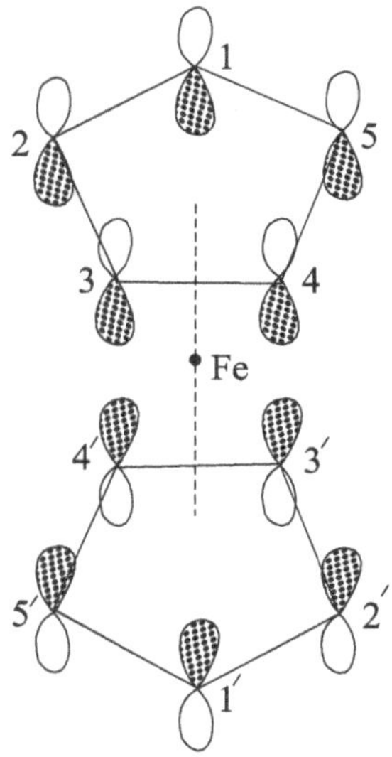

Fig. 9.10: Two cyclopentadiene rings in staggered orientation. Fe atom is in between the rings on the C_5-axis.

our treatment into two parts, one concerning the formation of $\pi-$group orbitals of the two cyclopentadiene rings by pursuing stages (i) to (iii) and following it up, in stage (iv), by suggesting the atomic orbitals of the metal atom capable of combining with the $\pi-$group orbitals to form $\pi-$mo's. The molecule belongs to the point group D_{5d} and is represented in Fig. 9.10.

The two cyclopentadiene rings are situated in two parallel planes in perfectly staggered orientation with respect to each other. At the ten C-atoms are set ten cartesian axial systems which are right-handed in the lower ring and left handed in the upper one. One more additional feature of the coordinate systems so set up is that the positive z-directions of the upper ring all point downwards and of the lower ring upwards. These then represent the directions $\phi_1\ \phi_2 ... \phi_5\ \phi'_1\ \phi'_2 ... \phi'_5$ which are shaded in the figure.

π-Group orbitals of the cyclopentadiene rings :

Stage (i). Using the ten p_z−orbitals as the basis functions and noting the shift or nonshift of the basis set, we write down the characters of the reducible representation.

D_{5d}	E	$2C_5$	C_5^2	C_5^3	$5C'_2$	$5\sigma_d$	$2S_{10}$	$2S_{10}^3$	S_{10}^5	$= i$
χ_{red}	10	0	0	0	0	2	0	0	0	

Stage (ii). The reducible representation contains the following IR's (application of Theo. 5.8),

$$\Gamma^{red} = A_{1g} + A_{2u} + E_{1g} + E_{1u} + E_{2g} + E_{2u} \qquad (9.57)$$

Stage (iii). This step is one of forming SALC's. We may, however, simplify the step by noting our results already obtained for cyclopentadienyl radical (Example 9.5). The π−orbitals for the individual rings will be similar to those of relations (9.50), viz., $\psi\pi(a)$, $\psi\pi(e_1a)$, $\psi\pi(e_1b)$, $\psi\pi(e_2a)$ and $\psi\pi(e_2b)$. Since two rings are involved, we combine these preserving normalisation and ensuring g and u characters of the IR's in (9,57). The π−group orbitals of the two rings are, therefore,

$$\psi\pi(a_{1g}) = \frac{1}{\sqrt{2}}\left[\psi_u\pi(a) + \psi_l\pi(a)\right], \text{ where u refers to upper ring and l to lower one.}$$

$$= \frac{1}{\sqrt{10}}[\phi_1 + \phi_2 + \phi_3 + \phi_4 + \phi_5 + \phi'_1 + \phi'_2 + \phi'_3 + \phi'_4 + \phi'_5]$$

$$\psi\pi(a_{2u}) = \frac{1}{\sqrt{2}}\left[\psi_u\pi(a) - \psi_l\pi(a)\right]$$

$$= \frac{1}{\sqrt{10}}\left[\phi_1 + \phi_2 + \phi_3 + \phi_4 + \phi_5 - \phi'_1 - \phi'_2 - \phi'_3 - \phi'_4 - \phi'_5\right]$$

$$\psi\pi(e_{1g}a) = \frac{1}{\sqrt{2}}\left[\psi_u\pi(e_1a) + \psi_l\pi(e_1a)\right]$$

$$= \frac{1}{\sqrt{5}}\left[(\phi_1 + \phi'_1)\cos w + (\phi_2 + \phi'_2)\cos 2w + \right.$$
$$\left. +(\phi_3 + \phi'_3)\cos 2w + (\phi_4 + \phi'_4)\cos w + (\phi_5 + \phi'_5)\right]$$

$$\text{where w} = \frac{2\pi}{5}$$

Likewise we have

$$\begin{aligned}
\psi\pi(\mathrm{e_{1g}b}) &= \frac{1}{\sqrt{5}}\big[(\phi_1+\phi_1')\sin \mathrm{w} + (\phi_2+\phi_2')\sin 2\mathrm{w} \\
&\quad -(\phi_3+\phi_3')\sin 2\mathrm{w} - (\phi_4+\phi_4')\sin \mathrm{w}\big] \\
\psi\pi(\mathrm{e_{1u}a}) &= \frac{1}{\sqrt{5}}\big[(\phi_1-\phi_1')\cos \mathrm{w} + (\phi_2-\phi_2')\cos 2\mathrm{w} \\
&\quad +(\phi_3-\phi_3')\cos 2\mathrm{w} + (\phi_4-\phi_4')\cos \mathrm{w} + (\phi_5-\phi_5')\big] \\
\psi\pi(\mathrm{e_{1u}b}) &= \frac{1}{\sqrt{5}}\big[(\phi_1-\phi_1')\sin \mathrm{w} + (\phi_2-\phi_2')\sin 2\mathrm{w} \\
&\quad -(\phi_3-\phi_3')\sin 2\mathrm{w} - (\phi_4-\phi_4')\sin \mathrm{w}\big] \\
\psi\pi(\mathrm{e_{2g}a}) &= \frac{1}{\sqrt{5}}\big[(\phi_1+\phi_1')\cos 2\mathrm{w} + (\phi_2+\phi_2')\cos \mathrm{w} \\
&\quad +(\phi_3+\phi_3')\cos \mathrm{w} + (\phi_4\phi_4')\cos 2\mathrm{w} + (\phi_5+\phi_5')\big] \\
\psi\pi(\mathrm{e_{2g}b}) &= \frac{1}{\sqrt{5}}\big[(\phi_1+\phi_1')\sin 2\mathrm{w} - (\phi_2+\phi_2')\sin \mathrm{w} \\
&\quad (\phi_3+\phi_3')\sin \mathrm{w} - (\phi_4+\phi_4')\sin 2\mathrm{w}\big] \\
\psi\pi(\mathrm{e_{2u}b}) &= \frac{1}{\sqrt{5}}\big[(\phi_1-\phi_1')\cos 2\mathrm{w} + (\phi_2-\phi_2')\cos \mathrm{w} \\
&\quad +(\phi_3-\phi_3')\cos \mathrm{w} + (\phi_4-\phi_4')\cos 2\mathrm{w} + (\phi_5-\phi_5')\big] \\
\psi\pi(\mathrm{e_{2u}b}) &= \frac{1}{\sqrt{5}}\big[(\phi_1-\phi_1')\sin 2\mathrm{w} - (\phi_2-\phi_2')\sin \mathrm{w} \\
&\quad +(\phi_3-\phi_3')\sin \mathrm{w} - (\phi_4-\phi_4')\sin 2\mathrm{w}\big]
\end{aligned}$$

Stage (iv). Having thus formed the ring group orbitals, it is now necessary to know how the valence shell atomic orbitals 3d, 4s, 4p of Fe atom transform under D_{5d} symmetry operations. Character table shows the following distribution:

Orbitals	IR
s, d_{z2}	A_{1g}
(d_{xz}, d_{yz})	E_{1g}
d_{xy}, $d(x^2-y^2)$	E_{2g}
p_z	A_{2u}
(p_x, p_y)	E_{1u}

We can thus foretell

1. Iron 4s and $3d_z^2$ orbitals will linearly mix with $\psi\pi(a_{1g})\pi-$group orbital to yield 3 mo's $\phi(a_{1g})$, $\phi'(a_{1g})$ and $\phi''(a_{1g})$.

2. $\psi\pi(a_{2u})$ and iron p_z orbital will combine to give two mo's $\phi(a_{2u})$, $\phi'(a_{2u})$.

3. $\psi\pi(e_{1g}a)$, $\psi\pi(e_{1g}b)$ and metal d_{xz} d_{yz} orbitals will linearly combine to give two e_{1g} orbitals i.e., four mo's in all.

4. $\psi\pi(e_{1u}a)$, $\psi\pi(e_{1u}b)$ and the Fe p_x, p_y orbitals will yield four mo's of symmetry e_{1u}.

5. $\psi\pi(e_{2g}a)$, $\psi\pi(e_{2g}b)$ and the metal $d_{xy}d_{(x^2-y^2)}$ orbitals mix to give two pairs of orbitals of e_{2g} symmetry, i.e., four mo's in all.

6. $\psi\pi(e_{2u}a)$ and $\psi\pi(e_{2u}b)$ of the ring group orbitals mix amongst themselves to form two mo's with no contribution coming from the metal orbitals.

Thus from symmetry we can come up to a point which reveals only the qualitative compositions of the mo's of the sandwich compounds. The actual values of the coefficients in the linear combinations have to be obtained from diagonalisation of the Fock matrix which yields the energies of the orbitals simultaneously. This is beyond the purview of symmetry and is purely a quantum mechanical job.

Unlike the previous illustrations of the $\pi-$systems (Examples 9.6, 9.7), where diagonalisations were rendered easy by virtue of Huckel parametrisations, the diagonalisation in the case of ferrocene is beset with formidable difficulties. Although in principle self consistent field methods do exist for solutions of Roothaan equations, these are scarcely completely solved for big molecules. instead, a number of simplifying assumptions in regard to matrix elements, spectroscopic data, ionisation potentials values are taken into consideration to construct an approximate energy correlation diagram of ferrocene molecule. The diagram is given below (Fig. 9.11) to convey an idea of the relative ordering of the energy levels without any quantitative scaling.

In actual practice, the entire 19×19 secular determinant splits into a number of smaller dimensional determinantal equations. It is further

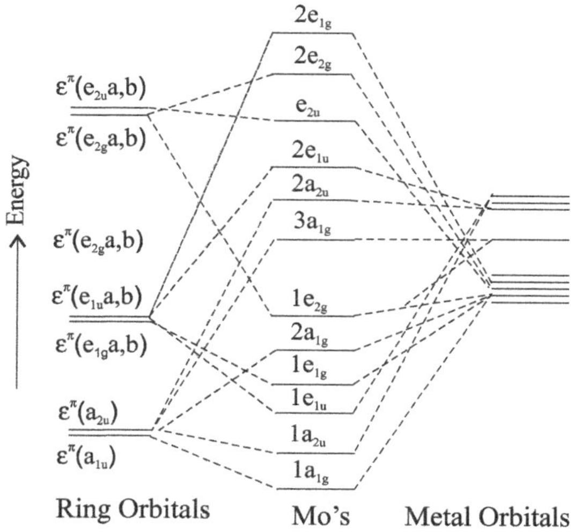

Fig. 9.11: Approximate energy diagram of ferrocene.

assumed that the g and u levels of the ring orbitals of the same gross symmetry have the same energy, e.g.,

$\epsilon\pi(a_{1g}) = \epsilon\pi(a_{2u})$, $\epsilon\pi(e_{1g}a) = \epsilon\pi(e_{1g}b) = \epsilon\pi(e_{1u}a) = \epsilon\pi(e_{1u}b)$ etc. The same value of energy has been attributed to a lack of direct interaction between the sets of ring orbitals on account of the distance between the two rings.

In the figure, the placement of most of the energy levels of the π−group orbitals below those of the metal atom orbitals is based on ionisation potential values of the electrons in the ring orbitals and in the valence orbitals of the Fe atom. The eighteen electrons (10 p_z electrons of the two cyclopentadiene rings and 6 electrons of 3d and 2 electrons of 4s of Fe atom) are accommodated in the low lying mo's, viz., $1a_{1g}$, $1a_{2u}$, $1e_{1u}$, $1e_{1g}$, $2a_{1g}$ and $1e_{2u}$. The orbitals of e symmetries accommodate four electrons each whereas those of a_{1g} and a_{2u} symmetries house two apiece. The electronic configuration is thus
$(1a_{1g})^2$ $(1a_{2u})^2$ $(1e_{1u})^4$ $(1e_{1g})^4$ $(2a_{1g})^2$ $(1e_{2g})^4$.

The treatment of dibenzene chromium is almost identical. Since the benzene π−orbitals are known, it is easy to form the π−group orbitals paying due regard to g and u characters. Finally these are mixed with matching orbitals of chromium atom and the energy diagram can be constructed.

D. MO's Of Simple Molecules.

The mo's of molecules, such as H_2O, NH_3, CH_4 etc. may be found out in two stages. Beginning with a symmetry based approach, we first arrive at the group orbitals of the ligand atoms. These then can be combined, with matching of symmetries, with the atomic orbitals of the central atom, e.g., O, N or C of our illustrative molecules. Symmetry principles thus lead only to qualitative compositions of the mo's. The final forms of mo's and the values of orbital energies are accessible to quantum mechanical methods as in the case of ferrocene. We briefly describe here the mo's of methane molecule.

Example 9.9.

To find the nature of mo's in methane molecule.

The molecule belongs to the point group T_d. The four 1s orbitals $(\phi_1\phi_2\phi_3\phi_4)$ of the four H atoms and the valence orbitals, 2s and the three 2p's $(\phi_5\phi_6\phi_7\phi_8)$ of the C atom combine to give the different mo's. One may start with the sp^3 hybridised orbitals of the C atom or even, quite generally, from the pure 2s and the 2p orbitals. As in the case of ferrocene molecule, we first find the group orbitals of ϕ_1, ϕ_2, ϕ_3, ϕ_4 conforming to the different IR's of the T_d point group.

The characters of the reducible representation with the basis set of $(\phi_1\phi_2\phi_3\phi_4)$ under the group operations of T_d are

T_d	E	$8C_3$	$3C_2$	$6S_4$	$6\sigma_d$
χ_{red}	4	1	0	0	2

This reducible representation can be expressed as (Theo. 5.8 and the character table of T_d)

$$\Gamma^{red} = A_1 \oplus T_2$$

To constitute the SALC's of A_1 and T_2 symmetries, we shun the projection operator technique here as its use will involve much labour which, however, is avoidable. Instead we proceed to construct these by mental analysis and inference.

The group orbital of A_1 symmetry is just

$$\psi(a_1) = \frac{1}{2}(\phi_1 + \phi_2 + \phi_3 + \phi_4) \qquad (9.58)$$

It is to be observed that $(p_x p_y p_z)$ orbitals form a basis of the IR, T_2, of the T_d point group. Hence ϕ_1, ϕ_2, ϕ_3, ϕ_4 are to be so combined that these yield the characteristic p-orbital lobes, Thus from the Fig. 9.4,

$$\left[(\phi_1 + \phi_4)\cos\left(\frac{\pi}{4}\right)\sqrt{\frac{2}{3}} - (\phi_2 + \phi_3)\cos\left(\frac{\pi}{4}\right)\sqrt{\frac{2}{3}}\right]$$

will have the lobe characteristics similar to that of p_x. In other words $\frac{1}{\sqrt{3}}(\phi_1 - \phi_2 - \phi_3 + \phi_4) \sim p_x$. Similarly, the other orbitals are $\frac{1}{\sqrt{3}}(-\phi_1 + \phi_2 - \phi_3 + \phi_4)$ and $\frac{1}{\sqrt{3}}(\phi_1 - \phi_2 + \phi_3 - \phi_4)$. Writing these in normalised forms and remembering that these group orbitals belong to T_2 symmetry we have

$$\left.\begin{aligned} \psi(t_2a) &= \tfrac{1}{2}(\phi_1 - \phi_2 - \phi_3 + \phi_4) \\ \psi(t_2b) &= \tfrac{1}{2}(-\phi_1 + \phi_2 - \phi_3 + \phi_4) \\ \psi(t_2c) &= \tfrac{1}{2}(\phi_1 - \phi_2 + \phi_3 - \phi_4) \end{aligned}\right\} \tag{9.59}$$

Having found the group orbitals (Eqs. 9.58, 9.59), we now assign symmetries to the 2s and the 2p orbitals of the C atom under the point group T_d. The distribution is (cf. the character table).

Orbitals	IR
$s(\phi_5)$	A_1
$p_x, p_y, p_z(\phi_6, \phi_7, \phi_8)$	T_2

The composition of the mo's are just the following:

1. $\Phi(a_1)c_1\psi(a_1) + c_2\phi_5$ giving rise to two mo's $(1a_1)$ and $(2a_1)$ corresponding to two sets of values of c_1 and c_2.

2. $\Phi(t_2) = d_1p_x + d_2p_y + d_3p_z + d_4\psi(t_2a) + d_5\psi(t_2b) + d_6\psi(t_2c)$ From symmetry of the system, it can be readily surmised that $d_1 = d_2 = d_3$ and similarly $d_4 = d_5 = d_6$. This will lead to six pairs of values of d_1 and d_4 corresponding to two triply degenerate states,. viz., $(1t_2)$ and $(2t_2)$. This is almost all that lie within the province of symmetry. For more quantitative informations as to the coefficients (eigenvectors) and the energies, one has to solve the secular equation (diagonalisation of the Fock matrix) by one of the self consistent field procedures. The energy diagram appears in Fig. 9.12.

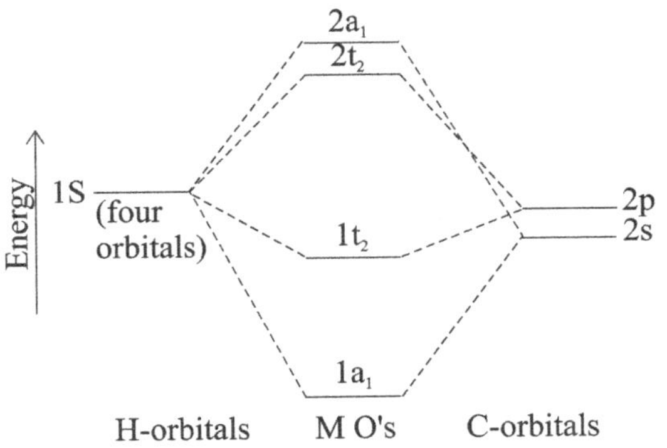

Fig. 9.12: Simple energy correlation diagram of methane molecule.

Example 9.10.

Localized MO's of Formaldehyde.

As a final example we shall consider the problem of finding the mo's of formaldehyde. This molecule, HH=C=0, is planer, belongs to C_{2v} point group and has the C_2 axis lying along the C-O bond in the plane of the paper. We implant a right handed coordinate system with the origin at the C-atom with the positive direction of X-axis pointing upward. A left handed system of coordinate axes is imagined to exist at the O-atom so that the positive Z-axis points toward the C-atom.

The valence shell basis orbitals for constructing the mo's are ten in number. These are two 1s-orbitals of H-atom represented by ϕ_1 and ϕ_2, four orbitals of C-atom of which three are sp^2-hybridised ϕ_3, ϕ_4, ϕ_5 and the fourth ϕ_6 is the $2p_x$-orbital perpendicular to the paper plane. The remaining four orbitals, provided by the O-atom, are ϕ_{2s}, ϕ_{2px}, ϕ_{2py} and ϕ_{2pz}. In the light of axial set-up, ϕ_{2pz} and ϕ_{2py} are in the paper plane and ϕ_{2px} is along the vertical axis.

Since the C and the O atoms lie on all the symmetry elements of formaldehyde molecule, their individual atomic orbitals form bases of the IR's of the molecular point group C_{2v}. A look at the character table suggests the following distribution of the orbitals over the IR's: (The behaviour of the sp^2 hybrids will be similar to the s-orbitals under the group operations)

$\phi_3,\ \phi_4,\ \phi_5,\ \phi'_{2s}\ \phi_{2p_z}$	A_1
ϕ_6 and ϕ_{2p_x}	B_1
ϕ_{2p_y}	B_2

We next form group orbitals with ϕ_1 and ϕ_2 of the two H-atoms. Without going through the formal steps, as explained in earlier examples, we can easily write out these in normalised forms.

$$\begin{aligned} G_1 &= \frac{1}{\sqrt{2}}(\phi_1 + \phi_2) \text{ of } A_1 \text{ symmetry} \\ \text{and } G_2 &= \frac{1}{\sqrt{2}}(\phi_1 - \phi_2) \text{ of } B_1 \text{ symmetry.} \end{aligned}$$

To form the completely delocalised mo's, the problem becomes somewhat unwieldy. Instead the simpler localised approximate molecular orbitals may be formed.

1. Five mo's formed from combinations of G_1 (i.e., $\phi_1,\ \phi_2$), $\phi_3,\ \phi_4$, ϕ_5 and ϕ_{2p_z} of which three are of bonding type (A_1) and two of antibonding (A_1) type. Effectively the lower three orbtals turn out to be localised mo's representing $\sigma-$bonds formed from overlaps of $\phi_1,\ \phi_2$ and ϕ_{2p_z} with the three sp^2 hybrids $\phi_3,\ \phi_4$ and ϕ_5 respectively.

2. ϕ_{2s}, the atomic orbital of O-atom, essentially turns out to be an mo (A_1 symmetry) localised over O-atom.

3. ϕ_6 and ϕ_{2p_x} linearly combine to give π and π^* orbitals, both of B_1 symmetry, of which π is of bonding and π^* antibonding types.

4. ϕ_{2p_y} turns out to be a nonbonding mo effectively belonging to oxygen. It is of symmetry B_2 in which only a negligible mixing of ϕ_{22p_y} with G_2 group orbital takes place. The antibonding b_2^* orbital results from their mixing.

The 12 bonding electrons of formaldehyde are accommodated in she six low-lying levels. The configuration of the ground state being.

$$(a_1 = \phi_{2s})^2(a_1')^2(a_1'')^2(a_1''')^2(b_1 = \pi)^2(b_2 = \phi_{2p_y})^2$$

In the above configuration a_1', a_1'' and a''' are the three σ−bonding mo's formed from the three sp^2 hybrids and ϕ_1, ϕ_2 and ϕ_{2p_z}. The HOMO (highest occupied molecular orbital) is the non-bonding ϕ_{2p_y}, i.e., the b_2 orbital. The LUMO (lowest unoccupied molecular orbital) is the π^*.

The feasibilities of some common excitation processes of formaldehyde molecule are considered in the next chapter (Sec. 10.7). The molecular orbital energy diagram of formaldehyde is given below.

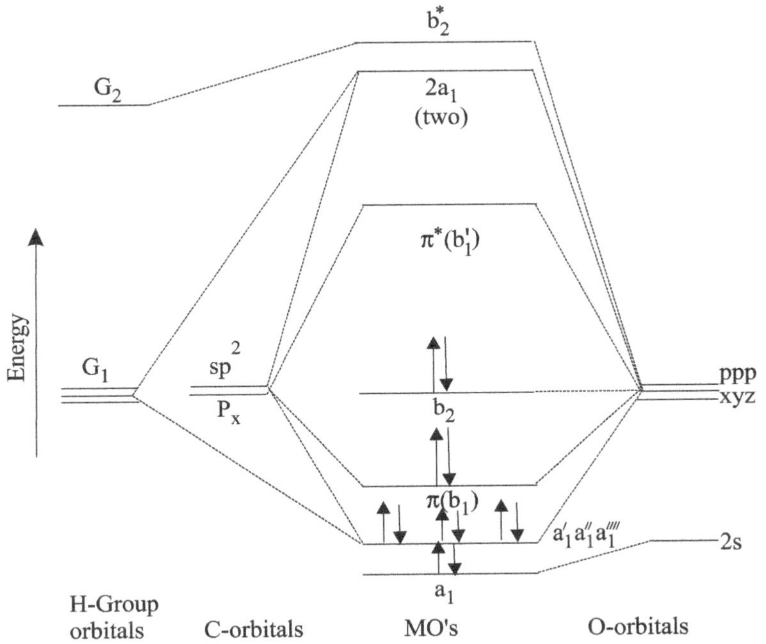

Fig. 9.13: Molecular orbital energy correlation diagram of formaldehyde molecule.

Chapter 10

Symmetry Principles And Transition Metal Complexes

10.1 General Remarks

The chemistry of transition metal complexes is a subject of engrossing interest to the experimental and theoretical inorganic chemists. The topic includes within its periphery preparation, stereochemistry, stability, thermodynamic properties, optical activity, reaction mechanisms, magnetic properties, spectral absorption and electronic structures of the complexes. Many aspects of these properties are basically understandable as consequences flowing from the splitting of degenerate energy levels of the d-orbitals of the metal ion. This splitting and the consequences are predictable directly by an electrostatic theory (Crystal Field Theory) or indirectly in a more sophisticated manner by the Molecular Orbital Theory of the complexes. We must take cognizance of two essential things: (i) cause of splitting of energy levels and (ii) effect of splitting. The theories which have been formulated and pressed into service are (a) Crystal Field Theory (CFT), (b) Adjusted Crystal Field Theory (or Ligand Field Theory (ACFT or LFT) and (c) Molecular Orbital Theory (MO theory), According to some chemists, however, the whole set of modifications comprising ACFT and the semiquantitative or quantitative MO theory are grouped under the general heading of ligand field theory (LFT). The idea of symmetry and the application of its principles help us very effectively only in some selected zones of coordination chemistry. These are in

(i) splitting of energy levels
(ii) construction of energy correlation diagrams } both by CFT or ACFT

(iii) forming LCAO MO of coordination complex
(iv) construction of MO energy diagram } both by MO theory

(v) electronic spectral transition
(involving d-d levels, i.e., vibronic coupling

(vi) vibronic polarisation, dichroism

(vii) spin orbit coupling, crystal field states and double groups.

It is only to these selective aspects of coordination complexes that we shall pay our attention. This may appear somewhat disjointed from the standpoint of the entire gamut of the transitional metal complexes, but it cannot be helped since only symmetry constitutes our focal point of interest in this book. As a preparatory measure for all these, we consider in the next section some basic ideas upon which the theories of CFT or ACFT are founded. We shall also make a grazing reference to the Russell Saunders' states and interelectron repulsion parameters of the transition metal ions assuming these to be already known in detail to the readers.

10.2 Basic Principles

The crystal field theory has its origin in the works of Bethe who suggested that the degeneracy of the d-orbitals of a metal in the environment of the groups of atoms or ions in a crystal would split. The idea was transplanted in the domain of transition metal complexes by Van Vleck who argued that similar splitting would also occur in such compounds. The CFT is based upon the concept that the metal ion at the centre has no orbital overlap with the ligand ions or molecules surrounding it. The ligands, if ions, are considered as merely negative point charges and, if molecules, as only point dipoles with their negative pole turned toward the metal ion. The ligands, whether ions or dipoles, thus simply act as dimensionless point sources of electric field exerting repulsions on the d-orbital electrons of the metal ion. This essential electrostatic concept involving point charges and point dipoles, though divorced from reality, enables one to treat the metal ion as a separate entity subjected to perturbation of the environmental field-a kind of internal stark effect. To elaborate, the hamiltonian, H_f of a transition metal atom, denuded of its 4s electrons, is given on neglecting the spin-orbit interaction by

$$H_f = -\frac{\hbar^2}{2m}\sum_i \nabla i^2 - \sum_k^{\text{core electrons}} \frac{ze^2}{r_k} - \sum_l^{\text{d electrons}} \frac{ze^2}{r_l} +$$

$$\sum_{k}\sum_{<j}^{\text{core electrons}}\frac{e^2}{r_{kj}}+\sum_{l<m}^{d-\text{electrons}}\frac{e^2}{r_{lm}}+\sum_{k,\,l}\frac{e^2}{r_{kl}} \tag{10.1}$$

Here i represents all electrons, k, j the core electrons, l, m the d-electrons. The last term in (10.1) indicates core electron d-electron repulsion terms. The potential energies due to nucleus-electron attraction and interelectorn repulsions have thus all been split into terms pertaining to core electrons (k,j) and 3d electrons (l, m). It can be shown that all the terms excepting $\sum_{k,\,l}\frac{e^2}{r_{kl}}$ and $\sum_{l<m}\frac{e^2}{r_{lm}}$ are central field[13] terms. If we substitute the $\sum\frac{e^2}{r_{kl}}$ by an almost equivalent central field term, the relation (10.1) becomes

$$H_f \approx H_f^{cent}+\sum_{l<m}\frac{e^2}{r_{lm}} \tag{10.2}$$

It is thus seen from (10.2) that the inter d-electron repulsion terms act as a perturbation on an otherwise central field hamiltonian of the unperturbed metal ion.

Now when the metal ion is placed in the environment of the ligands so as to form the metal ion complex then, in terms of the CFT, the ligand point charges or point dipoles exert an electrical force on the metal electrons. The hamiltonian of the metal ion in the complex can then be written as

$$H_{m(complex)} = H_f^{cent}+\sum_{l<m}\frac{e^2}{r_{lm}}+V_{cryst} \tag{10.3}$$

where V_{cryst} is the crystal field potential energy. Relation (10.3) admits of two ways of writing it depending on whether $\sum_{l<m}\frac{e^2}{r_{lm}} \gg V_{cryst}$ or just the reverse. If $\sum_{l<m}\frac{e^2}{r_{lm}}$ is very large, then, remembering (10.2), it is seen

$$H_{m(complex)} = H_f + V_{cryst} \tag{10.4}$$

Thus for the calculations of energy and wavefunctions, we can start from the free ionic term inclusive of interelectron repulsion as the zero order state (unperturbed state) with the hamiltonian H_f. The crystal field hamiltonian, V_{cryst}' is treated as a perturbation giving rise to new states with new energies. Secondly, in the other extreme, when

$V_{cryst} \gg \sum_{l<m} \frac{e^2}{r_{lm}}$, relation (10.3) can be moulded as

$$\begin{aligned} H_{m(complex)} &= \left(H_f^{cent} + V_{cryst}{}^{\propto}\right) + \sum_{l<m} \frac{e^2}{r_{lm}} \\ &= H_{m(cryst)}{}^{\propto} + \sum_{l<m} \frac{e^2}{r_{lm}} \end{aligned} \qquad (10.5)$$

The metal ion in the limit of very strong (infinite but hypothetical) crystal field of the ligands can, therefore, be deemed as the initial zero order state to which are added the interelectron repulsions $\sum_{l<m} \frac{e^2}{r_{lm}}$, as perturbations in the real complex yielding new states with new energies.

We thus have two extreme anchor spots-first the free ion terms as the unperturbed states and the second, viz., the infinitely large crystal field states as the initial states-from either of which symmetry-based approaches can be made to find the new perturbed states (Tables 10.3 and 10.4) and correlations of the states at the two extreme ends can be effected (Sec. 10.4, Fig. 10.3).

The hamiltonians (10.2), (10.4) and (10.5) apply to d^n metal ion complexes with n greater than 1. In other words, these hamiltonians apply when there exist inter *d*-electron repulsion. For d^1 system, with no inter *d*-electron repulsion, the hamiltonians are respectively

$$H_f = H_f^{cent} \qquad (10.2a)$$

$$H_{m(complex)} = H_f^{cent} + V_{cryst} \qquad (10.4a)$$

$$\text{and } H_{m(complex)} = H_{m(cryst)}{}^{\propto} \qquad (10.5a)$$

From the foregoing description we thus have two primary functions theoretically in dealing with the CFT, viz., (i) to find the perturbed (crystal field) states in the limits of zero and very strong crystal fields and (ii) to correlate the two sets of states energetically. While the first task is purely one of symmetry, the second job is one of quantum mechanics where symmetry ideas (such as direct product representation and others) are invoked as aids.

The Adjusted Crystal Field Theory (ACFT)

The adjusted crystal field theory tries to remove the incongruities in the assumptions of point charges and point dipoles by allowing a certain degree of overlap of the metal ion orbitals and ligand orbitals. It also allows possible polarisations of the ligands by the central charge of the metal ion. These flexibilities are allowed for only indirectly in the form of adjusted parameters-such as interelectron repulsion parameters (*Racah*) B and C-in crystal fields of different strengths. Moreover, in the limit of weak crystal field, relation (10.4) states that in CFT the anchorpoint is the free ion term with the hamiltonian, $H_{m(complex)} = H_f$. But in ACFT, the anchor point is the metal ion term as situated in the environment of the ligands exuding zero field. In this case the initial hamiltonian $H_{m(complex)} = H_f' = \alpha H_f$ where α is around 70%. In fact the modifications (there are many ramifications of these) are so incorporated empirically that these permit of numerical calculations executable with much less difficulty than are possible even in a simplified MO treatment. The smaller separations between the Russell Saunders' states (vide infra), of the metal ion in the complex than in the free ion states are due to smaller inter d-electron repulsions of the ion in the complex. This smaller interelectron repulsion, therefore, means the charge cloud of the d-electrons is more extended permitting overlap[14] with the ligand charge cloud. This view helps in interpreting the experimentally observed nepheleauxetic effect.

Russell Saunders States. Interelectron Repulsion Parameters.

The electronic configurations of the incompletely filled subshells of any polyclectronic atom or ion give rise to different spectroscopic states depending on the relative magnitudes of the electrostatic interelectron repulsions and of magnetic (spin-orbit) interactions. Russell Saunders' states are the ones that prevail when electrostatic interactions very much outweigh the magnetic interactions. Assuming the basic principles to be all known to the readers and also the importance of Pauli exclusion principle in the determination of such states, it may be pointed out that the number of permissible Russell Saunders' or LS states increases very rapidly with the number of electrons in the incompletely field subshells. The same remark is true of the transition metal ions of d^n electronic configurations with, however, a small departure. Here the LS terms for the transition metal ions go on increasing from d^l to d^5 and afterward decrease in number from d^6 to d^9. This is understandable since a d^4 electron configuration giving rise to terms would be formally equivalent (though with opposite energy sequence) to a d^4-positron configuration. Since a positron is equivalent to a disappearance of an electron, i.e., equivalent to a hole, a d^{10} electron system is a zero hole system. Hence a d^4 electronic configuration≡a d^4 positron system ≡d^4 hole system≡ d^{10-4} i.e., d^6 electronic configuration. Similarly d^1 and d^9, d^2 and d^8 and d^3 and d^7 can be pairwise bracketed as equivalent with opposite ordering of energy within a given term.

The general process of writing down the Pauli permitted microstates corresponding to a d^n configuration before deciphering the actual terms is a laborious and exacting task. The process is ludicrously simplified by using a branching rule technique[15] non-violative of the Pauli exclusion principle. The terms so obtained are given in table 10.1 for the d^n systems.

Table 10.1

LS Terms for d^n Electronic Configurations

Configuration	Terms
d^1, d^9	^2D
d^2, d^8	^3F ^3P ^1G ^1D ^1S
d^3, d^7	^4F ^4P ^2P ^2D ^2F ^2G ^2H ^2D
d^4, d^6	^5D ^3P ^3D ^3F ^3G ^3H ^3P ^3F ^1S ^1D ^1F ^1G ^1I ^1S ^1D ^1G
d^5	^6S ^4D ^4G ^4P ^4F ^2S ^2D ^2F ^2G ^2I ^2P ^2D ^2F ^2G ^2H ^2D

The perturbational correction to the energy value of a central field system due to $\sum_{l<m} \frac{e^2}{r_{lm}}$ of relation (10.2), i.e., inter d-electron repulsions is given in terms of certain integrals[16], F_n and C_n, called Slater Condon parameters. The specific forms of such integrals involve integration of the radial parts of the d-orbitals of a pair of electrons besides other radially dependent components and also over triple product of spherical harmonics. The actual magnitudes are adjusted by referring to the spectral data of the atoms. In later works, specially for d^n and f^n systems, the Slater Condon parameters have been replaced by what are called Racah parameters B and C. These are certain linear combinations of Slater Condon parameters $B=F_2-5F_4$ and $C=35F_4$. These latter parameters are the ones usually used in CFT or ACFT theories. The molecular orbital theory, however, has its own theoretical way of calculating the replusion and exchange integrals independently which are not identifiable with B and C.

10.3 Symmetry and Splitting of Energy Levels

10.3.1 Crystal Field Effect on p^1, d^1 and f^1 Systems

The crystal field theory demands that the atomic orbitals of the metal ion in a complex be treated as an independent set of basis functions. In the absence of environmental ligands all the d-orbitals of the metal ion are degenerate and since the metal ion has a spherical symmetry, these d-orbitals form the basis of an IR of the spherical symmetry group. The same remark applies to the set of p or f orbitals. Let us in the first place consider a d^1 metal ion and an octahedral environment constituted by the ligands. Placement of the free metal ion within the environment of the ligands will mean a switching on of a perturbation and the symmetry of the system (i.e., the metal ion) will be lowered form its original spherical symmetry. One may recall Eq. (6.9) of Sec. 6.3 and the discussion therein including the Example 6.3 to be convinced of how the d-orbitals split, into a doubly degenerate e_g and a triply degenerate t_{2g} orbitals (Fig. 10.1). In demonstrating this dichotomy of the originally

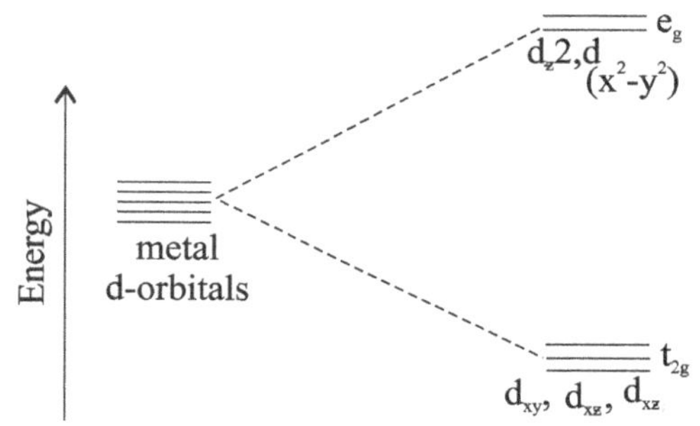

Fig. 10.1: Splitting of energy levels of the *d*-orbitals in octahedral symmetry.

degenerate energy levels, two methods have been employed-firstly formal transformation of the basis set under group operations and secondly inference from the stereo dispositions of the orbitals (Example 6.3).

We shall employ another method here to show which IR's should result when perturbational fields of different symmetries are allowed to play on the centrally placed metal ion. The analytic form of a single-electron orbital (exclusive of spin wavefunction) is

$$\psi = R(r)\theta_{lml}(\theta)\Phi_{ml}(\phi) \tag{10.6}$$

Here R=normalized radial function invariant under symmetry operation, θ_{lml}=normalized associated Legendre's polynomials

$$\theta_{\mathrm{lml}} = \frac{(-1)^{\mathrm{l}}}{2^{\mathrm{l}}l!}\sqrt{\frac{(2\mathrm{l}+1)}{2}\frac{(\mathrm{l}-\mathrm{m_l})!}{(\mathrm{l}+\mathrm{m_l})!}}\sin^{\mathrm{ml}}\theta\frac{\mathrm{d}^{\mathrm{l}+\mathrm{m_l}}}{(\mathrm{d}\cos\theta)^{\mathrm{l}+\mathrm{m_l}}}(\sin^{2\mathrm{l}}\theta)$$

and $\Phi_{\mathrm{ml}} = \frac{1}{\sqrt{2\pi}}\mathrm{e}^{\mathrm{iml}\phi}$ where l=orbital angular momentum quantum number of the electron and ml=z-component orbital angular momentum quantum number capable of taking values $-\mathrm{l} \leq \mathrm{m_l} \leq \mathrm{l}$

Temporarily we narrow down our general perspective to consider the d-electron orbitals. Here with l=2 and $\mathrm{m_l} = 2,\ 1,\ 0,\ -1$ and -2 one gets five degenerate complex wavefunctions d_2, d_1, d_0, d_{-1} and d_{-2}. All these involve the polar angle θ defined with respect to the quantising axis (polar axis) in the configuration space. (This is called quantising axis because the $\mathrm{m_l}$ quantum numbers are defined with respect to the orbital angular momentum component along this axis). Any rotation of the d-wavefunction about this quantising axis will leave θ unchanged. Hence R(r) and $\theta_{\mathrm{l,\ ml}}(\theta)$ will be unaffected but Φ_{ml} will change since ϕ, the azimuthal angle, will change under rotation. Let us consider an octahedral metal complex where we know that different rotational axes of symmetries, such as $\mathrm{C_4}$, $\mathrm{C_2}$, $\mathrm{C_3}$, $\mathrm{C_2}$' etc., do exist. Using, as the basis set, the five d-wavefunctions defined with respect to $\mathrm{C_4^z}$ symmetry axis as the quantising axis in the configuration space, the transformations of the basis functions will be as follows (cf. Sec. 4.4.)

$$\mathrm{C_4}\psi_{22} = \mathrm{R}\theta_{22}\frac{1}{\sqrt{2\pi}}\mathrm{e}^{2\mathrm{i}(\phi-\frac{2\pi}{4})} = \mathrm{R}\theta_{22}\Phi_2\mathrm{e}^{-2\mathrm{i}.\frac{2\pi}{4}} \tag{10.7}$$

Similarly,

$$\begin{aligned}\hat{\mathrm{C}}_4\psi_{21} &= \mathrm{R}\theta_{21}\Phi_1\mathrm{e}^{-\mathrm{i}.1.\frac{2\pi}{4}}; \qquad & \hat{\mathrm{C}}_4\psi_0 = \mathrm{R}\theta_{20}\Phi_0\mathrm{e}^{\mathrm{i}.0.\frac{2\pi}{4}}\\ \hat{\mathrm{C}}_4\psi_{2,\ -1} &= \mathrm{R}\theta_{2,\ -1}\Phi_{-1}\mathrm{e}^{\mathrm{i}(1)\frac{2\pi}{4}} \qquad & \hat{\mathrm{C}}_4\psi_{2,\ -2} = \mathrm{R}\theta_{2,\ -2}\Phi_{-2}\mathrm{e}^{\mathrm{i}(+2)\frac{2\pi}{4}}\end{aligned}$$

The transformation matrix, scooped from Eqs. (10.7), is a diagonal one in the basis set $\mathrm{d}_2\ \mathrm{d}_1\ \mathrm{d}_0\ \mathrm{d}_{-1}\ \mathrm{d}_{-2}$ having a particular trace value, Tr $\mathrm{C_4}$.

$$\mathbf{D}(\mathrm{C_4}) = \begin{bmatrix} \mathrm{e}^{-\mathrm{i}.2.\frac{2\pi}{4}} & & & & \\ & \mathrm{e}^{-\mathrm{i}.\frac{2\pi}{4}} & & & \\ & & \mathrm{e}^{\mathrm{i}.0.\frac{2\pi}{4}} & & \\ & & & \mathrm{e}^{+\mathrm{i}.\frac{2\pi}{4}} & \\ & & & & \mathrm{e}^{+\mathrm{i}.2.\frac{2\pi}{4}} \end{bmatrix} \tag{10.8}$$

It will be readily perceived that the representation matrix with the same basis set for C'_2 will be nondiagonal and cannot be *so easily constructed* since it will simultaneously involve changes in θ (quantising angle) and ϕ under rotations. Nevertheless if the matrix is constructed, it will lead to some trace value, viz., Tr C'_2. Similar considerations will apply to C_3.

If instead one uses the set of d-wavefunctions defined with respect to C'_2−symmetry axis as the quantising axis, then $\mathbf{D}(C'_2)$ will turn out to be diagonal, viz.,

$$\mathbf{D}(C'_2) = \text{Diag}\left(e^{-i.2.\frac{2\pi}{2}},\ e^{-i.\frac{2\pi}{2}},\ e^{i.0.\frac{2\pi}{2}},\ e^{+i.\frac{2\pi}{2}},\ e^{+i.2.\frac{2\pi}{2}}\right) \tag{10.9}$$

However, $\mathbf{D}(C_4)$ with this newly defined quantising axis will be, unlike (10.8), a nondiagonal matrix. But since the first and the second sets of the d-wavefunctions are the same function space although with different axial set-ups in configuration space, these matrices must be related through a similarity transformation and hence the trace values of $\mathbf{D}(C_4)$, $\mathbf{D}(C'_2)$ will each be invariant (Theo. 4.2). This will be true for all the rotational operators of the octahedral group of the present instance. We thus seize this unique advantage of finding the trace (character) values of the matrices of rotational operators by constructing the corresponding diagonal matrices with the basis set of *d*-wavefunctions defined each time with respect to different symmetry axes as the quantising axes. The characters so collected for the typical class elements of the group can then be utilised, by employing Theo. 5.8, to find the IR's spanned by the *d*-wavefunctions.

From our self restricted vista of d-electron orbitals we expand the perspective to consider the basis set of (2l+1) single-electron orbitals with quantum number l. Any rotation operator C_n in this basis of (2l+1) orbitals, each consisting of a radial and a spherical harmonic part, can be represented, with a suitable choice of quantising axis, by the diagonal matrix

$$D(C_n) = \text{Diag}\left\{e^{-i.l.\frac{2\pi}{n}} e^{-i(l-1)\frac{2\pi}{n}} \ldots e^{i.0.\frac{2\pi}{n}} e^{+i.\frac{2\pi}{n}} \ldots e^{+i(l-1)\frac{2\pi}{n}} e^{+i.l.\frac{2\pi}{n}}\right\} \tag{10.10}$$

$\chi(C_n) = \sum\limits_{p=-l} e^{ip\frac{2\pi}{n}}$. This is a G.P. series with a common ratio $e^{i.\frac{2\pi}{n}}$

$$\chi(C_n) = \frac{\sin\left(l+\frac{1}{2}\right)\beta}{\sin\frac{\beta}{2}} \tag{10.11}$$

where $\beta = \frac{2\pi}{n}$ and relation is valid for $\beta \neq 0$.

The case $\beta = 0$, occurs when one considers the character of the E (identity) operator. It is absolutely clear from the general matrix (10.10) that $\chi(E) = 2l + 1$, being the sum of $(2l + 1)$ number of e° terms.

We now revert to our original problem of a d^1 transition metal ion situated in an octahedral environment of the ligands. The gross symmetry of an $O_\hbar$ point group can be realised by considering the O-subgroup having the symmetry operators (one of each separate class) E, C_4, C_2, C_3 and C_2'. For single-electron d-orbital with l=2, it is evident $\chi(E)=5$ and using the relation (10.11) one gets $\chi(C_4) = -1$, $\chi(C_2) = 1$, $\chi(C_3) = -1$ and $\chi(C_2') = 1$. Employing the relation $n_i = \frac{1}{g} \sum_R \chi_i^*(R)\chi^{red}(R)$ of Theo. 5.8 and the character table of the O point group, it is found

$$\begin{aligned} n_E &= \frac{1}{24}[2.5 + 3.1.2 + 8.(-1)(-1)] = 1, \\ n_{T2} &= \frac{1}{24}[3.5 + 6.(-1)(-1) + 3.(-1)(+1) + 6.1.1] = 1 \end{aligned}$$

and the other representations are absent. We conclude, therefore, that the set of five d-wavefunctions thus splits into a doubly degenerate e and a triply degenerate t_2 base functions having, the transformation properties of the IR's E and T_2 of the O-point group. Remembering that the d-wavefunctions are gerade (g) in character these five d-wavefunctions will split into e_g and t_{2g} basefunctions in $O_\hbar$ symmetry group.

Proceeding in a similar vein we may also consider p^1 or f^1 system in an environment of O-point group. For the three singlelectron p wavefunctions acting as the basis set, relations (10.10 and 10. 11) yield $\chi(E) = 3$, $\chi(C_4) = 1$, $\chi(C_3) = 0$, $\chi(C_2) = -1$ and $\chi(C_2') = -1$. This gives only one representation T_1. The energy level of the original triply degenerate p wavefunctions remains unsplit and degenerate (though of different magnitude) even when the octahedral perturbing environment is there. Since the p-wavefunctions are of the ungerade type, the IR will be T_{1u} in $O_\hbar$- symmetry group.

An exercise with the basis set of seven f wavefunctions will display a splitting of the originally degenerate level of the metal ion into non-degenerate a_{2u} and triply degenerate t_{1u} and t_{2u} orbitals in octahedral environment.

Enough has been said in the present section and in Secs. 5.8 and 6.3

(together with the illustrative examples 5.11, 5.12, 6.1, 6,2, 6.3 and 6.4) to permit us draw the following conclusions.

1. The degenerate single-electron orbitals or wavefunctions of an atom undergo splitting or remain unsplit when the atom is placed in environments of different symmetries, e.g., $O_{\hbar}$, T_d, $D_{4\hbar}$ etc. Symmetry principles alone help us arrive at this conclusion.

2. The techniques that are applied to show whether the degeneracy will be lifted or not and to apportion the orbitals to different symmetries (in case the degeneracy is lifted) are:

   (i) full transformation properties of the degenerate set of basis functions under symmetry operations of the group (Ex. 5.11, 5.12) followed by finding the IR subspaces.

   (ii) inference based on stereo disposition of the orbital lobes (Example 6.3). This eliminates the necessity of applying SALC method for determining the identities of the split levels.

   (iii) transformation of the spherical harmonics about suitably chosen quantisting axis.

3. The splitting of the energy levels in environments of lower symmetries $D_{4\hbar}$, C_{4v}, D_3 can be inferred by applying the techniques listed under 2 above or by applying the principle of descent in symmetry (explained adequately in Sec. 6.3 and in the illustrations there).

10.3.2 Crystal Field Effect (Splitting). Multielectron Configurations

Having known from the previous description the nature of, splittings of p^1, d^1 and f^1 systems, our natural inquiry is what will happen to p^n, d^n or f^n systems. For these multielectron configurations n may range from 2 to 5 for an unfilled p^n, 2 to 9 for an unfilled d^n and 2 to 13 for the incompletely filled f^n systems.

Taking d^2 as an example, we ought to know what would be the total wavefunction of the metal atom or ion before we can subject this to transformations. A follow-up of Sec. 10.1 (Table 10.1) indicates that the Russell Saunder's states are ^3F, ^3P, ^1G, ^1D, ^1S with the total L being 3, 1, 4, 2 and O respectively. The commutation properties of angular momentum operators and the consequences flowing therefrom tell us that for any given L, there will be (2L+1) M_L integral values ranging from $-L \leq M_L \leq L$ and that the given state will be (2L+1) fold degenerate. Hence there would have to be, omitting spin factors, (2L+1) number of atomic (or ionic) degenerate wavefunctions each of which will be differentiated from the other within the fold by the inclusion of one of the different permissible values of M_L. The M_L values are just the different possible sums of the single electron m_l values of the multielectron systems. Let us ask ourselves what atomic wavefunctions we can write for the ^3F state of a d^2 system. For each M_L value, this will be a Slater determinant or sum of Slater determinants. The elements of such a Slater determinant of the ^3F state of a d^2 system are just the products of two single-electron spin orbitals of two d-electrons. Leaving out the spinpart, a general element will be

$$\begin{aligned}(D)_{ij} &= R_{nl}(r_1)\theta_{lml_1}(\theta_1)\Phi_{ml1}(\phi_1)R_{nl}(r_2)\theta_{lml2}(\theta_2)\Phi_{ml2}(\phi_2) \quad (10.12)\\ &= R_{n2}(r_1)\theta_{2ml1}(\theta_1)\Phi_{ml1}(\phi_1).R_{n2}(r_2)\theta_{2ml2}(\theta_2)\Phi_{ml2}(\phi_2)\\ &= (m_{l1})(m_{l2})\end{aligned}$$

Each bracketed term here represents a corresponding single-electron orbital. For M_L=3 of the ^3F state (d^2 system) the Slater determinants wavefunction is

$$\psi_{^3F(M_L=3)} = \frac{1}{(2!)^{\frac{1}{2}}}\begin{vmatrix}(m_{l1})^1 & (m_{l1})^2\\ (m_{l2})^1 & (m_{l2})^2\end{vmatrix} \qquad (10.13)$$

with actual m_{l1} and m_{l2} values in the product function being 1 and 2 to make M_L=3. In the determinant (10.13), the numerical factor $\frac{1}{(2!)^{\frac{1}{2}}}$ is the normalising factor and the superscripts associated with the elements represent electrons 1 and 2 respectively.

The state, $\psi_{^3F(M_L=1)}$, can be obtained with single electron m_l values as

$m_{l1}m_{l2} = 2 + (-1) = 1 = M_L$ or $m'_{l1} + m'_{l2} = 1 + 0 = 1 = M_L$. Hence, there may be written two possible determinants

$$D_1 = \frac{1}{(2!)1/2}\begin{vmatrix}(m_{l1})^1 & (m_{l1})^2\\ (m_{l2})^1 & (m_{l2})^2\end{vmatrix} \text{ and } D_2 = \frac{1}{(2!)1/2}\begin{vmatrix}(m'_{l1})^1 & (m'_{l1})^2\\ (m'_{l2})^1 & (m'_{l2})^2\end{vmatrix}$$

$$\text{Therefore, } \psi_{^3F(M_L=1)} = a\,D_1 + bD_2. \qquad (10.14)$$

which is a linear combination of determinants.

[It should be remembered that the actual determinantal wave -functions should also include spin and, as such, are really different from what appears in relation (10.13) or in (10.14). But to compute symmetry effect, one can make do with these hypothetical wave-functions comprising the space part only.]

Our main concern is, however, to know the transformation property of the determinantal wave-functions under group operations. Fixing our attention on $\psi_{^3F(M_L=3)}$ of determinant (10.13) we find the diagonal term is $\left(\text{dropping the numerical factor} \frac{1}{(2!)^{\frac{1}{2}}}\right)$

$$
\begin{aligned}
(m_{l1})^1(m_{l2})^2 &= \left(R_{n2}(r_1)\theta_{21}(\theta_1)\frac{1}{\sqrt{2\pi}}e^{i.1\phi_1}\right) \\
&\quad \left(R_{n2}(r_2)\theta_{22}(\theta_2)\frac{1}{\sqrt{2\pi}}e^{i.2\phi_2}\right)
\end{aligned}
$$

Adopting the respective symmetry axis as-the quantising axis as described earlier, a C_4 operation on $(m_{l1})^1(m_{l2})^2$ will be

$$
\begin{aligned}
\hat{C}_4(m_{l1})^1(m_{l2})^2 &= \{R_{n_2}(r_1)R_{n2}(r_2)\theta_{21}(\theta_1)\theta_{22}(\theta_2)\}\,. \\
&\quad \left\{\frac{1}{\sqrt{2\pi}}e^{1.i}\left(\phi_1 - \frac{2\pi}{4}\right).\frac{1}{\sqrt{2\pi}}\,e^{2i}\left(\phi_2 - \frac{2\pi}{4}\right)\right\} \\
&= (m_{l1})^1(m_{l2})^2e^{-i\frac{2\pi}{4}}.e^{-i.2.\frac{2\pi}{4}} \\
&= (m_{l1})^1(m_{l2})^2e^{-i(1+2)\frac{2\pi}{4}} \\
&= (m_{l1})^1(m_{l2})^2e^{-iM_L\frac{2\pi}{4}} \qquad (10.15)
\end{aligned}
$$

where $M_L = 3$. Transformation (10.15) of the product function is just similar to that of the single-electron spherical harmonics. Similar results will appear when the transformation is effected of any other term obtainable on splitting the determinant (10.13). If one uses the basis set of seven degenerate functions of the ^3F state, the extreme two of which are $\psi^3F(M_L = 3)$ and $\psi^3F(M_L = -3)$ one can easily construct the seven-dimensional matrix representation of the symmetry operator C_4 and compute the character value.

Leaving aside all mathematical detail, it will do well to remember that a multielectron atomic wavefunction of an F state has a form for its angular part similar to that of a corresponding single-electron f-wavefunction. Analogously the D and the P states have similar angular parts as of a single-electron d and p wavefunction respectively. This conclusion follows when one tries to derive the general form[17] of the angular part of a multielectron wavefunction from the principles and properties of the total orbital angular momentum operator, L, step-up and step-down operators, L_+ and L_-, respectively.

A general table for the splitting of the levels of a few d^n multielectron states of an atom or ion placed at the centre of different environmental symmetries of perturbing points (i.e. ligands) is given below.

Table 10.2

Russell Saunders states	$O_\hbar$ Symmetry	T_d Symmetry	$D_{4\hbar}$ Symmetry	D_3 Symmetry
		IR's of the split levels		
S	A_{1g}	A_1	A_1	A
P	T_{1g}	T_1	A_{2g}, E_g	A_2' E
D	T_{2g}, E_g	T_2, E	A_{1g}, B_{1g}, A_{2g}, E_g	A_1, E(2)
F	$A_{2.g}$ T_{1g}, T_{2g}	A_2, T_1, T_2	A_{2g}, B_{1g}, B_{2g}, $E_g(2)$	A_1, $A_2(2)$, E2

The numerals within the parenthesis of certain symmetries in the table represent the number of such symmetry states when the levels are split.

While discussing the splitting of levels, we have been silent on the spin multiplicities of the states. Since the perturbing centres (i.e., the ligands) do not affect the spin state of an atom or ion, the split terms will have the same spin multiplicity as of the states from which these are derived.

10.4 Energy of Split Levels. Energy Diagram.

10.4.1 Principles:

To find the energies of the split levels is really a job of quantum mechanics and does not fall within the purview of symmetry. All the same some symmetry considerations are involved in the form of direct product representation and in drawing the energy correlation diagram.

Confining our attention to a d^1 system, the perturbation effect of the set of five degenerate d-orbitals of a metal ion is obtainable from solutions of the determinantal equation.

$$\begin{vmatrix} H_{11-\epsilon} & H_{12} & \cdots\cdots & H_{15} \\ H_{21} & H_{22-\epsilon} & & H_{25} \\ \vdots & & & \vdots \\ H_{51} & H_{52} & & H_{55-\epsilon} \end{vmatrix} = 0 \qquad (10.16)$$

where $H_{12}=H_{21}=< m_{l1} \mid V_{cryst} \mid m_{l2} >$ and similarly for the others. The kets $\mid m_{l1} >$, $\cdots\cdots \mid m_{l5} >$, are the d-wavefunction, viz., d_0, d_{+2}, d_{-2}, d_{+1} and d_{-1} respectively. The term V_{cryst} (cf. Eq. 14a) the potential energy operator due to crystal field, depends on the nature of the environment. Since it is electrostatic in nature, $V_{cryst} = \sum_i \frac{z'e^2}{r_{ij}}$, remembering there is one d-electron only and the suffix j stands for it. The suffix represents the point ligands each carrying a negative charge $-z'e$. An expression of $\frac{1}{r_{ij}}$ in terms of spherical harmonics[18] defined with respect to an origin located at the metal ion shows that for an $O_\hbar$ environment

$$V_{cryst} = V_{sp} + V_0$$

that is, the perturbation of the environment is the sum total of

(i) a spherical perturbation (V_{sp}) on the degenerate set of d-orbitals due to a uniform smear of $-6_{z'e}$ units of charges on the surface of a sphere of suitable radius (metal-ligand distance).

(ii) an additional perturbation V_0 due to charges gathered in clots of −z'e units at the six apices of an octahedron circumscribed about the sphere from the previously smeared charges on its surface. This gathering in process will involve no work, since this happens on an equipotential surface [nonalteration of baricentre achieved in state (i)].

The expression for V_{cryr} may be obtained either by detailed calculation or mostly by using symmetry consideration.

Thus the effect of spherical perturbation (V_{sp}) is to raise the original energy level E of the degenerate d-wave functions by an amount

ε_g (total energy E+ε_g) which is still five-fold degenerate. The remant perturbation (V_0) brings about energy changes relative to ε_g and lifts the degeneracy. The original perturbation determinant (10.16) may be simplified by considering the spherically perturbed state as the initial one. Thus the determinants equation is

$$\begin{vmatrix} H_{11}' - \varepsilon_0' & H_{12}' \cdots\cdots H_{15}' \\ H_{21}' & H_{22}' - \varepsilon_0' \cdots H_{25}' \\ \vdots & \vdots \\ \vdots & \vdots \\ H_{51}' & H_{52} \cdots\cdots H_{55}' - \varepsilon_0' \end{vmatrix} = 0 \qquad (10.17)$$

where $H_{11}' - \varepsilon_0' =< m_{l1} \mid V_0 \mid m_{l1} >= \varepsilon_0' = H_{22} < m_{l2} \mid V_0 \mid m_{l2} >$ etc. $H_{12}' = H_{21} =< m_{l1} \mid V_0 \mid m_{l2} >$ and so also for others. The same type of calculations can be made where real d-o0rbitals, instead of complex d-wavefunctions, are used.

The solution of Eq. (10.17) with requisite input of data leads to two values of ε_0', viz., x which is doubly degenerate and −y, this level being triply degenerate. The corresponding zero order orbital sets used are (i) the base functions d_{x^2} and $d_{(x^2-y^2)}$ of Eg symmetry and (ii) the base functions d_{xy}, d_{xz}, d_{yz} of T_{2g} symmetry or the five complex d-wavefunctions. A thorough calculation on the basis of CFT leads to results which are qualitatively correct but quantitatively untenable. This has led to introducing modifications of the ACFT which incorporate empirical parameters. In any practical calculation, however, the energy gap between t_{2g}, and e_g levels of a single d-electron is determined spectroscopically and symbolised by 10 Dq units.

$$x - (-y) = x + y = 10Dq \qquad (10.18a)$$

A d^{10} free ion is in ^1S state. When such a d^{10} ion is placed in octahedral environment of perturbation $V_{cryst} = V_{sp} + V_o$, its energy will be raised by ε_s^0 under V_{sp}, but V_0 will no longer be able to split it any further. In terms of product wavefunctions of the single-electron d-orbitals of this S-state these 10 electrons would be distributed as $(d_{z2})^2(d_{x^2} - y^2)^2(d_{xy})^2(d_{xz})^2(d_{yz})^2$ under V_{sp} and as $(e_g)^4(t_{2g})^6$ under V_0. Hence, we may write for the preservation of the baricentre

$$4\varepsilon_0'(e_g) + 6\varepsilon_0'(t_{2g}) = 0$$

$$\text{i.e.,}\quad 4x - 6y = 0 \tag{10.18}$$

Eqs. (10.18a and 10.18b) lead to

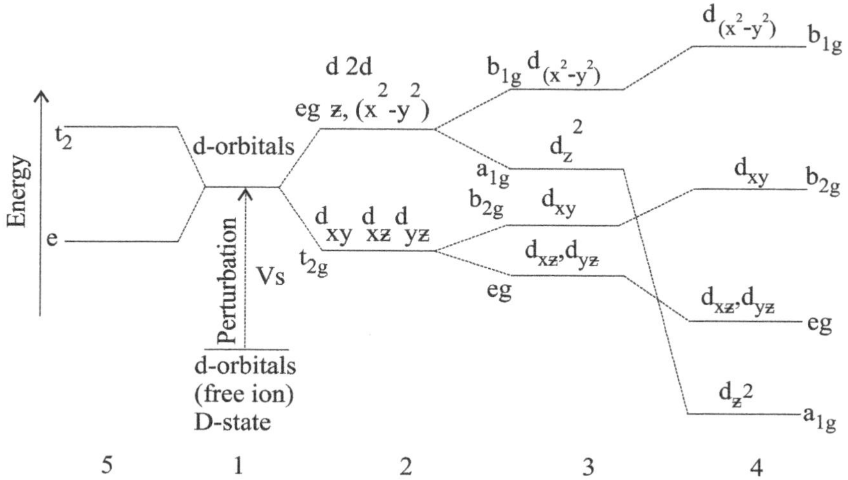

10.2: Nature of splitting of D state of d-electron systems. 1. Spherically perturbed d-shell. 2. Octahedral perturbation. 3. Tetragonal perturbation obtainable from (1) or by distortion (elongation along z-axis) of O_h symmetry. 4. Squareplanar perturbation. 5. Tetrahedral perturbation.

$$\begin{aligned} x &= 6\ \text{Dq} \\ \text{and } y &= 4\ \text{Dq} \end{aligned} \tag{10.19}$$

The foregoing process of CFT described for finding the relative energies in octahedral complexes of d^1 systems may be applied with equal facility for tetrahedral and other complexes. For tetrahedral systems the degenerate d-orbitals split into upper t_2 and lower e orbitals. This is just a reversal of octahedral field effect. Fig. 10.2 describes the splitting of d^1 system in different environments. This will also be true for a D-state of the metal ion of d^n system.

10.4.2 Energy Correlation Diagram

It is to be noted that in a transition metal ion complex there are two perturbing effects on the electrons in the metal ion d-orbitals. The first is the inter d-electron repulsions leading to correlation (Russell Saunders' Coupling). The second one is the interaction between metal d-electrons and ligand point charges leading to crystal field effect (splitting or non-splitting of levels). When the crystal field effect is weak relative to correlation effect, the initial state is the free ion term and the final are the crystal field terms. Thus for O_h symmetry, the terms appear in Table 10.3.

Table 10.3 Weak Field Effect (O_h environment)

Configuration	Initial term (Russell Saunders' Term)	Final term (Crystal field Symmetry Term)	Corresponding Crystal field orbitals
d^1	2D	2E$_g$	e_g
		2T$_{2g}$	t_{2g}
d^2	^3F	3A$_{1g}$	a_{1g}
		$^3T_{1g}$	t_{1g}
		$^3T_{2g}$	t_{2g}

and so on for other configurations.

But when the crystal field is infinitely strong, couplings due to electrical repulsions completely break down and the electrons are forced to obey the mandate of the crystal field and occupy orbita4s of the lowest energy upwards with spin pairing, if necessary. Table 10.4 sums up the situation.

Table 10.4 Strong Field (O_h environment)

Configuration	Crystal field orbital configurations (infinitely strong crystal field)	Final crystal field terms when electrical interactions are not negligible but still relatively small.
d^1	$(t_{2g})^1$–ground	
	$(e_g)^1$–excited	
		To be decided
d^2	$(t_{2g})^2$–ground	for different
	$(t_{2g})^1(e_g)^1$–excited	crystal field states
	$(e_g)^2$–excited	
d^3	$(t_{2g})^3$–ground	
	$(t_{2g})^2(e_g)^1$–excited	
	$(t_{2g})^1(e_g)^2$–excited	
	$(e_g)^3$–excited	

The task before us is two-fold-(i) to find the crystal field terms in the third column of Table 10.4 and (ii) to correlate the weak crystal field terms (Table 10.3) which are readily available to us (Sec. 10.3). with those of the strong field both spinwise and symmetry-wise. This task, when completed, yields the energy correlation diagram and demonstrates how a weak field term changes with the increasing strength of the crystal field. To make things clear we specifically consider a d^2–metal ion in an octahedral environment of the ligands. The respective Russell Sannders' states of the free ion are split (Table 10.1), in the weak field of $O_\hbar$ symmetry, into the crystal field terms, viz.,

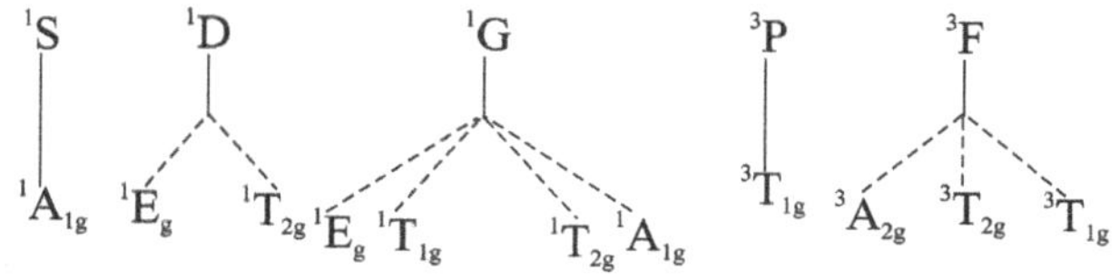

The same metal ion etc., under infinitely strong crystal field has three possible orbital configurations (Table 10.4). The wavefunction of the $(t_{2g})^2$ configuration is a product wavefunction of two single-electron wavefunctions each of symmetry T_{2g}. Hence, when the electric repulsions are just beginning to be relatively felt, the representation of the group $O_\hbar$ with the net wavefunction as the basis is the same as the direct product representation with the individual electronic wavefunctions as the bases. Hence we may Write (of. Example 5.10)

$$\left.\begin{array}{c}
\text{for } (t_{2g})^2 \text{ state, } \Gamma^{(T_{2g}\otimes T_{2g})} = T_{2g} \otimes T_{2g} \\
= A_{1g} \oplus E_g \oplus T_{1g} \oplus T_{2g} \\
\text{For the } (t_{2g})^1(e_g)^1 \text{ and } (e_g)^2 \text{ state} \\
T_{2g} \otimes E_g = T_{1g} \oplus T_{2g} \\
\text{and finally } \Gamma^{(E_g\otimes E_g)} = A_{1g} \oplus A_{2g} \oplus E_g
\end{array}\right\} \qquad (10.20)$$

Unlike the weak field states, one does not know the spin states of the strong field terms immediately. This has to be found out indirectly. One must not loose sight of the fact that the configuration $(t_{2g})^1(e_g)^1$ can have two electrons in both spin paired and unpaired conditions. Therefore, considering the two possibilities of triplet and singlet states of $(t_{2g})^1(e_g)^1$ configuration, the outcome is evident (see figure and cf.

Eq. 10.20). Spin assignments for the rest of the strong crystal field terms may be done by arguments (Also general techniques exist for asscribing the spin multiplicities). All the weak field terms with their respective

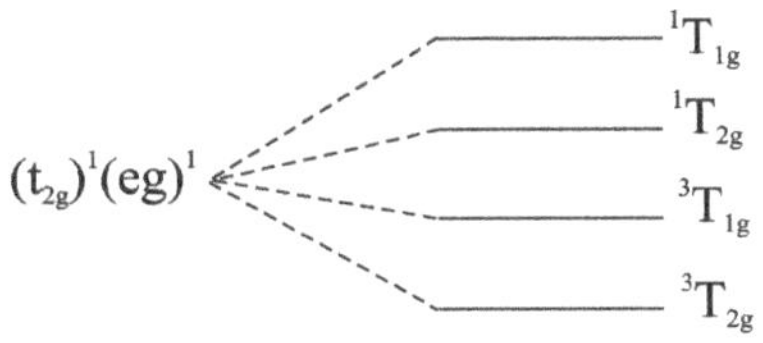

spin multiplicities must also reappear as the strong field terms with preservation of the multiplicities. It is noted further that

(i) all the A_{1g}, E_g weak field terms are singlets, the A_{2g} term is triplet.

(ii) there are three weak field T_{2g} terms of which two are singlets and one triplet

(iii) there are three weak field T_{1g} terms of which two are triplets and one singlet.

Taking note of the terms of $(t_{2g})^1(e_g)^1$ and of the above three observations we can write, paying attention to relations (10.20),

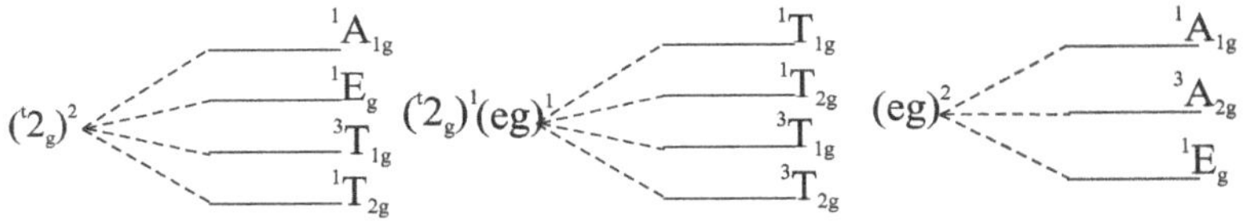

We thus establish the perturbational terms near the two anchor points of the crystal field states, i.e., on the sides of very low and very high fields respectively. Using two principles, viz.,

(i) There exists one-to-one correspondence between the low and high crystal field states,

(ii) States of the same symmetry and spin multiplicity do not cross.

we can join the different crystal field states to obtain the correlation diagram (Fig. 10.3). In doing this one should initially known how to place the free ion terms in increasing order of energies from their analytic energy expressions involving Slater Condon parameters and supplementary spectroscopic data. As regards the energy sequence of very strong crystal field terms, a modified form of Hund's ground state rule may be followed.

(i) States of highest mutiplicity of a given symmetry lie lowest.

(ii) For given multiplicity, states with higher degeneracy (T > E > A) tend to lie lower.

But these rules arc only tentative and not always valid for the crystal field states. Fig. 10.3 depicts the energy correlation diagram of the crystal field states and shows qualitatively how with increasing field strength of the ligands the energies vary for a d^2 metal ion forming a complex of $O_\hbar$ symmetry.

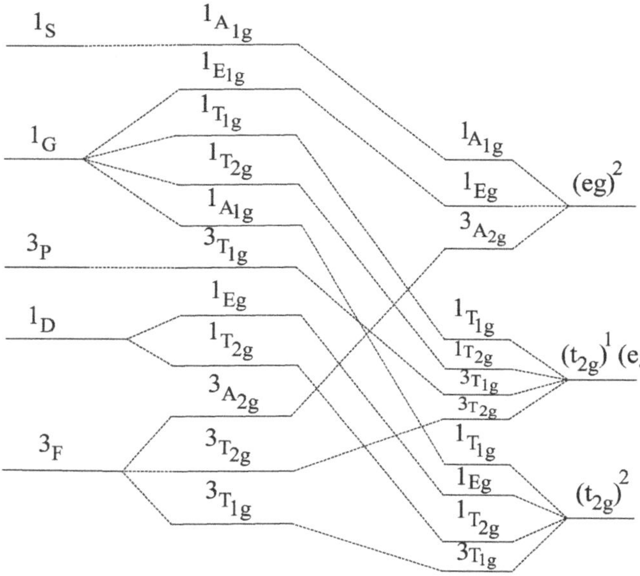

10.3: Strong and weak fields correlation diagram for d^2 metal ion in octahedral field.

The principles developed for the d^2-system in $O_\hbar$ environment can similarly be extended to other environments. Also similar correlation diagrams can be developed for d^n systems with $n>2$. Many additional features are revealed including changes from 'high spin' weak field ground state to 'low spin' strong field ground state in some d^n systems of $O_\hbar$

symmetry. In tetrahedral fields, there is no reversal or change of ground crystal field terms.

10.5 Molecular Orbital Theory of Transition Metal Complexes

Principles: A natural follow-up of crystal field treatment of coordination complexes from symmetry viewpoint is their molecular orbital treatment. The *mo* theory of transition metal complexes is quite similar to what has been described in Chapter 9 under molecular orbitals.

In an *mo* treatment one should know essentially (or assume using intrinsic chemical sense) the metal atom orbitals and the ligand orbitals which are to combine linearly to form the mo's. For example in $[CoF_6]^{3-}$, nine atomic orbitals of cobalt, viz., $3d_{z^2}$, $3d_{(x^2-y^2)}$, $3d_{xy}$, $3d_{yz}$, $3d_{zx}$, 4s, $4p_x$, $4p_y$, $4p_z$ and four from each ligand (one 2s and three 2p orbitals), involving a total of 33 atomic orbitals, participate in yielding the mo's. In $[PtCl_4]^{2-}$, it is 25 and in $[Cr(NH_3)_6]^{3+}$, the total number of atomic orbitals involved is fifteen only.

Secondly, instead of setting up the secular determinant in the basis set of the atomic orbitals, it is advantageous to use the metal atomic orbitals and certain linear combinations of the ligand orbitals as the basis set. This technique affords factoring the determinantal equation into blocks-a principle which has also been used by using symmetry orbitals (or group orbitals) to construct the mo's of Chapter 9.

These specific linear combinations of ligand orbitals are the group orbitals of the ligands and are obtainable totally from symmetry principles.

The stage of combining the group orbitals and atomic orbitals of the metal atom (ion) to form the mo's depends on the symmetries of the combining orbitals. The quantitative evaluation of the coefficients, however, is a quantum mechanical job.

Finally an approximate mo energy diagram can be set up.

Example 10.1. MO's and energy diagram of $[Cr(NH_3)_6]^{3+}$

The ion has the symmetry of O_h point group. The actual orbitals that participate in forming the mo's are $3d_{z^2}$, $3d_{(x^2-y^2)}$, $3d_{xy}$, $3d_{xz}$,

$3d_{yz}$, 4s, $4p_x$, $4p_y$ and $4p_z$ of chromium and the hybrids h_1, h_2, h_3, h_4, h_5, h_6 of the N atoms in the six NH_3 ligands. The ligands are

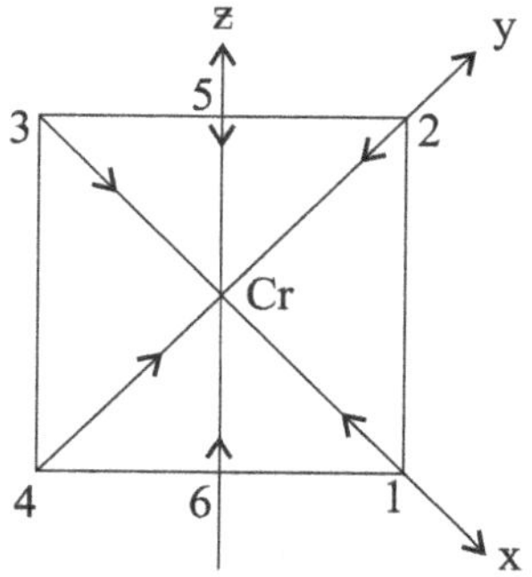

Fig. 10.4: Axial system in $[Cr(NH_3)_6]^{3+}$

numbered 1 to 6 and the hybrids are poised for $\sigma-$bond formation with the chromium orbitals.

(i) We work with the axial system as shown in Fig. 10.4. The main cartesian axes X Y Z are set up with Cr-atom at the orgin. The $h_1 \cdots\cdots h_6$ hybrids of N atoms of the six NH_3 ligands have their positive lobes, indicated by arrow heads, turned toward the Cr-atom for $\sigma-$bond formation.

(ii) To solve the Roothaan equation (Eq. 9.28) the secular determinant $| \mathbf{F} - \varepsilon\mathbf{S} |$ will be 15×15 dimensional in the basis set of 15 atomic orbitals chosen. But the following factoring device is possible by exploiting the symmetry aspects of the complex ion.

(iii) It is to be noted that under group operations of $O_\hbar$, the ligand orbitals exchange positions amongst themselves and not with the metal atom obitals. Hence if any reducible representation, Γ^{red}, be generated with the 15 orbitals as the basis set, it should turn out to be a direct sum of two representations, viz.,

$$\Gamma^{red} = \Gamma_M \oplus \Gamma_L$$

where Γ_M is the reducible representation of the point group in the basis set of the metal atomic orbitals (d_{z^2}, $d_{(x^2-y^2)}$, d_{xy}, d_{xz}, d_{yz}, s, p_x, p_y, p_z) and Γ_L, the representation in the basis set of ligand orbitals (h_1 h_2 h_3 h_4 h_5 h_6).

(iv) It is not necessary to carry out the reduction of Γ_M here since the results are already known from piecemeal treatments in chapters 5 (Examples) and 9 as also from the table 10.2.

$$\Gamma_M = A_{1g} \oplus E_g \oplus T_{2g} \oplus T_{1u} \text{ where}$$

s orbital transforms as A_{1g},
d_{z^2}, $d_{(x^2-y^2)}$ transform as E_g,
(d_{xy}, d_{yz}, d_{zx}) transform as T_{2g},
and $(p_x\ p_y\ p_z)$ transform as T_{1u} (not T_{1g} as may be inferred from table 10.2)

This information is also obtainable from the character table of the $O_\hbar$ point group.

(v) To reduce Γ_L, we find the characters of the reducible representation first, the basis set being $(h_1 \cdots\cdots h_6)$.

	$\hat{E}$	$6\hat{C}_4$	$3\hat{C}_2$	$6\hat{C}'_2$	$8\hat{C}_3$	$\hat{i}$	$3\hat{\sigma}_\hbar$	$6\hat{S}_4$	$8\hat{S}_6$	$6\hat{\sigma}_d$
χ^{red}	6	2	2	0	0	0	4	0	0	2

On using Theo. 5.8 along with the character table of $O_\hbar$, we have

$$\Gamma_L = A_{1g} \oplus E_g \oplus T_{1u}$$

(vi) We next form the group orbitals (symmetry orbitals) having symmetries of A_{1g}, E_g and T_{1u}. Avoiding the general method of projection operators, we may also rely on the easier process of observation and inference (cf. Example 9.9)

$$\psi^\hbar(a_{1g}) = \frac{1}{\sqrt{6}}(h_1 + h_2 + h_3 + h_4 + h_5 + h_6) \tag{10.21}$$

The pair of e_g orbitals must mirror the symmetries of d_{z^2} and $d_{(x^2-y^2)}$
Since $d_{z^2} = d(2z^2 - x^2 - y^2) = d(z^2 - x^2 + z^2 - y^2)$. we can write

$$\left.\begin{aligned} \psi^\hbar(e_g^1) &= \frac{1}{\sqrt{12}}(h_5 + h_6 - h_1 - h_3 + h_5 + h_6 - h_2 - h_4) \\ &= \tfrac{1}{\sqrt{12}}(2h_5 + 2h_6 - h_1 - h_2 - h_3 - h_4) \\ \psi^\hbar(e_g^2) &= \tfrac{1}{2}(h_1 - h_2 + h_3 - h_4) \end{aligned}\right\} \tag{10.22}$$

The triply degenerate T_{1u} orbitals should mirror the symmetry of p_x p_y p_z under $O_\hbar$.

$$\left.\begin{aligned}\psi^\hbar(t_{1u}1) &= \tfrac{1}{\sqrt{2}}(h_1 - h_3) \sim p_x\\ \psi^\hbar(t_{1u}2) &= \tfrac{1}{\sqrt{2}}(h_2 - h_4) \sim p_y\\ \psi^\hbar(t_{1u}3) &= \tfrac{1}{\sqrt{2}}(h_5 - h_6) \sim p_z\end{aligned}\right\} \tag{10.23}$$

(vii) Now that the group orbitals have been obtained, we are-ready to combine these with metal orbitals of the same symmetry. What we have achieved is that if we replace the basis set of the original 15 atomic and hybrid orbitals by a basis set of 9 metal atom orbitals and the six combinations of ligand orbitals (viz., the group orbitals of Eqs. 10.21, 10.22 and 10.23), the determinant will be-factored into blocks because of vanishing integral rule (cf. Chapter 6). Thus

(a) s and $\psi^\hbar(a_{1g})$ will mix to yield two mo's $1a_{1g}$ and $2a_{1g}$ obtainable from a (2×2) block.

(b) d_{z^2} and $\psi^\hbar(e_{g1})$ combine to yield two e_g orbitals. These are obtained from two equivalent 2×2 blocks. The mo's are $1e_g$ and $2e_g$ each being a degenerate pair.

(c) From three equivalent (2×2) blocks, we get mo's from combination of p_x and $\psi^\hbar(t^1_{1u})$, p_y and $\psi^\hbar(t^2_{1u})$, p_z and $\psi^\hbar(t^3_{1u})$. The final yield is a set of six mo's divided into two triply degenerate orbitals $1t_{1u}$ and $2t_{1u}$.

(d) The metal atom orbitals d_{xy}, d_{yz}, d_{xz} do not mix with the ligand orbitals and form just a triply degenerate t_{2g} mo.

Actual eigen vectors and energy eigenvalues are to be found out by-quantum mechanical methods following some self consistent field. procedure or some approximate mo method. Even without this information, it is feasible to draw the energy correlation diagram as has been done for dicyclopentadienly iron (ferrocene) and methane (Chapter 9).

It will be noted that the ground state electronic configuration of $[Cr(NH_3)_6]^{3+}$ ion, excluding the core part of Cr, N and N-H bonding, is $(1a_{1g})^2$ $(1t_{1u})^6$ $(1e_g)^4$ $(t_{2g})^3$ accounting for the twelve bonding and

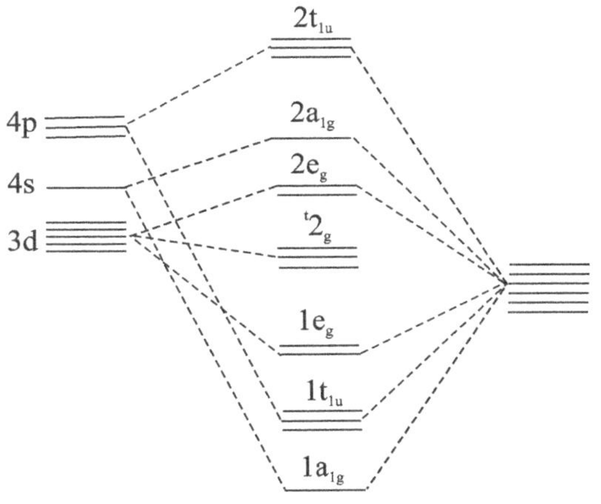

Fig. 10.5: Molecular Orbital energy correlation diagram of $[Cr(NH_3)_6]^{3+}$ ion.

three upaired electrons of the complex ion. While t_{2g} is a nonbonding mo, $2e_g$, $2a_{1g}$ and $2t_{1u}$ are antibonding mo's. It is seen that even in mo treatment, there occur t_{2g} and $2e_g$ mo's with the d-orbital energy as their baricentre. Although $2e_g$ is not purely a split d-orbital, the contributions of the ligand orbitals are very small compared to those coming form d_{z^2} and $d(x^2 - y^2)$. This is confirmed by actual calculation. It is in this sense that an mo treatment confirms the prediction of crystal field theory on the splitting of degenerate d-orbitals into e_g and t_{2g} under $O_\hbar$ point group symmetry.

The complex ion $[Cr(NH_3)_6]^{3+}$ provides no scope for π bonding with the metal atom orbitals. The ligand orbitals are used up in forming $\sigma-$bonds. We shall now dwell upon a case where both σ and $\pi-$bondings occur. As the previous example was dealt with in detail, the next one of the mo's of the coordination complex is treated with pertinent elaboration of only the newer points of the problem.

Example 10.2. Molecular Orbitals of $[PtCl_4]^{2-}$ ion.

The ion belongs to $D_{4\hbar}$ point group. The participating atomic orbitals are the usual nine metal atom orbitals (5d, 6s, 6p) and sixteen ligand atom orbitals consisting of one 3s and three 3p's from each Cl atom.

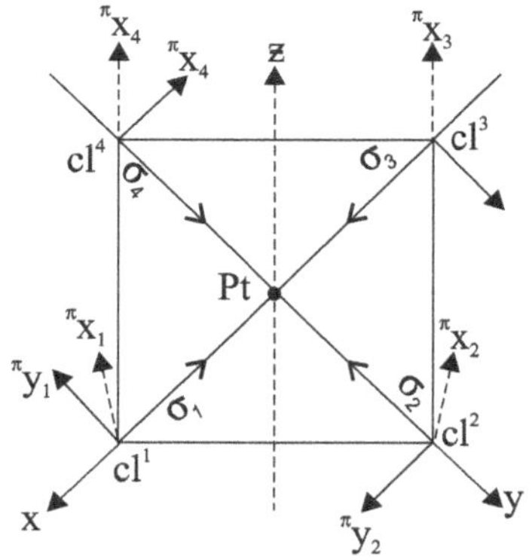

Fig. 10.6: Coordinate system in $[PtCl_4]^{2-}$. Right handed at Pt and left handed at the ligands.

A right handed cartesian frame is set up at the Pt-atom and four other ligand coordinate axial sets, all left handed, are installed at the Cl-atoms (Fig. 10.6). The positive four lobes of the ligand p_z-orbitals, indicated by σ_1, σ_2, σ_3, σ_4 point toward the Pt atom. The set of ligand p_x and p_y orbitals, denoted by $\pi_{xi}.\pi_{yi}$(i=1 to 4) are arranged following the left handedness of the coordinate system. The positive lobes of the ligand p_x orbitals (π_{xi}) all point upwards from the molecular plane and the ligand π_{yi}'s lie in the complex ion plane. Arrow heads indicate the positive lobes.

It would be noted that under the symmetry operations of $D_{4\hbar}$ there occur no interchanges of the metal atom orbitals and the-ligand orbitals. Consequently $\Gamma^{red} = \Gamma_M \oplus \Gamma_L$.

The assignment of symmetries to metal orbitals in $D_{4\hbar}$ point group is as follows.

Orbitals	S	$p_x p_y$	p_z	d_{z^2}	$d_{(x^2-y^2)}$	d_{xy}	$(d_{xz} d_{yz})$
IR	A_{1g}	E_u	A_{2u}	A_{1g}	B_{1g}	B_{2g}	E_g

Reduction of Γ_L:

Before proceeding to reduction, we note that under the group-operations of $D_{4\hbar}$

(i) the ligand s orbitals (s_1 s_2 s_3 s_4) do not interchange with the σ_i set, π_{xi} set or the π_{yi} set.

(ii) the σ_i set orbitals do not exchange position with any of the s_i, π_{xi} and π_{yi} sets.

(iii) the π_{xi} set orbilals do not trade places with the s_i, σ_i or π_{yi} sets.

(iv) the π_{yi} set orbitals also behave similarly.

It is thus at once clear that Γ_L can be expressed as a direct sum, viz.,

$$\Gamma_L = \Gamma_{si} \oplus \Gamma_{\sigma i} \oplus \Gamma_{\pi xi} \oplus \Gamma_{\pi yi}$$

where each representation is a reducible one in the respective basis sets. To find the characters in these three basis sets, we draw up-the following:

	E	$2C_4$	C_2	$2C_2'$	$2C_2''$	$\sigma_\hbar$	i	$2S_4$	$2\sigma_d'$	$2\sigma_d''$
χ^{si}	4	0	0	2	0	4	0	0	2	0
$\chi^{\sigma i}$	4	0	0	2	0	4	0	0	2	0
$\chi^{\pi xi}$	4	0	0	−2	0	−4	0	0	2	0
$\chi^{\pi yi}$	4	0	0	−2	0	4	0	0	−2	0

Employing Theo. 5.8 and the relevant character table, we have

$$\begin{aligned}
\Gamma_{si} &= A_{1g} \oplus B_{1g} \oplus E_u \\
\Gamma_{\sigma i} &= A_{1g} \oplus B_{1g} \oplus E_u \\
\Gamma_{\pi xi} &= A_{2u} \oplus B_{1u} \oplus E_g \\
\Gamma_{\pi yi} &= A_{2g} \oplus B_{2g} \oplus E_u
\end{aligned}$$

Group Orbitals

The ligand group orbitals of the required symmetries, as shown above, are now to be formed from the corresponding basis function sets, viz., s_i, σ_i, π_{xi} and π_{yi}. All these, excepting B_1u, may be formed by inspection and inference using as a mental aid the graphical form of a suitable base function transforming as the relevant IR. The B_{1u} group orbital may be obtained by trial or by method of projection operator. We thus have

$$\left.\begin{aligned}
\psi^{si}(a_{1g}) &= \tfrac{1}{2}(s_1 + s_2 + s_3 + s_4)\ \text{reflectingthesymmetryofsorbital} \\
\psi^{si}(b_{1g}) &= \tfrac{1}{2}(s_1 - s_2 + s_3 - s_4) \cdots\cdots d_{(x^2-y^2)} \\
\psi^{si}(e_u^1) &= \tfrac{1}{\sqrt{2}}(s_1 - s_3) \cdots\cdots p_x \\
\psi^{si}(e_u^2) &= \tfrac{1}{\sqrt{2}}(s_2 - s_4) \cdots\cdots p_y
\end{aligned}\right\} \tag{10.24}$$

Similarly

$$\left.\begin{aligned} \psi\sigma_{\mathrm{i}}(\mathrm{a}_{1\mathrm{g}}) &= \tfrac{1}{2}(\sigma_1+\sigma_2+\sigma_3+\sigma_4) \\ \psi\sigma_{\mathrm{i}}(\mathrm{b}_{1\mathrm{g}}) &= \tfrac{1}{2}(\sigma_1-\sigma_2+\sigma_3-\sigma_4) \\ \psi\sigma_{\mathrm{i}}(\mathrm{e}_{\mathrm{u}}^1) &= \tfrac{1}{\sqrt{2}}(\sigma_1-\sigma_3) \\ \psi\sigma_{\mathrm{i}}(\mathrm{e}_{\mathrm{u}}^2) &= \tfrac{1}{\sqrt{2}}(\sigma_2-\sigma_4) \end{aligned}\right\} \tag{10.25}$$

$$\left.\begin{aligned} \psi\pi\chi_{\mathrm{i}}(\mathrm{a}_{2\mathrm{u}}) &= \tfrac{1}{2}(\pi_{\mathrm{x}1}+\pi_{\mathrm{x}2}+\pi_{\mathrm{x}3}+\pi_{\mathrm{x}4}) \quad \text{mirroring } \mathrm{p_z} \text{ symmetry} \\ \psi\pi\chi_{\mathrm{i}}(\mathrm{b}_{1\mathrm{u}}) &= \tfrac{1}{2}(\pi_{\mathrm{x}1}-\pi_{\mathrm{x}2}+\pi_{\mathrm{x}3}-\pi_{\mathrm{x}4}) \quad \text{(by projection operator method)} \\ \psi\pi\chi_{\mathrm{i}}(\mathrm{e}_{\mathrm{g}}^1) &= \tfrac{1}{\sqrt{2}}(\pi_{\mathrm{x}1}-\pi_{\mathrm{x}3}) \text{ mirroring the symmetry of } \mathrm{d_{xz}} \\ \psi\pi\chi_{\mathrm{i}}(\mathrm{e}_{\mathrm{g}}^2) &= \tfrac{1}{\sqrt{2}}(\pi_{\mathrm{x}2}-\pi_{\mathrm{x}4}) \cdots\cdots \mathrm{d_{yz}} \end{aligned}\right\} \tag{10.26}$$

and finally,

$$\left.\begin{aligned} \psi^{\pi}\mathrm{yi}(\mathrm{a}_{2\mathrm{g}}) &= \tfrac{1}{2}(\pi_{\mathrm{y}1}+\pi_{\mathrm{y}2}+\pi_{\mathrm{y}3}+\pi_{\mathrm{y}4}) \text{ symmetry of } \mathrm{R_z} \text{ vector} \\ \psi^{\pi}\mathrm{yi}(\mathrm{b}_{2\mathrm{g}}) &= \tfrac{1}{2}(-\pi_{\mathrm{y}1}+\pi_{\mathrm{y}2}-\pi_{\mathrm{y}3}+\pi_{\mathrm{y}4}) \cdots \mathrm{d_{xy}} \text{ orbital} \\ \psi^{\pi}\mathrm{yi}(\mathrm{e}_{\mathrm{u}}^1) &= \tfrac{1}{\sqrt{2}}(\pi_{\mathrm{y}2}-\pi_{\mathrm{y}4}) \cdots\cdots \mathrm{p_x} \\ \psi^{\pi}\mathrm{yi}(\mathrm{e}_{\mathrm{u}}^2) &= \tfrac{1}{\sqrt{2}}(\pi_{\mathrm{y}3}-\pi_{\mathrm{y}1}) \cdots\cdots \mathrm{p_y} \end{aligned}\right\} \tag{10.27}$$

The table 10.5 summarises the above results for possible formation of molecular orbitals from group orbitals and Pt orbitals.

Table 10.5 Possible Group Orbitals for Combinations with Pt orbilals

Symmetry	Pt orbitals	Ligand group orbitals for $\sigma-$ bonding	Ligand group orbitals for $\pi-$ bonding
A_{1g}	s, d_{z2}	$\frac{1}{2}(s_1 + s_2 + s_3 + s_4)$; $\frac{1}{2}(\sigma_1 + \sigma_2 + \sigma_3 + \sigma_4)$	
B_{1g}	s, $d(x^2 - y^2)$	$\frac{1}{2}(s_1 - s_2 + s_3 - s_4)$; $\frac{1}{2}(\sigma_1 - \sigma_2 + \sigma_3 - \sigma_4)$	
E_u	p_x p_y	$\frac{1}{\sqrt{2}}(s_1 - s_3)$; $\frac{1}{\sqrt{2}}(\sigma_1 - \sigma_3)$ $\frac{1}{\sqrt{2}}(s_2 - s_4)$; $\frac{1}{\sqrt{2}}(\sigma_2 - \sigma_4)$	$\frac{1}{\sqrt{2}}(\pi_{y2} - \pi_{y4})$ $\frac{1}{\sqrt{2}}(\pi_{y3} + \pi_{y1})$
A_{2u}	p_z		$\frac{1}{2}(\pi_{x1} + \pi_{x2} + \pi_{x3} + \pi_{x4})$
B_{1u}			$\frac{1}{2}(\pi_{x1} - \pi_{x2} + \pi_{x3} - \pi_{x4})$
E_g	d_{xz} d_{yz}		$\frac{1}{\sqrt{2}}(\pi_{x1} - \pi_{x3})$ $\frac{1}{\sqrt{2}}(\pi_{x2} - \pi_{x4})$
A_{2g}			$\frac{1}{2}(\pi_{y1} + \pi_{y2} + \pi_{y3} + \pi_{y4})$
B_{2g}	d_{xy}		$\frac{1}{2}(-\pi_{y1} + \pi_{y2} - \pi_{y3} + \pi_{y4})$

The foregoing table suggests the compositions of the molecular orbitals. The net results are:

(i) Four mo's of A_{1g} symmetry, $1a_{1g}$, $2a_{1g}$, $3a_{1g}$, $4a_{1g}$.

(ii) Three mo's of B_{1g} symmetry, $1b_{1g}$, $2b_{1g}$, $3b_{1g}$.

(iii) Four different degenerate pairs of mo's of E_u symmetry. viz., $1e_u$, $2e_u$, $3e_u$ and $4e_u$.

(iv) Two mo's of A_{2u} symmetry, $1a_{2u}$, $2a_{2u}$.

(v) One mo of B_{1u} symmetry, b_{1u} identifiable with a group orbital.

(vi) Two degenerate pairs of mo's of Eg symmetry, $1e_g$, $2e_g$.

(vii) One mo of A_{2g} symmetry, a_{2g} identifiable with a group orbital.

(viii) Two mo's of B_{2g} symmetry, $1b_{2g}$, $2b_{2g}$.

It may be noted that the mo's of (iv), (vi) and (viii) are formed through lateral overlaps ($p\pi - p\pi$, $d\pi - p\pi$ bondings). The mo's of (iii) involve both σ and π overlaps. The π−mo's of (v) and (viii) are concentrated over the ligand regions.

The two examples provided for the molecular orbital treatment of the transition metal ion complexes clearly demonstrate how powerful is the tool of symmetry and to what an amazing extent it simplifies the procedure before it is handed over to a quantum chemist to carry on the task to completion.

10.6 Spectral Properties. Vibronic Coupling, Vibronic Polarisation.

We shall see in the present section how symmetry principles continue to be applied in interpreting the electronic transitions in transition metal complexes. It is known that two general selection rules, provided there are no secondary complicating effects, govern electronic transitions.

(i) Transition takes place between states of the same spin multiplicities.

(ii) For centrosymmetric systems, transitions are governed by Laporte rules (Sec. 8.3).

The electronic spectra of transition metal complexes generally exhibit three types of absorption which are (in the concepts of CFT) (a) d-d transitions, i.e, transition from the ground crystal field state to an upper one. (b) Charge transfer transitions which involve transfer of electrons between metal and ligand orbitals. (c) Inter-ligand transitions. Besides there may be some spin-forbidden transitions of low intensity. We shall just restrict ourselves to d-d transitions. The complex metal ion may be centrosymmetric (symmetry $O_\hbar$, $D_{4\hbar}$. etc.) or a non-centrosymmetric one (T_d, D_3 etc.) depending on the possession or the lack of an inversion centre respectively.

A. Centrosymmetric Complexes. Vibronic Coupling

The crystal field states of complexes of $O_{\hbar}$, $D_{4\hbar}$ are all of gerade type and hence transitions between g and g states demand that $d - d$ transitions be of zero intensity. Contrarily, however, d–d trnsitions of reasonable intensities do occur. Within the context of Born Oppenheimer Approximations, it is assumed that the total wavefunction $\Psi = \psi_{\mathrm{el}}\ \psi_{\mathrm{v}}\ \psi_{\mathrm{r}}$, i.e., a product of electronic, vibrational and rotational wavefunctions. It is further assumed that for most of the spectral absorption properties, electronic motions may be treated independently of the vibrational and rotational motions of the nuclei. This means ψ_{el}, ψ_{v} and ψ_{r} are mutually separable. In this context the electric dipole transition moment integral, $< \psi'_{\mathrm{el}}\, |e(\overrightarrow{x} + \overrightarrow{y} + \overrightarrow{z})|\, \psi_{\mathrm{el}} >=< g|u|g >$ will vanish where ψ_{el}'s are gerade crystal field states. But physically it is understandable that a crystal field state (say of $O_{\hbar}$, symmetry) may not in reality be cent percent gerade. The different normal modes of vibrations of the atoms in the complex lead in some cases to some average effect which may distort the gerade nature to a small extent. Leaving out the rotational wavefunction ψ_{r}, the role of which is not of any major or minor consequence here, we may deem the remnant part of the total wavefunction as due to a coupling of the vibrational and electronic motions and call it a vibronic wavefunction $\psi_{\mathrm{el}}\ \psi_{\mathrm{v}}$, wherefrom ψ_{v} is no longer separable from ψ_{el}. The transition moment integral would thus be $< \psi'_{\mathrm{el}}\ \psi'_{\mathrm{v}}\, |e(\overrightarrow{x} + \overrightarrow{y} + \overrightarrow{z})|\, \psi_{\mathrm{el}}\ \psi_{\mathrm{v}} >$. The direct product representation with the integrand (Secs. 5.7., 6.4) as the basis function may be split into set of IR's which may contain a totally symmetric A_1g component along with others. This hints at the feasibility of a d-d transition in a centrosymmetric complex. We may plunge into examples right now to clarify the concepts. Consider, as an example, $[V(H_2O)_6]^{3+}$ complex ion which is a d^2-system. The qualitative correlation diagram for d^2 shows that there ought to be three d-d transitions from the ground state of this octahedrally symmetrical complex ion (Sec. 10.4, Fig. 10.3). These are ${}^3T_{2g}(F) \leftarrow^3 T_{1g}(F)$, ${}^3T_{1g}(P) \leftarrow^3 T_{1g}(F)$ and ${}^3A_{2g} \leftarrow^3 T_{1g}(F)$. The letters P, F within parentheses represent the original free ion terms from which the respective crystal field terms have been obtained. The ${}^3T_{1g}(F)$

is the ground term.

For the first transition one may write

$$< \psi^3T_{2g}\psi'_v\, |e(\vec{x} + \vec{y} + \vec{z})|\, \psi^3T_{1g}\psi_v >$$

If the molecule or ion be in its ground vibrational state initially, $\psi_v = \prod_i^{3N-6} \Phi_i(Q_i)$ (Sec. 8.2 Chap. 8), is always symmetric, i.e., of A_{1g} symmetry. Leaving this out the symmetry of the rest of the integrand, viz., $e[\psi^3T_{2g}\psi'_v(\vec{x}+\vec{y}+\vec{z})\psi^3T_{1g}]$ is to be investigated. To find whether the direct product representation contains an A_{1g} symmetry component or not is equivalent to investigating whether the direct product representation of $[\psi^3T_{2g}(\vec{x} + \vec{y} + \vec{z})\psi^3T_{1g}]$ contains an IR characteristic also ψ'^*_v (Sec. 6.4). Since (x y z) form a basis for T_{1u} under $O_\hbar$ symmetry, the direct product representation, on resolution can be expressed as

$$\Gamma^{(T_{2g}\otimes T_{1u}\otimes T_{1g})} = A_{1u} \oplus A_{2u} \oplus 2E_u \oplus 3T_{1u} \oplus 4T_{2u}$$

To investigate the symmetry of ψ'_v, one considers a simpler form of the complex $[V(H_2O)_6]^{3+}$. It holds to reason if we suppose that the vibrations of the metal ion and of its immediate neighbouring atoms of the ligands (viz., the six oxygen atoms) are the ones which influence the electronic motions. In other words, these vibrations are responsible for vibronic coupling. One can, in this light, regard the vibrations of the complex as of the (hypothetical) octahedral complex $[VO_6]^{3+}$ having a total of $(3 \times 7 - 6)$, i.e., 15 normal modes of vibrations. The IR's of the normal modes of vibrations of such an octahedral complex (cf. Example 8.3) are

$$A_{1g},\ Eg,\ 2T_{1u},\ T_{2g} \text{ and } T_{2u}$$

We thus detect, by comparing these IR's with those obtained from the direct product representation, that the final vibrational states ψ'_v, corresponding to the normal modes of T_{1u} and of T_{2u} symmetries, can vibronically couple leading to nonzero value of electric dipole transition integral. This permits a d-d transition band (one vibronic band per vibronic state, i.e., different vibrational states coupled with one electronic state $^3T_{2g}$).

Similar conclusions on the feasibilities of the other two d-d transitions, viz., $^3T_{1g}(P) \leftarrow^3 T_{1g}(F)$ and $^3A_{2g} \leftarrow^3 T_{1g}(F)$ of $[V(H_2O)_6]^{3+}$ are qualitatively supported by symmetry arguments based on vibronic

coupling. Experimentally one really comes across two such d-d transition bands and this is quantitatively explained with the help of Tanabe Sugano correlation diagrams[19] (not discussed in this book) accounting for transitions ${}^3T_{2g} \leftarrow {}^3T_{1g}$ and ${}^3T_{1g}(P) \leftarrow {}^3T_{1g}(F)$. The existence of the third band
${}^3A_{2g} \leftarrow {}^3\Gamma_{1g(F)}$, predictable from symmetry arguments as explained above, is not negated, however, though not observed. Experimentally only a strong charge transfer band overlaps this d-d transition and prevents it from being directly discerned.

Example 10.3.

A d^6 low spin metal ion complex of octahedral stereochemistry gives a correlation diagram which suggests the following transitions
${}^1T_{1g} \leftarrow {}^1A_{1g}$ and ${}^1T_{2g} \leftarrow {}^1A_{1g}$. Can these be supported as being due to vibronic transitions?

Transition ${}^1T_{1g} \leftarrow {}^1A_{1g}$: Here we first find the direct product representation. $\Gamma^{(T_{1g} \otimes T_{1u} \otimes A_{1g})}$ indicative of the symmetry of $\{\psi^{*1}T_{1g}(\overrightarrow{x} + \overrightarrow{y} + \overrightarrow{z})\psi^1 A_{1g}\}$ which occurs as part of the total integrand of the transition moment integral. The representation works out as $A_{1u} + E_u + T_{1u} + T_{2u}$. The normal modes of vibration in octahedral symmetry found out before (Sec. 8.3) are A_{1g}, Eg, $2T_{1u}$ T_{2g} and T_{2u}, Hence it is seen that when the upper vibration state ψ_v' is achieved by excitation along T_{1u} or T_{2u} modes, this will vibronically couple giving rise to the transition.

Transition ${}^1T_{2g} \leftarrow {}^1A_{1g}$: Here $\Gamma^{(T_{2g} \otimes A_{1g} \otimes T_{1u})}$ splits into A_{2u}, E_2, T_{1u} and T_{2u}. It is thus evident, for reasons already stated, that transition can really take place if there is a vibrational excitation along T_{1u} or T_{2u}.

B. Non centrosymmetric Complex. *pd* Mixing. Vibronic Coupling

We may briefly relate here that in complexes lacking an inversion centre, e.g., tetrahedral complexes, pure d-d electronic transitions [transition moment $= \langle \psi'_{el} \mid \overrightarrow{\mu} \mid \psi_{el} \rangle$] are theoretically possible (though not always) since neither ψ'_{el} nor ψ_{el} is of gerade nature. Experimentally, however, these electronic bands are wide enough to suggest that vibrational transitions, theoretically not always essential, also accompany these. Moreover, d_{xy}, d_{xz}, d_{yz} orbitals of a metal atom or ion in T_d

symmetry form bases of IR, T_2, which is also spanned by p_x, p_y and p_z orbitals. As a consequence the triply degenerate t_2 orbitals of a tetrahedral complex do not comprise only the pure d_{xy}, d_{xz} and d_{yz} orbitals, It is a set consisting of a mixed character of d and p orbitals. This dp mixing augments the magnitude of the transition probability integral. However, in some instances, electronic transitions become feasible because of vibronic interactions. As an illustration consider the ion $[CoCl_4]^{2-}$, a d^7 high spin system of T_d symmetry. Correlation diagram for d^7 system suggests a d-d transition ${}^4T_2 \longleftarrow {}^4 A_2(F)$ in addition to two others. Experimentally also this transition is observed at 5800 cm^{-1}. If it were a pure electronic transition, we then have to consider the direct product representation, viz., $\Gamma^{(T_2 \otimes T_2 \otimes A_2)}$. This splits as $A_2 \oplus E \oplus T_1 \oplus T_2$ without any symmetric component A_1. It at once eliminates the possibility of its being a pure electronic transition. On the other hand if it be a vibronic band, then the transition moment integral will look like $e < \psi^4 T_2 \psi'_v \,|(\overrightarrow{x} + \overrightarrow{y} + \overrightarrow{z})|\, \psi^4 A_2 \psi_v >$. Since the ground state ψ_v is symmetric (Chap. 8), it is necessary that the direct product $\Gamma^{(T_2 \otimes T_2 \otimes A_2)}$ should have an IR in common with Γ'_v. Analysis of the normal modes of vibration of a tetrahedral molecule yields the IR's A_1, E, $2T_2$. Thus the representations E and T_2, being common to both the normal modes and the split components of the direct product representation mentioned above, lead to the conclusion that the transition is really a vibronic one.

C. Vibronic Polarisation. Dichroism

Certain transition metal complexes in their crystalline states possess the property of exhibiting different absorptions of polarised light depending on the orientation of the crystal with respect to the plane of polarisation. This phenomenon is termed dichroism.

If a polarised light be incident on a single crystal (say, a tetragonally symmetrical complex compound) such that the z-direction coincides with the plane of polarisation, polarised absorption in the z-direction will differ form that in the x, y directions. This differential behaviour in polarised absorptions can be explained in terms of symmetry arguments and vibronic coupling.

Consider a crystal of trans-dichlorobisethylene diamine cobalt chloride, trans-$[Co(en)_2Cl_2]Cl$. The ion $[Co(en)_2Cl_2]^+$ has the symmetry

$D_{4\hbar}$ and it is a d^6 metal ion system. Since the ion has an inversion centre, the d-d transitions in it will be vibronic in character.

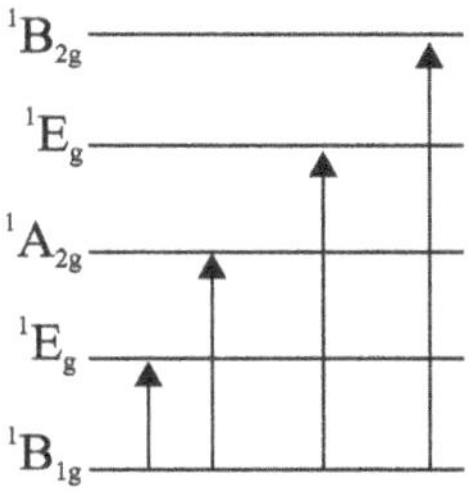

Fig. 10.7: Polarised absorptions by trans-$[Co(en)_2Cl_2]^+$

The possible d-d transitions from the ground $^1A_{1g}$ state, shown in the Fig. 10.7, are

(a) $^1A_{2g} \leftarrow ^1A_{1g}$
(b) $^1B_{2g} \leftarrow ^1A_{1g}$
(c) $^1E_g \leftarrow ^1A_{1g}$
(d) $^1E_g \leftarrow ^1A_{1g}$.

The selection rule for vibronic transition as we by now know, is that the direct product representation of the product of initial wave function, final wavefunction and the dipole moment operator must include an IR which is spanned by one or the other normal mode of vibrations of the species. We thus tabulate the ingredients of the selection rule to arrive at the result.

Table 10.6

Polarised Absorptions By Trans-$[Co(en)_2Cl_2]Cl$ Crystals

Transitions	Direct Product representation of the product of electronic wave function	Symmetry of Dipole Operator		Final Direct product representation	Symmetry of Normal modes vibrations.	Polarised absorptions
		z component	x, y component			
(a) $^1A_{2g} \leftarrow ^1A_{1g}$	A_{2g}	A_{2u}	E_u	A_{1u} E_u	A_{2u} B_{2u}	(x,y) polarised
(b) $^1B_{2g} \leftarrow ^1A_{1g}$	B_{2g}	A_{2u}	E_u	B_{1u} E_u	E_u	(x,y) polarised
(c & d) $^1E_{2g} \leftarrow ^1A_{1g}$	E_g	A_{2u}	E_u	$A_{1u}+A_{2u}$ $B_{1u}+B_{2u}$		z polarised (x,y) polarised

Experimental confirmation of this symmetry based result has been obtained. Of course there are some approximations involved in the assumption of the structure subjected to symmetry analyses. Many interesting and complicated instances of complex compounds showing dichroism have also besn successfully handled from the viewpoint of polarised vibronic coupling.

10.7 Electronic Transitions. Selection Rules and Polarisation

We have discussed previously the symmetry principles governing the infra red and Raman transitions in Chapter 8. In the present one, the electronic transitions between different crystal field states and the justifiability of some 'so called' forbidden transitions due to vibronic coupling have been treated. It will be convenient here to lump together the selection rules for electronic transitions in symmetry terminology although this may mean straying into fields not fully covered by the specific caption of the present chapter.

It may be recalled that the quantum mechanical transition moment integral between two states of the same spin multiplicity is $< \psi_m \mid \mu \mid \psi_n >$. For non-vanishing values of the integral, the symmetries of the states involved are given for linear molecules and nonlinear polyatomic molecules.

A. Linear molecules of the point group $C_{\propto v}$

The following points may be noted.

(i) The $\overrightarrow{z}$ vector of the dipolar operator transforms as $\sum^+$.

(ii) The $(\overrightarrow{x}, \overrightarrow{y})$ vectors transform as Π

(iii) Both the Π and Δ doubly degenerate states contain a symmetric and an antisymmetric components with respect to rotation about the principal axis.

(iv) The direct products $\sum^+ \otimes \sum^+$, $\sum^- \otimes \sum^-$, $\Pi \otimes \Pi$ and $\Delta \otimes \Delta$ contain $\sum^+$ as the result or as one of the components on reduction.

With the foregoing points in view, it is necessary that the direct product of the IRs, of which ψ_m and ψ_n are bases, comprises on reduction, the IR for z (or x, y). The selection rules and the nature of polarisations are, therefore, the following:

z−polarization	(x, y) polarization
$\sum^+ \longleftrightarrow \sum^+$(symmetric to symmetric)	$\sum^+ \longleftrightarrow \Pi$
$\sum^- \longleftrightarrow \sum^-$(Antisymmetric to antisymmetric)	$\sum^- \longleftrightarrow \Pi$
$\Pi \longleftrightarrow \Pi$	$\Pi \longleftrightarrow \Delta$
$\Delta \longleftrightarrow \Delta$	

The significances of symmetric to symmetric and of antisymmetric to antisymmetric transitions are discussed under $D_{\sigma\hbar}$, linear molecules.

B. Linear Molecules of the point group $D_{\propto\hbar}$

It may be noted that

(i) the z-vector of the dipole operator transforms as $\sum_u^+$

(ii) the (x, y) vectors belong to Π_u.

For the non-zero value of the transition moment integral, the direct product of the IR's, of which ψ_m and ψ_n form bases respectively, should contain the IR, $\sum_u^+$ (or Π_u). The Π_u, involves both the symmetric and antisymmetric components with respect to rotation.

For the sake of simplicity, instead of thinking in terms of the IR's, one may note that the product function $\psi_m^* \psi_n$ should invert under i, because the z-vector (and x, y vectors) do so Hence, if the ψ_m is g, ψ_n must be u (or vice versa). The transition should, therefore, be from g $\longleftrightarrow$u.

Secondly since the dipole vector is invariant under rotation (symmetric) the product function ψ_m ψ_n should behave as a symmetric function under rotation. Hence ψ_m and ψ_n should either be both symmetric or antisymmetric.

The following transition rules and polarisations will, therefore, be valid.

z−polarization	(x, y)−polarization
$\sum_g^+ \longleftrightarrow \sum_u^+$	$\sum_g^+ \longleftrightarrow \Pi_u$
$\sum_g^- \longleftrightarrow \sum_u^-$	$\sum_g^- \longleftrightarrow \Pi_u$
$\Pi_g \longleftrightarrow \Pi_u$	$\sum_u^+ \longleftrightarrow \Pi_g$
$\Delta_g \longleftrightarrow \Delta_u$	$\sum_u^- \longleftrightarrow \Pi_g$
	$\Pi g \longleftrightarrow \Delta_u$
	$\Pi u \longleftrightarrow \Delta_g$

In general terms, the selection rules are (for the same spin states)

(a) g $\longleftrightarrow$ u
(b) symmetric $\longleftrightarrow$ symmetric
(c) antisymmetric $\longleftrightarrow$ antisymmetric

} under rotation

C. Polyatomic Molecules

One has to resolve the direct product of the IR's spanned by ψ_m and ψ_n respectively. The reduction should involve the IR's representing transformations of any or some of the vectors x, y and z or the translational vectors T_x, T_y and T_z.

Electronic Transitions: nπ^* and $\pi\pi^*$

The spectral investigations of aldehydes and ketones are replete with data for nπ^* and $\pi\pi^*$ transitions. We shall consider the feasibilities of such transitions in formaldehyde from symmetry viewpoint. As already described (Chap. 9, Example 9.10) the highest two occupied orbitals in the ground electronic state of formaldehyde are a π mo and a nonbonding mo of B_2 symmetry. The wavefunction has A_1 symmetry.

If an nπ^* excitation takes place,...$(\pi)^2(b_2)^1(\pi^*)' \leftarrow ...(\pi)^2(b_2)^2$, the excited state wavefunction has the symmetry given by the direct product $B_1 \otimes B_2 = A_2$ in the C_{2v} point group. The transition moment integral $< (\pi^*)/\mu/(b_2) >$ spans the component IR's obtainable on reduction of the direct product $A_2 \otimes \Gamma_\mu \otimes A_1$. Remembering that the dipolar operator has the IR's A_1, B_1 and B_2, the direct product

$$\left.\begin{array}{r} A_2 \otimes \Gamma_\mu \otimes A_1 = A_2 \otimes \Gamma_\mu = A_2 \otimes A_1 \\ \text{or } A_2 \otimes B_1 \\ \text{or } A_2 \otimes B_2 \end{array}\right\} \text{for the } C_{2v} \text{ point group}$$

The resulting symmetries are A_2 or B_2 or B_1 respectively. Since the direct product does not contain an A_1 component, nπ^* transition is theoretically forbidden. But in spite of this forbiddenness, this transition occurs in practice. This is mainly due to vibronic coupling.

The $\pi\pi^*$ transitions are theoretically permitted since the direct product representation is $B_1 \otimes \Gamma_\mu \otimes B_1$ which comprises an A_1 component on resolution.

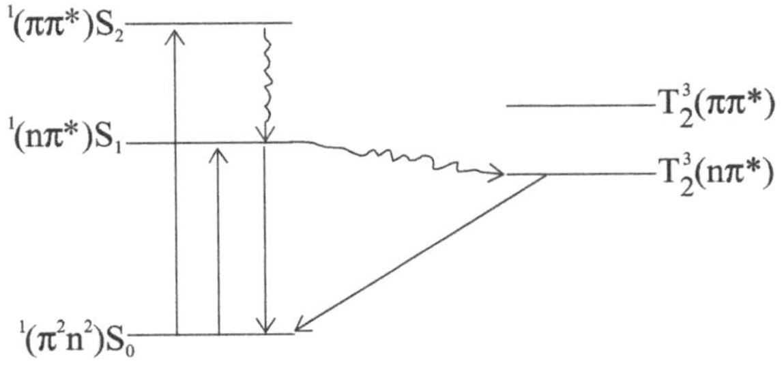

The different absorption and the de-excitation processes of vibronically coupled nπ^* and of the $\pi\pi^*$ transitions are meaningfully expressed in the following Jablonski's diagram. Here n represents the nonbonding orbital (b_2) and the wavy arrows indicate nonradiative transitions of internal conversion and inter-system crossing.

10.8 Double Groups. Spin Orbit Coupling And Crystal Field States.

Introduction : The previous treatment of the crystal field effect on the Russell Saunders' free ion or atomic term is based upon an implicit assumption that the total orbital angular momentum quantum number L is a good quantum number. Additionally also the LS (spin orbit) interaction leads to multiplets which are closely spaced and the term splittings are relatively small compared to those of the crystal field states. It is in this sense that the perturbation hamiltonian comprised V_{cryst} only (Sec. 10.2) in the weak field case.

The symmetry effects in all such cases were considered only upon orbitals or wavefunctions characterised by L. Spin was completely excluded from such wavefunctions or orbitals.

Three queries now become uppermost in our mind.

(a) How will the crystal field states of our previous acquisition , split into further states if now the perturbation in the forms of spin orbit interaction be imposed?

(b) If the LS interaction of an orbital state defined by L leads to multiplet term separations which are relatively large compared to crystal field separations (in the weak field case), spin orbit interactions become the first order perturbation hamiltonian. How do then these multiplets decompose in the crystal field of the ligands? This is just the converse of the situation described under (a).

(c) In the case of j-j coupled ionic or atomic term, L is no longer a good quantum number but J=L+S is. The wave-function of an atom or ion is now to be described in a form characterized by J. How does a j-j coupled term split when the atom or ion is placed in a crystal field?

The answers to all such queries can be given in terms of the concepts and properties of double groups which will presently engage our attention.

A. Double Groups. Their Necessity and Properties.

It is known that the atomic orbitals of an atom or ion form, for all values of M_L for a given L, a basis for the representation of of any finite rotational group (C_n, D_n, O). The representation is diagonal (cf. Sec. 10.3) with a character $\chi(C_\alpha) = \frac{\sin(L+\frac{1}{2})\alpha}{\sin\frac{\alpha}{2}}$. This representation and the character values are valid only when (2L+I), giving the total number of M_L, is an odd integral number. But when spin is included, the atomic states are given in terms of orbital functions where there occur (2J+1) number of M_J values for any given J. This J may be half integer or integer depending on the odd or even number of electrons present. The determination of the character values of the symmetry elements of a rotational group with this degenerate basis set, when J is half integer [and hence (2J+1) is even], presents some difficulty to be pointed out shortly. This difficulty can be resolved in a very general way by having re-course to 'spinors' which can handle the continuous rotation groups (see chap 11). In an artificial way, however, this difficulty can be circumvented by Bethe's suggestion of double groups. We turn to the difficulty first before prescribing Bethe's remedy.

When any rotation α is substituted by $(\alpha+2\pi)$, it means an identical position. Hence it is imperative that $\chi(C_\alpha) = \chi(C_\alpha + 2\pi)$. This is borne out by the formula $\chi(C_\alpha) = \frac{\sin(L+\frac{1}{2})\alpha}{\sin\frac{\alpha}{2}}$, where (2L+1) is odd integer, for

$$\begin{aligned}\chi[C_{(\alpha+2\pi)}] &= \frac{\sin(L+\frac{1}{2})(\alpha+2\pi)}{\sin(\alpha/2+\pi)}\\ &= \frac{\sin[(L+\frac{1}{2})\alpha + (\text{odd integer})\pi]}{-(\sin\alpha/2)}\\ &= \frac{\sin(L+\frac{1}{2})\alpha}{\sin\alpha/2} = \chi(C_\alpha)\end{aligned}$$

But when L is substituted by J and J is half integer [i.e., (2J+1) is even],

$$\begin{aligned}\chi[C(\alpha+2\pi)] &= \frac{\sin[(J+\frac{1}{2})\alpha + (\text{even integer})\pi]}{-(\sin\alpha/2)}\\ &= \frac{-\sin(J+\frac{1}{2})\alpha}{\sin\alpha/2} = -\chi(C_\alpha)\end{aligned}$$

It thus appears that the character is double-valued and not unique. Let us, after Bethe, adopt the fiction that a rotation of (2π) is not an identity operation, but a new operation represented by symmetry element R. A rotation of 4π, i.e., R^2 is the identity operation E. While the second assumption does not conflict with reality, the first one does and should only be considered as a mathematical artifice without any physical significance.

With this assumption a rotation group acquires twice as many group elements as in the original group. Such an artificially contrived group is termed a double group. For example, if D_3 be the original point group with six elements, viz., E, C_3, C_3^2, C_2', C_2'', C_2''', the double group D_3' will consist of twelve elements E, C_3, C_3^2. C_2', C_2'', C_2''', R, RC_3, RC_3^2, RC_2', RC_2'' and RC_2'''. Similarly the double group D_4' will comprise sixteen elements which are twice as many as are present in D_4. An O′ group has 48 elements. It may be mentioned incidentally that in each double group $RC_n^m = C_n^m R$.

We have now to consider several properties of the double groups before we turn to the queries raised in the introduction of this section.

(1) **Classes:**

The classes in a double group consist of the following

(i) E forms a single element class.

(ii) R forms a single element class.

(iii) C_2' and RC_2' belong to one class. Any other C_2' and RC_2' pair in which C_2' is normal to the first C_2' will also belong to the same class as the first pair (cf. D_4).

(iv) The pair C_n^n and RC_n^{n-p} belong to a class. Thus in the group D_3', C_3 and RC_3^2 form a class.

Classwise distributions of elements in D_4' and O′ are as follows

$D_4' \rightarrow$	E	R	$C_4RC_4^3$	C^2RC_2	C_4^3, RC_4	$2C_2'$, $2RC_2'$	$2C_2''$,	$2RC_2''$
O′	E	R	$3C_4$, $3RC_4^3$	$3C_4^3$, $3RC_4$	$3C_2$, $3RC_2$	$4C_3$, $4RC_3^2$	$4C_3^2$, $4RC_3$	$6C_2'$, $6RC_2'$

(2) **Characters:**

The characters of the different elements of the double groups in the basis set of the harmonics with J= half integral or integral are to be obtained by applying the formula $\chi(C_\alpha) = \frac{\sin(J+\frac{1}{2})\alpha}{\sin\ \alpha/2}$. Thus $\chi(E) = \frac{0}{0}$. An evaluation of this using the special method of limits leads to $\chi(E) = (2J+1)$.

$$\text{Again } \chi\ (R) = \frac{\sin(2J+1)\pi}{\sin\ 2\pi/2} = \begin{array}{l}(2J+1) \text{ for } J = \text{integer and} \\ -(2J+1) \text{ for } J = \text{half integer.}\end{array}$$

$$\chi(C_2) = \chi(RC_2) = 0$$

The other character values can be found out also by using the well' established formula.

(3) **Theorems of IR's, Characters and Nomenclature of IR's.**

The ordinary theorems of the IR's and of the characters are all applicable to the double groups. The symbols for the IR's are also, similar to those of the ordinary point groups except in respect of some minimum unavoidable modifications.

To exemplify, we take the double group O', i.e., the augmented point group of the cubic class. We find, as recorded already, that there are eight classes and there ought to be eight IR's. Utilizing the dimensionality theorem and remembering that there are 48 elements we may write.

$$l_1^2 + l_2^2 + l_3^2 + l_4^2 + l_5^2 + l_6^2 + l_7^2 + l_8^2 = 48$$

The set of values l, 1, 2, 3, 3, 2, 2 and 4 for the l's satisfies the above requirement. It is thus noted that there are two one-dimensional, one two-dimensional, two three-dimensional IR's as in the original point group and additionally two more two-dimensional and one four-dimensional new IR's, On examining the full character table of the O'-double group all the IR's of the O:-point group (vide Appendix III) are detected in the assembly of the IR's of the double group. Whatever new IR's occur additionally these are 'evenorder'-dimensional and hence degenerate. The symbols for the IR's are either numerically subscripted prime Γ' (gamma) symbols of Bethe, viz., Γ'_1, Γ'_2, Γ'_3, Γ'_4, Γ'_5, Γ'_6, Γ'_7 and Γ'_8 or the primed Mulliken symbols A'_1, A'_2, E'_1, T'_1, T'_2, E'_2, E'_3 and G'. The symbols are so written that the IR's of the unaugmented point group O are presented first followed by the new finds.

The generalities of our discussion in relation to the double group O' will also apply to D'_4 or the other double-groups.

(4) **Reducibility of Representations and Direct Products**

The ideas of decomposing a reducible representation into a set of IR's and of reducing the direct product of the IR's also hold in the case of double groups. This will be demonstrated in stages with the help of a few examples.

Example 10.4

What IR's of the D'_4 double group are spanned by the spin-orbit fine structure wave functions of the ^2D state of a d^1 ion ?

The ^2D state of a single d-electron ion splits into two fine structures having J values 5/2 and 3/2. The wavefunction describing the upper fine structure is one of six spin orbitals* i.e., kets $|\alpha j M_j > |\alpha >$ with M_j values ranging from 5/2 to $-5/2$. Similarly the lower fine structure level is described by spin orbitals of the form. $|ajM_j > |\beta >$ with M_j taking values 3/2, 1/2, $-1/2$ and $-3/2$. α and β represent the spin wavefunctions of the electron. In the set of the six degenerate spin orbitals the characters of the reducible representation of D'_4 is first found out using the formula for $\chi(C_\alpha)$. The same exercise is also undertaken using the four basis functions corresponding to the lower fine structure level of J = 3/2.

The $\chi(E)$, $\chi(R)$ and $\chi(C'_2)$ values are all known and the rest can easily be worked out. Some sample calculations appear below:

$$\text{For J} = \frac{5}{2},\ \chi(C_4) = \frac{\sin\left(\frac{5}{2}+\frac{1}{2}\right)\pi/2}{\sin\ \pi/4} = \frac{\sin 3\frac{\pi}{2}}{\sin\ \pi/4} = -\sqrt{2}$$

$$\text{For J} = \frac{5}{2},\ \chi(C_4^3) = \frac{\sin\left(\frac{5}{2}+\frac{1}{3}\right)\frac{3\pi}{2}}{\sin\ \pi/4} = \frac{\sin 9\frac{\pi}{2}}{\sin\ \pi/4} = \sqrt{2}$$

$$\text{For J} = \frac{3}{2},\ \chi(C_4) = \frac{\sin\left(\frac{3}{2}+\frac{1}{2}\right)\pi/2}{\sin\ \pi/4} = \frac{\sin \pi}{\sin\ \pi/4} = 0$$

The reducible representation thus leads to the following

D'_4	E	R	$(C_4RC_4^3)$	(C_2RC_2)	$(C_4^3RC_4)$	$(2C'_2RC'_2)$	$(2C''_22RC''_2)$
$\chi J=5/2$	6	−6	$-\sqrt{2}$	0	$\sqrt{2}$	0	0
$\chi J=3/2$	4	−4	0	0	0	0	0

Theo. 5.8 and the character table D'_4 group are now used for reduction.

$$\text{Reduction of } \Gamma\left(J=\frac{5}{2}\right)$$

$$\begin{aligned} nE'_2 &= \frac{1}{16}\left[6\times 2+6\times 2+2\times\sqrt{2}\times(-\sqrt{2})+2\times\sqrt{2}\times(-\sqrt{2})\right] \\ &= 1 \\ nE'_3 &= \frac{1}{16}\left[6\times 2+6\times 2+2\times(-\sqrt{2})\times(-\sqrt{2})\right. \\ &\qquad \left.+2\times(+\sqrt{2})\times(+\sqrt{2})\right] \\ &= 2 \\ \Gamma\left(J=\frac{5}{2}\right) &= E'_2+2E'_3,\ \textit{i.e.},\ \Gamma'_6+2\Gamma'_7 \end{aligned}$$

Reduction of $J=\frac{3}{2}$
$nE'_2 = 1$ and $nE'_3 = 1$, so that

$$\Gamma\left(J=\frac{5}{2}\right)=E'_2+E'_3,\ \textit{i.e.},\ \Gamma'_6\oplus\Gamma'_7$$

The interpretation of the results obtained in this Example is that an ion with one d-electron has its spin orbit fine structure terms further split into a number of crystal field terms when the system is placed in an environment of D_4 or D_h symmetry. Diagrammatically.

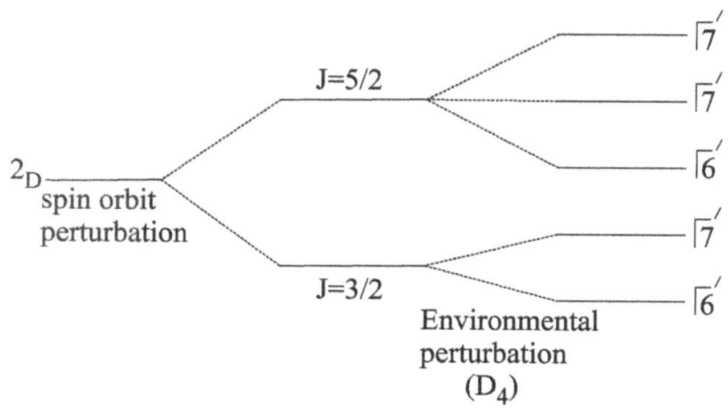

Example 10.5

What IR's of the O′ double group are spanned by the electronic spin wavefunctions α and β ?

The spin angular momentum of the electron is $\sqrt{s(s+1)}\hbar$ and the spin quantum number $s = \frac{1}{2}$ =J. To find the characters of the elements of the O′ double group, we may use the conventional formula. Thus

$$\begin{aligned}
\chi(C_4) &= (RC_4^3) = \frac{\sin\left(\frac{1}{2}+\frac{1}{2}\right)\frac{\pi}{2}}{\sin\ \pi/4} = \sqrt{2} \\
\chi(C_4^3) &= (RC_4) = \frac{\sin\ 3\pi/2}{\sin\ \pi/4} = -\sqrt{2} \\
\chi(C_3) &= (RC_3^2) = \frac{\sin\ 2\pi/3}{\sin\ 2\pi/6} = 1 \\
\chi(C_3^2) &= (RC_3) = \frac{\sin\ 4\pi/3}{\sin\ 4\pi/6} = -1
\end{aligned}$$

The rest of the χ's being known, we can draw up the following table.

0′	E	R	($3C_4$, $3RC_4^3$)	($3C_2$, $3RC_2$)	($3C_4^3$, $3RC_4$)	($4C_3$, $4RC_3^2$)	($4C_3^2$, $4RC_3$)	($6C_2'$, $6RC_2'$)
$\chi J=\frac{1}{2}$	2	−2	$\sqrt{2}$	0	$-\sqrt{2}$	1	−1	0

Thus $\Gamma(J=\frac{1}{2})=\Gamma_6'=E_2'$ (cf. the character table for O′ group). Thus the pair of spin wavefunctions span a degenerate IR.

If we had considered an atom or ion with a single 1 s−electron, the ^2S state would then be characterized by a J value$=\frac{1}{2}$. The spin orbitals of the 2S state will, therefore, have the same symmetry behaviour as that of the pair of spin wavefunctions.

It may be rioted that D'_4 double group is a subgroup of O' and Γ'_6 (i.e., E'_2) of O' group corresponds to Γ'_6 (i.e., E'_2) of D'_4. Hence the pair of spin wavefunctions will similarly span the IR, Γ'_6, only of D'_4.

We already know a ^{2}D stale splits into e_g and t_{2g} orbitals in octahedral symmetry. If we use as the degenerate basis set the product functions, viz., the spin orbitals $d_z^2\alpha$, $d(x^2-y^2)\alpha$, $d_z^2\beta$, $d(x^2-y^2)\beta$, the IR's of O', which will be spanned, are given by a resolution of the direct product representations. Similar arguments will apply to the product (unctions of the basis set of T_{2g} and the spin wave functions. These will be illustrated in the following example. The reduction of the direct product representation is just identified with the spin orbit perturbation effect on the crystal field terms.

Example 10.6

Find the effect of the spin orbit interaction on the crystal field terms of ^{2}D state in $O\hbar$-symmetry.

When spin is considered, the crystal field orbitals are further split and the results are given by reduction of the direct product of the component IR's in O' group. Now E_g symmetry of O_h corresponds to E of the group O which in turn correlates with E'_1 (i.e., Γ'_3) of O' double group. This is because dz^2, $d(x^2-y^2)$, which form a basis of E_g of $O_\hbar$ and E of O, are characterized by an L=2. Hence E symmetry of the group O is the same as E'_1 of O' since 2L+1 = even. Again $\Gamma j = \frac{1}{2} = E'_2$ (i.e., Γ'_6) under Γ' (Example 10.5)

The direct product that we need to consider is

$$E'_1 \otimes E'_2 = \Gamma'_3 \otimes \Gamma'_6$$

1	E	R	$4C_3$	$4C_3R$	$3C_2$	$3C_4$	$3C_4R$	$6C'_2$
$\chi E'_1 \times \chi E'_2$	4	-4	-1	1	0	0	0	0

which is just G', i.e., Γ'_8

e_g orbital changes to Γ'_8 under spin orbit perturbation. A similar set of arguments leads to a consideration of the direct product $T'_2 \otimes E'_2 =$

$\Gamma'_5 \otimes \Gamma'_6$ to ascertain the splitting of t_{2g}.

Now $\Gamma'_5 \otimes \Gamma'_6 = \Gamma'_7 \oplus \Gamma'_8$. The final result is thus conveyed by the previous diagram.

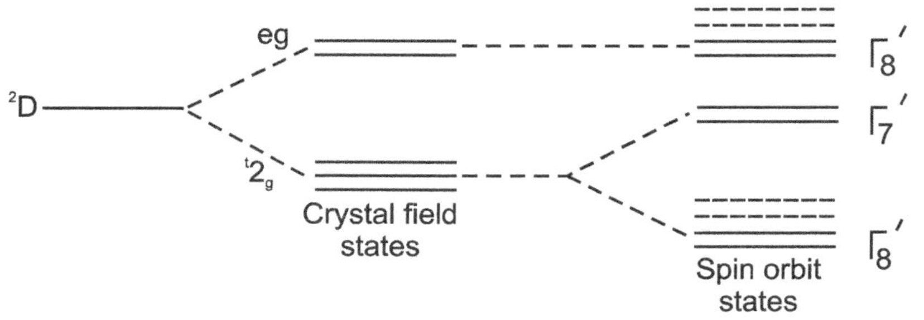

It is now time to turn our attention to the three queries raised at the beginning of Sec. 10.8. The first query is adequately answered by the worked out Example 10.6. An answer to the second query-is reflected in the specific case treated in Example 10.4. A little thought will enable one to realize that the answer to the third query is the same as the second one. The only difference is that one does not start with the Russell Saunders' free ion term as the unperturbed state. Each of the j-j coupled term plays the role of the unperturbed term and the environmental effect constitutes the first order perturbation for each $j - j$ term.

Chapter 11

Atomic Symmetry and Quantum Mechanical Problems. R(2), R(3) SU(2) and R(4) Lie Groups

We preface our study of the topic by a worthwhile: quote

Lie groups were studied by the Norwegian mathematician Sophus Lie at the end of the 19th century. Lie was interested in solving equations. A typical tool was to make a clever change of variables which would make one of the variables drop out of the equations. Lie's basic insight was that when this happened, this was due to an underlying symmetry of the equations and that underlying the symmetry was what is now called a Lie group.

Lie groups lie at the intersection of two fundamental fields of mathematics: algebra and geometry. A Lie group is first of all a group. secondly it is a smooth manifold which is a specific kind of geometric object. The circle and the sphere are examples of smooth manifolds. Informally, a Lie group is a group of symmetries where the symmetries are continuous. A circle has a continuous group of symmetries: you can rotate the circle an arbitrarily small amount and it looks the same.

11.1 Lie Group of Transformation

Uptil now we have dealt with finite molecular point groups excepting two, *viz*, $C_{\propto v}$ and $D_{\propto h}$. Infinite groups containing infinite number of group elements may principally be of two categories;
(i) infinite groups with discrete group elements
(ii) infinite groups with continuous set of group elements containing one or more parameters capable of taking up a continuous set of values within a certain range or ranges.

Examples of these two types are provided by (i) the infinite set of 0 and +ve and -ve integers with the group combination law of arithmetic addition. (ii) rotation of a vector in the XY plain with the angular parameter ω varying within the range O$\leq \omega \leq 2\pi$. This admits of an infinite number of transformations, successively continuous, of the tip of the rotating vector. All these transformations, collectively considered, satisfy the definition of a group. The individual member of the group, viz the respective magnitudes of the transformations, are indicated by the corresponding parameter value of ω

The group element with ω=0, (i.e., zero value of angular transformation of the tip of the vector) is the identity element E, of the group. The infinitesimal values for ω, viz, $\delta\omega_1$, $\delta\omega_2$, $\delta\omega_3$.... represent group elements (i.e., the angular transformations in the neighborhood of the identity element). Similarly there are more distant group elements with ω values, say, $\frac{\pi}{2}$, $1.5\frac{\pi}{2}$ or 2π.

Most of the continuous groups with parameter p, their important results and consequences can be theoretically treated by a method, originally devised by Sophus Lie, a Norwaygean mathematician, called Lie group of transformations.

11.2 Classification of Linear Transformations:

All linear transformations of a vector in m- dimensional linear space belong to a group called general linear transformation group symbolised by GL (m). The subgroup of GL (m) consisting of orthogonal formations with real matrix elements is (i.e., transformation matrix $\underline{\mathbf{A}} = \widetilde{\underline{\mathbf{A}}} = \underline{\mathbf{A}}^{-1}$) is the orthogonal group and is denoted by O(m). Special orthogonal group SO (m), a subgroup of O(m), consists of all the unimodular orthogonal transformations (i.e., det $\underline{\mathbf{A}}$=1) Amongst these transformations are pure rotations in the m- dimensional space and are indicated simply as R(m). Thus the real orthogonal unimodular rotations in 2-, 3-, 4-dimensional spaces are indicated by group symbols R(2), R(3) and R(4) respectively.

Corresponding to the transformations of the SO group, the parallel set viz the unimodular unitary transformations (rotations) of complex vector are grouped under special unitary group, SU(m) with transformation matrix $\underline{\mathbf{A}}$ being such that $\mathbf{A}^{\neq} = \widetilde{\underline{\mathbf{A}}}^{*} = \underline{\mathbf{A}}^{-1}$ and that det $\underline{\mathbf{A}}$ is + 1. Thus we have groups SU(2), SU(3) with unimodular unitary rotations of complex vector in two- and three dimensions respectively with an additional restriction that $|$ det $\underline{\mathbf{A}}$ $|$ be +1.

11.3 Lie Groups: Number of Parameters and General Process of Treatment.

The groups R(2), R(3), R(4), SU(2) are all exemplified, amongst others by rotational motions of real or complex vectors respectively in their proper dimensional plains. The corresponding representative matrices (reducible or irreducible forms notwithstanduig) are unimodular and orthogonal or unitarily unimodular with the additional requirement det $\underline{\mathbf{A}}=+1$

We now try to ascetain the number of independent parameters that can charactarise the representative two, three and four dimensional motions of R(2), R(3), R(4) and SU(2)

R(2): Let the matrix be $\begin{pmatrix} a\ b \\ c\ d \end{pmatrix}$ the elements being all real numbers. Since it is to be unimodular and orthogonal, the rows are to behave as normalised ones and mutually orthogonal. Two restrictions for normalisation and one for orthogonalization i.e., an overall of 3 restrictions are to be satisfied by the 4 elements of the matrix. Thus remains only one independent parameter for R(2) Lie group of transformations.

R(3): There are 9– matrix elements. The orthonormality conditions together impose (3+3) i.e., 6 restrictions. Hence R(3) group is a 3 parameter Lie group.

R(4): Total matrix elements =16 Number of restrictions (4+6) i.e., 10. Hence R(4) is a 6- parameter Lie group.

For rotations of complex vectors, the matrix elements are in general complex numbers

SU(2): Of the four complex elements $\begin{pmatrix} a\ b \\ c\ d \end{pmatrix}$, each is associated with a real and imaginary part. There thus exist 8 choices.

Number of restrictions:

(i) normality conditions 2 (viz $aa^*+bb^*=1$) and similarly ($cc^*+dd^*=1$)

(ii) Orthogonality conditions =2

$$\begin{aligned} \text{viz} \quad ac^* + bd^* &= 0 \\ \& \quad a^*c + b^*d &= 0 \end{aligned}$$

(iii) additional restriction $| \det \underline{\mathbf{A}} |= 1$.
Therefore SU (2) is (8-5) i.e., 3- parameter Lie group of transformations. Arguing in a like manner, it can be concluded that SU (3) is an 8- parameter group.

11.4 General Steps in Lie Group Treatment:

The process is based on the assumption that if a group element is obtained with a parameter value p=c, and this group element can also be arrived at by successive multiplication with parameter values p=a and then p=b of two group elements starting from the identity element (with p=0) for the first (p=a) and using the latter for further transformation, with p=b, then the parameter c=ϕ(a,b).

Such an assumption is also valid when the transformations involved are infinitesimals say (δp_1, δp_2) in the neighbourhood of the identity element E(p=0). The process consists of

Step 1. Carrying out infinitesimal transformations in the vicinity of the identity element, one finds out a set of infinitesimal operators IO's, satisfying certain commutation rules.

These commutation rule (or rules) constitute the crux of the Lie algebra of the corresponding Lie group. These IO's define a linear operator space.

With the help of IO's, operators for a macro transformation (parameter p =large) are formed.

Step 2. Suitable linear combinations of IO's are made to constitute basic operators sustaining the linear space and reformulation of the CR's (commulation rules).

Step 3. Representations of the basic operators are derived giving a representation of Lie algebra.

Step 4. Irreducible representations of the Lie group operations are effected.

Step 5. Ancillary relations or related quantum mechanical problems are attended to. The general and original method of treatment of Lie groups is rather hard to follow, Sophus Lie showed how generators could be defined without mentioning representation at all. We shall follow a very simplified method stressing on representation.

11.5 The Group R (2)

The simplest example of R (2) is the picture of all possible angular pushes in the xy plain given to the tip P of a vector fixed at one end at the origin. The different pushes comprise continuous set of values of ω in the range $0 \leq \omega \leq 2\pi$ All these angular transformations together form the group R(2) with the identity clement E having ω=0. Let P, the tip of a vector, have the coordinates (x, y) at the initial moment when the game starts with the angular push $\omega = 0$ given to the tip P leaving the latter unmoved. Let it then suffer an infinitesimal transformation $\delta\omega$, corresponding to a group element in the neighbourhood of the identity element E(ω=0) in the group element space. The new rotational tip position P′ has coordinates (x′ y′). Since the transformation is infinitesimal and $\delta\omega$ is a rotational magnitude we can write

$$\left.\begin{aligned} x' &= x + dx = x\cos\delta\omega - y\sin\delta\omega \\ y' &= y + dy = x\sin\delta\omega + y\cos\delta\omega \end{aligned}\right\} \tag{11.1}$$

As $\delta\omega$ is infinitesimal $\cos\delta\omega = 1$ and $\sin\delta\omega \approx \delta\omega$, Eq (11.1) leads to

$$\left.\begin{aligned} dx &= -y\delta\omega \\ dy &= x\delta\omega \end{aligned}\right\} \tag{11.2}$$

Since $\delta\omega$ is an infinitesimal angular rotation, it can be assumed $\delta\omega = d\omega$, the differential of ω. Now, if -F be any orbitrary function of x, y there

$$\begin{aligned} dF = \frac{\partial F}{\partial x}dx + \frac{\partial F}{\partial y}dy &= \left(dx\frac{\partial}{\partial x} + dy\frac{\partial}{\partial y}\right)F \\ &= \partial\omega\left(-y\frac{\partial}{\partial x} + x\frac{\partial}{\partial y}\right)F \qquad \text{(from 11.2)} \\ &= \partial\omega\left(x\frac{\partial}{\partial y} - y\frac{\partial}{\partial x}\right)F \\ &= \partial\omega XF, \qquad \text{where} \\ X, \text{ an infinitesimal operator} &= \left(x\frac{\partial}{\partial y} - y\frac{\partial}{\partial x}\right) \end{aligned} \tag{11.3}$$

and is quite independent of $\delta\omega$ i.e., $d\omega$

Hence

$$\frac{dF}{d\omega} = XF \tag{11.4}$$

X being rate of change (of a function) per infinitesimal angular variation. Writing F=R(ω)=rotation through macro angle ω, which is also a group element of R(2), we have from (11.4)

$$\frac{dR(\omega)}{d\omega} = XR(\omega), \quad \text{so } R(\omega) = e^{\omega X} \tag{11.5}$$

$$\begin{aligned} \text{Further} \qquad X &= \left(x\frac{\partial}{\partial y} - y\frac{\partial}{\partial x}\right) \\ &= \frac{i}{\hbar}\cdot\frac{\hbar}{i}\left(x\frac{\partial}{\partial y} - y\frac{\partial}{\partial x}\right) = \frac{i}{\hbar}L_z. \end{aligned} \tag{11.6}$$

where l_z is the z dimensional angular momentum operator. Additionally, it is to be noted that X sustains an one- dimensional operator space in which the commutator $[l_z, l^2]=0$

It may be noted that X, an infinitesimal rotational operator is directly linked, through i, to the angular momentum operator l_z, when the latter is expressed in units of $\hbar$ (eq. 11.6). It is to be realised that it is valid only when we put the identity element

$$E = g\ (\omega)|_{\omega=0} = 0 \tag{11.7}$$

where g(ω) the group element is indicated by ω parameter.

We now turn our attention to finding the IR's of the operator, X and of the group element.

The group elements of $R(2)$ are such as to satisfy commutation property in their binary product, viz

$$\begin{aligned} \delta(\omega_1)\delta(\omega_2) &= \delta(\omega_2)\delta(\omega_1) = \delta(\omega_p) \\ \text{where} \qquad \delta\omega_p &= \delta\omega_1 + \delta\omega_2 = \delta_{\omega_2} + \delta_{\omega_1}. \end{aligned}$$

This commutation holds for every binary pair of group elements. Hence the R(2) group is abelian and is characterized by one- dimensional irreducible representations.

It is known that the spherical harmonics, $\psi_{l,\ k}$'s from the bases of the representations of the rotational operators. All the individual members of ψ's, for any given l, are individual eigenfunctions of l operator and hence of the infinitesimal operator X. Let us choose any one such function ψ_{lm}

$$X\psi_{lm} = \frac{i}{\hbar}l_z\psi_{lm} = \frac{i}{\hbar}m\hbar\psi_{lm} = im\psi_{lm}$$

where im is the character (also eigenvalue) of the one dimensional (irreducible) representation (im) of the i.o X $=\mathbf{X}$
Now R(ω)= $e^{\omega \mathbf{X}}$.
Thus the representation of the group R(2) is known, since the IR of each such group element will be of the from

$$\underline{\mathbf{R}}(\omega) = e^{\omega \mathbf{X}} = e^{im\omega} = \exp(im) \tag{11.8}$$

Singlevaluedness property of the character demands that the X values of R(ω) and R($\omega + 2\pi$) be the same i.e., $e^{im\omega}=e^{im(\omega+2\pi)}$. This can be true only when m is a +ve or -ve integer including zero. The possible values of m lie in the range -l$\leq$m$\leq$l.

11.6 General Form of Generator of Lie Group

The method of obtaining a general expression for infinitesimal transformation operator, especially for those in the vicinity of the identity element E, is to resort to Taylor expansion of some function of α, representing the group elements, with the respect to α. Let g(α) represent the group and $\mathbf{D}_g(\alpha)$ its representation. If $\delta\alpha$ be any group element near the identity element,

$$\begin{aligned} \mathbf{D}_g(\delta\alpha) &= [D_g(\alpha)]_{\alpha=0} + \delta\alpha \left[\frac{\partial}{\partial\alpha} D_g(\alpha)\right] + \frac{\delta\alpha^2}{2!}\left[\frac{\partial^2}{\partial\alpha^2} D_g(\alpha)\right] + \ldots \\ &= \mathbf{D}_g(E) + \delta\alpha \left[\frac{\partial}{\partial\alpha}\mathbf{D}_g(\alpha)\right]_{\alpha=0}, \text{ neglecting higher order differentials} \\ &= \left[1 + i\delta\alpha\left[-i\frac{\partial}{\partial\alpha}\right]\mathbf{D}_g(\alpha)\right]_{\alpha=0} \end{aligned}$$

Putting in $\mathbf{X}$ representing $(-i\frac{\partial}{\partial a})$, the representation for infinitesimal transformation,

$$\left.\begin{aligned} D_g(\delta\alpha) &= (1 + i\delta\alpha\mathbf{X}) \\ &= (1 + id\alpha\mathbf{X}) \end{aligned}\right\} \tag{11.9}$$

If, however, the α parameter shows dependence on x, y and z dimensions separately through components α_1 α_2 and α_3 then the infinitesimal transformation operator (or generator)

$$\left(1 + \sum_{\mu=1,\ 2,\ 3} i(\delta\alpha_\mu \mathbf{X}_\mu)\right) \tag{11.10}$$

If we want the representation of a macro group element α where $\alpha =$ n$d\alpha$, (n=extremly large positive integer), it is obtained by repeated self multiplication of the generator (11.9) viz., the group multiplication law. Then

$$\begin{aligned} \mathbf{D}_{\mathrm{g}}(\alpha) = (1 + \mathrm{i}d\alpha\mathbf{X})^{n} &= \left[1 + \mathrm{i}\left(\frac{\alpha}{n}\right)\mathbf{X}\right]^{\mathrm{n}} \\ &= \mathrm{e}^{\mathrm{i}\alpha\mathbf{X}} \\ &\quad \text{when n = infinitaly large + ve integer} \end{aligned} \tag{11.11}$$

We have used a common symbol, X, both in the treatment of our infinitesimal operator and in the treatment of generator with, however, different implications for X. In the infinitesimal operator case, X is related to angular momentum through multiplication by i. In the generator case, the definition of X involves already a multiplication by i. Hence X, in the generator case when general rotation is involved, can be identified with the angular momentum operator associated with differential multiplicative factor.

11.7 The group R(3) i.e, SO(3) [sub group of the spinless Atomic Symmetry Group]

There are numerous examples of SO(3). All orthogonal transformations of linear functions Ax+By+Cz with varying parameters A, B, C subject to the condition $A^2+B^2+C^2$=constant, also the infinite set of 3×3 orthogonal matrices with three naturally independent parameters with group law of matrix multiplication are illustrative of group elements of SO(3).

If we ignore the improper rotations of a sphere but take into consideration the infinite number of distinct rotations about a unit vector axis $\vec{n}$ passing through the center of the sphere, this bunch of operations constitute a sub group R(3) of the complete spherical symmetry group. The complete spherical group can also be regarded as the symmetry group of the hypothetical spinless atom

Consider a sphere with XYZ axes meeting at the centre (origin) and also a unit vector $\vec{n}$ emerging from the origin. An infinitesimal rotational operation in the neighbourhood of the identity element will shift the coordinates of the rotating vector. These are

$$\left.\begin{aligned} x + dx &= (1 + p_{11})x + p_{12}y + p_{13}z \\ y + dy &= p_{21}x + (1 + p_{22})y + p_{23}z \\ z + dz &= p_{31}x + p_{32}y + (1 + p_{33})z \end{aligned}\right\} \tag{11.12}$$

where p_{ij}'s are all infinitesimals.

Reference to sec. (11.3) allows us to retain only 3 out of nine p_{ij} numerical quantitics. We retain three independent parameters viz.,

$$p_{12} = -p_{21} = \xi, p_{13} = -p_{31} = -\eta \qquad \text{and} \qquad p_{23} = -p_{32} = \zeta$$

and set the diagonal coefficients p_{jj}'s=0 each. Such a choice ensures infinitesimality of the change and the retention of the orthogonal nature of the transformation matrix.
The corpus of (eq. 11.12) is thus reduced to

$$dx = \xi y - \eta z; \quad dy = -\xi x + \zeta z; \quad dz = \eta x - \zeta y \tag{11.13}$$

If F be a function of x, y, z only

$$\begin{aligned} dF &= \frac{\partial F}{\partial x}dx + \frac{\partial F}{\partial y}dy + \frac{\partial F}{\partial z}dz \\ &= \left\{(\xi y - \eta z)\frac{\partial}{\partial x} + (-\xi x + \zeta z)\frac{\partial}{\partial y} + (\eta x - \zeta y)\frac{\partial}{\partial z}\right\}F, \quad \text{on using (11.13)} \\ &= \left\{\zeta\left(z\frac{\partial}{\partial y} - y\frac{\partial}{\partial z}\right) + \eta\left(x\frac{\partial}{\partial z} - z\frac{\partial}{\partial x}\right) + \xi\left(y\frac{\partial}{\partial x} - x\frac{\partial}{\partial y}\right)\right\}F \quad (11.14) \end{aligned}$$

Relation (11.14) provides us with three component infinitesimal linearly independent operators for effecting infinitesimal changes, viz

$$\underline{X}_1 = \left(z\frac{\partial}{\partial y} - y\frac{\partial}{\partial z}\right), \ \underline{X}_2 = \left(x\frac{\partial}{\partial z} - z\frac{\partial}{\partial x}\right) \ \text{and} \ \underline{X}_3 = \left(y\frac{\partial}{\partial x} - x\frac{\partial}{\partial y}\right) \tag{11.15}$$

It will be evident later that these become rotation operators when F is some rotation. Two characteristics of these infinitesimal operators are to be noted.

[a] X_1, X_2 and X_3 when multiplied by +i are just the same as the corresponding orbital angular momentum component operators viz l_x, l_y and l_z all expressed in units of $\hbar$.

$$X_1 = -il_x, \ X_2 = -il_y, \ X_3 = -il_z \tag{11.16}$$

[b] Using the operator forms, the following commutation relations can immediately be established.

$$[X_1,\ X_2] = X_3,\ [X_2,\ X_3] = X_1;\ [X_3,\ X_1] = X_2 \qquad (11.17)$$

These three commutation relations can be more compactly expressed as

$$[X_a,\ X_b] = \sum_{a,\ b,\ c=1}^{3} f^{c}_{a,\ b} X_c \qquad (11.18)$$

where a, b, c can assume any of the values 1, 2, and 3. All the coefficients f^{c}_{ab}'s are o's except the following

$$f^{3}_{12} = f^{1}_{23} = f^{2}_{31} = 1 \qquad (11.19)$$

for the R(3) group

[i] It is appareut, that the IO's, satisfying the closure properties, form the basis set for a 3- dimensional operator spece.

[ii] The binary products of the operators, occurring in the commutation relations (11.17), suggest a product space which is also closed.

[iii] Most importantly the relations (11.17) being strongly akin to the orbital angular momentum commutation relations $[l_x,\ l_y] = il_z$ (in unit of $\hbar$) etc, the IO's will follow and generate all the algebraic results that ensued from the commutation relations of the quantum mechanical orbital angular momentum operators. We can legitimately form the raising and lowering operators like $l_{i\pm}$, the basis set of eigenkets (degenerate) for a given X_3 just like l_z. All these algebraic follow- ups constitute what is called commutation algebra of the Lie group of rotations with three independent parameters, R(3). The only difference between X_i's and L_i's will be the presence or absence of i's in the corresponding algebraic consequences.

[iv] We thus have the luxury of using either X_i's or l_i's for revealing the properties of the Lie group and Lie algebra.

The tasks for us now is to find the IR's of the operators that span the three dimensional linear space of Lie algebra, of the Lie group elements and to solve any other incidental relevant problem.

From what has been described before we can use l_i's $l_{\pm}$, l^2, viz, the angular momentum operators, their eigenkets and their commutation properties, instead of the IO's to unravel the properties of the R(3) group. We know that the common eigenket set $\mid l, m>$'s, i.e., the spherical harmonics Y_{lm}'s from the basis for diagonal representations of l^2, l_z in a rigid rotator (free axis) system. Furthermore, we may regard the linear operator space, spanned by X_1, X_2 and X_3 be equivalently and alternatively spanned by l_x, l_y and l_z or even by l_z, l_+ and l_- since l_x and l_y are jointly related to l_+ and l_-. Again, with the following substitution

$$A = \frac{1}{\sqrt{2}} l_z, \; E_+ = \frac{1}{2} l_+ \text{ and } E_- = \frac{1}{2} l_- \; ,$$

the linear operator space may now be deemed as spanned by the basis operators (called the standand base operators) A, E_+ and E_-. We are now to use the basis set $\mid l, m>$'s for any given value of l to find the representations of A, E_+ and E_-. For a given value of l, there occurs a(2l+1) odd dimensional irreducible representation, symbolised by (l) where l= any positive integer including zero. The set of (2l+1) degenerate eigenkets (the $Y_{l,m}$'s) are all eigenkets of the operator A eg., (in units of $\hbar$)

$$\{A \mid \underline{l}\underline{l} >= \underline{l} \mid \underline{l}\underline{l} >, \; A \mid \underline{l}, \; \underline{l} - 1 >= (\underline{l} - 1) \mid \underline{l}, (\underline{l} - 1) >, \\ A \mid \underline{l}, m >= m \mid l, > m\}$$

Hence, these $(2l + 1)$ functions may be used as basis set for the representations of A, E^+ E^- operators of which $\mathbf{D}(A)$ will turn out to be a diagonal one.

Thus $\mathbf{A}^l$ (in units of $\hbar$) will be diagonal as shown below

$$\begin{aligned} & \hat{A}[\mid \underline{l}, l >, \; \mid l, (\underline{l} - l) > \ldots \mid \underline{l}, m >; \ldots \mid \underline{l}, -(\underline{l} - 1) >, \mid \underline{l}, -l >] \\ = \; & [\mid \underline{l}, l > \mid \underline{l}, (\underline{l} - 1) > \ldots \mid \underline{l}, m > \ldots \mid \underline{l}, -(\underline{l} - 1) > \mid \underline{l} - l >] \\ & \operatorname{diag}\left[\frac{1}{\sqrt{2}} \{\underline{l}, (\underline{l} - 1), (\underline{l} - 2), ..m, .. - (\underline{l} - 1), -\underline{l}\}\right] \end{aligned} \tag{11.20}$$

The matrix elements of (2l+1)-dimensional non-diagonal matrix of $\mathbf{D}^l(E_{\pm})$ are given by

$$< (l, m \pm 1) \mid E_{\pm} \mid (l, m) >= \frac{1}{2}\sqrt{l(l+1) - m(m \pm 1)} \tag{11.21}$$

Having represented the linear space of the Lie algebra, we now try to represent the group elements, i.e., the representation for a general rotation ω about an arbitrary unit- vector $\overrightarrow{n}$ passing through the origin.

The differential forms of X_1, X_2 and X_3 are familiar enough. These are the angular rates of differential rotations about X, Y and Z axes respectively each preceded by a minus sign, i.e.,

$$X_1 = -\left(\frac{d}{d\omega}\right)_X, \quad X_2 = -\left(\frac{d}{d\omega}\right)_Y \quad \text{and} \quad X_3 = -\left(\frac{d}{d\omega}\right)_Z$$

Now if F be any rotation $R(\omega)$ and we rotate the function about the Z-axis only, no contribution will result from the terns involving X_1 and X_2 of eq. (11.14). The contribution coming from the third part is $\xi X_3 R(\omega) = dF$. But since ξ is arbitrary and independent of x, y,z a second choice ξ would have fetched a value for $\xi_1 X_3 R(\omega) = dF_1$. So we can always conceive of an intrinsic net contribution $X_3R(\omega)$ toward dF and independent of further reduction or enhancement of the value by the ξ's. In view of this we can write

$$\begin{aligned} X_3 R(\omega) &= -\left(\frac{d}{d\omega}\right)_Z R(\omega) \\ -X_3 d\omega &= \frac{dR_z(\omega)}{R_z} \\ \text{i.e., } R_z(\omega) &= e^{-\omega X_3} \end{aligned} \qquad (11.22)$$

$$\text{In a like manner} \quad R_x(\omega) = e^{-\omega X_1} \text{ and } R_y(\omega) = e^{-\omega X_2} \qquad (11.23)$$

For a general rotation about arbitrary unit vector $\overrightarrow{n}$ axis

$$\begin{aligned} R_n^{(l)}(\omega) &= e^{-\omega n.X} \\ &= \exp\{-\omega(n_x + n_y + n_z).(X_1 + X_2 + X_3)\} \\ &= \exp\{i\omega(n_x l_x + n_y l_y + n_z l_z)\} \end{aligned} \qquad (11.24)$$

where n_x, n_y and n_z are the direction cosines of X, Y, Z laboratory fixed axes at the same point viz the origin as well as $\overrightarrow{n}$,

Now writing the part within the parenthesis as

$$\left(\frac{1}{2}n_x l_x + \frac{1}{2}n_x l_x + \frac{1}{2}n_y l_y + \frac{1}{2}n_y l_y + \frac{1}{2}n_y i l_x - \frac{1}{2}n_y i l_x + n_z l_z\right)$$

it follows that

$$R_n^l(\omega) = \exp\{i\omega\,[E_+(n_x - in_y) + E_-(n_x + in_y)] + n_z l_z\} \qquad (11.25)$$

The representation of their operation is, in principle, known since l_z, l_x and l_y are expressible in the matrix operator forms of $\sqrt{2}A$, E_+ and E_- of (11.20) and (11.21)

Thus the representative matrix of $R_z(\omega)$ in the basis set (2l+1) kets is

$$\mathbf{R}_z(\omega) = \exp(-\omega \mathbf{X}_3) = \exp(i\omega \mathbf{l}_z) = \exp(i\omega\sqrt{2}\mathbf{A}) \qquad (11.26)$$

The IR matrix of a group element provides a set of eigenvalues which add up to give the character χ of that group element. Having established (11.25) and (11.26) for a rotation ω, we now inquire about their character in the representation (l). In this case, the relevant matrices E_+, E_-, $\sqrt{2}A$ all occur in the exponential forms. Will the eigenvalues of say $\sqrt{2}A$ simply sum up to yield the $\chi[R_z^{(l)}(\omega)]$. To answer this, we have to look up some general properties of exponential matrix.

If F and G be two exponential matrices, then

[1] $e^{\mathbf{F}} = \mathbf{1} + \mathbf{F} + \frac{\mathbf{F}^2}{2!} + \frac{\mathbf{F}^3}{3!} +$

[2] The series is a convergent one

[3] $e^{\mathbf{F}}.e^{\mathbf{G}} = e^{(\mathbf{F}+\mathbf{G})}$, only if $[\mathbf{F},\mathbf{G}]=0$, otherwise not

[4] If λ_1, λ_2, λ_3.... be the eigenvalues of $\mathbf{F}$ matrix, then e^{λ_1}, e^{λ_2}, e^{λ_3} etc. are the eigenvalues of $e^{\mathbf{F}}$

[5] The character of $e^{\mathbf{F}}$ is $(e^{\lambda_1}+e^{\lambda_2}+e^{\lambda_3}+....)$

Now, rotations through the same amount ω about different axes passing through the centre of spherical symmetry, as in R(3), belong to the same class and have the same character. Hence $\chi[R_n^{(l)}(\omega)]$ and $\chi[R_z^{(l)}(\omega)]$ will be identical. We, therefore, have to find the character of the group element, $[Rz^{(l)}(\omega)]$.

Since $\sqrt{2}A = \mathrm{Diag}[\underline{l}, (\underline{l}-1), (\underline{l}-2), ... - (\underline{l}-1), -\underline{l}]$, using (11.26)

we can write

$$\begin{aligned}\chi[R_z^{(l)}(\omega)] &= e^{i\omega\underline{l}} + e^{i\omega(\underline{l}-1)} + e^{i\omega(\underline{l}-2)} .... + e^{-i\omega(\underline{l}-1)} + e^{-i\omega\underline{l}} \\ &= \frac{\sin\left(\underline{l}+\frac{1}{2}\right)\omega}{\sin\frac{\omega}{2}} \qquad (11.27)\end{aligned}$$

(The series is a G P and hence the summation result can immediately be written down)
l, the orbital angular momentum quantum numbers (0), (1), (2), (3), (4), (5), (6), (7), (8) etc. are notated in atomic symmetry by S, P, D, F, G, H, I, K, L etc. (note that J is not there)

The group R(3) contains all rotations about any axis passing through the centre. If we now include all roto-reflections (or roto- inversions), the spherical symmetry group, symbolized R′(3), describes the symmetry of a (hypothetical) "spinless" atom. For a real atom, having spin angular momenta, additional symmetry operators are feasible. These are taken care of both by SU(2) group and R*(3) double group.

The IR's of R′(3) are twice as many as those of R(3) group. The IR's are notated by S_g, S_u, P_g, P_u, D_g, D_u, F_g, F_u and so on. The subscripts g and u arise from the inversion effect stemming from roto- iuversion operations of R′(3).

Let us consider the R(3) subgroup of the total symmetry group of a spinless atom with two electrons with orbital moments l_1 and l_2. We can have separate individual irreducible representations of R(3). Since we know how to evaluate the χ values both in (l_1) and (l_2) representations, we can readily evaluate the character in the direct product representations of the IR's (l_1) and (l_2) [see Example 11.2]

An operator C, called Casimir operator, is often used in continuous groups. It is defined, depending on the Lie algebra of the group concerned. For the algebra, called B_1, of R(3) group,

$$\begin{aligned} C &= A^2 + E_+E_- + E_-E_+ \\ &= \frac{1}{2}l_z^2 + \frac{1}{2}(l_x^2 + l_y^2) \\ &= \frac{1}{2}(l_x^2 + l_y^2 + l_z^2) \\ &= \frac{1}{2}l^2 \end{aligned}$$

$$2C \mid \underline{l}, m> = \underline{l}(\underline{l}+1) \mid \underline{l}, m>$$

The step- up and step- down operators $l_\pm$ have been defined and used a little differently by several authors. This has caused a corresponding alteration in the expression of the normalising factors accompanying the step- up and step- down resulting eigenket. The variations are shown below.

$$
\begin{aligned}
l_{\pm} &= (l_x \pm \mathrm{i}l\mathrm{y}) && \text{Normalising factor } \sqrt{l(l-1)-\mathrm{m}(\mathrm{m}\pm 1)} \\
\mathrm{E}_{\pm} &= \frac{1}{2}l_{\pm} = \frac{1}{2}(l_x \pm \mathrm{i}l\mathrm{y}) && \text{Normalising factor } \sqrt{\frac{1}{2}l(l-1)-\mathrm{m}(\mathrm{m}\pm 1)} \\
l'_{\pm} &= \frac{1}{\sqrt{2}}l_{\pm} = \frac{1}{\sqrt{2}}(l_x \pm \mathrm{i}l\mathrm{y}) && \text{Normalising factor } \sqrt{\frac{l(l-1)-\mathrm{m}(\mathrm{m}\pm 1)}{2}}
\end{aligned}
$$

Example 11.1. find the commutator relations $[l_{ßz}, l'_{\pm}]$ and $[l'_{+} l'_{-}]$

$$
\begin{aligned}
[l_z, l'_+] &= \frac{1}{\sqrt{2}}\{l_z l_x + \mathrm{i}l_z l_y - l_x l_z - \mathrm{i}l_y l_z\} \\
&= \frac{1}{\sqrt{2}}[l_z, l_x] + \mathrm{i}[l_z, l_y] \\
&= \frac{1}{\sqrt{2}}\{\mathrm{i}l_y + (\mathrm{i})(-\mathrm{i})l_x\} \text{ in units of } \hbar \\
&= l'_+ \text{ in units of } \hbar
\end{aligned}
$$

In a similar manner, it can be shown

$$[l_z,\ l'_-] = -l'_- \text{ in units of } \hbar$$

$$
\begin{aligned}
[l'_+,\ l'_-] &= \frac{1}{2}\{l_+ l_- - l_- l_+\} = \frac{1}{2}\{(l_x + \mathrm{i}l_y)(l_x - \mathrm{i}l_y) - \\
&\qquad (l_x - \mathrm{i}l_y)(l_x + \mathrm{i}l_y)\} \\
&= \frac{1}{2}\{l_x^2 + l_y^2 - \mathrm{i}l_x l_y + \mathrm{i}l_y l_x - l_x^2 - l_y^2 - \mathrm{i}l_x l_y + \mathrm{i}l_y l_x\} \\
&= \mathrm{i}\{l_y l_x - l_x l_y\} = \mathrm{i}[l_y,\ l_x] \\
&= (\mathrm{i})(-\mathrm{i}l_z) \\
&= l_z
\end{aligned}
$$

Example 11.2 Evaluate the character in the direct product representations of the IR's (l_1) and (l_2), i.e., $\chi^{(l_1)\times(l_2)}$ of the subgroup R(3) of a hypothetical spinless atom with two electrons.

Since $\chi^{[(l_1)\times(l_2)]} = \chi^{(l_2)}.\chi^{(l_2)}$, we write out the diagonal matrix elements explicitly.

$$
\begin{aligned}
\chi^{(l_1)}.\chi^{(l_2)} &= \left(\mathrm{e}^{\mathrm{i}\omega l_1} + \mathrm{e}^{\mathrm{i}\omega(l_1-1)} + \ldots \mathrm{e}^{-\mathrm{i}\omega(l_1-1)} + \mathrm{e}^{-\mathrm{i}l_1\omega}\right) \times \\
&\quad \left(\mathrm{e}^{\mathrm{i}\omega l_2} + \mathrm{e}^{\mathrm{i}\omega(l_2-1)} + \ldots \mathrm{e}^{-\mathrm{i}\omega(l_2-1)} + \mathrm{e}^{-\mathrm{i}l_2\omega}\right).
\end{aligned}
$$

Let the two parentheses be indicated by A and B and let $l_1 > l_2$. Each term in A, on multiplying all the terms of B will generate $(2l_2+1)$ terms. Taking the first term $e^{i\omega l_1}$ and multiply with it all the terms in B including the highest power $e^{i\omega(l_1+l_2)}$, and serially down to $e^{i\omega . o}$, as well as some extraterms with other exponents. The first $(l_1+l_2+1)^{th}$ terms are included in a parenthesis C. The rest of the terms that are housed in C come from multiplying the first (l_1+l_2) terms of B by the lowest power term $e^{-i\omega l_1}$ of A. There will also be some extra terms. These along with the first set of extras are accommodated in a new parenthesis C′.

The two series of resultant product terms are

$$\begin{aligned} C &= e^{i\omega(l_1+l_2)} + e^{i\omega(l_1+l_2)-1} + ... + e^{i\omega o} + ... + e^{-i\omega(l_1+l_2-1)} + e^{-i\omega(l_1+l_2)} \\ C' &= e^{i\omega|l_1-l_2|} + e^{i\omega(|l_1-l_2|-1)} + ... + e^{i\omega o} + ...e^{-i\omega|l_1-l_2|} \end{aligned}$$

Repeating this exercise, by taking second highest power term in A and the penultimate lowest power term in A, and multiplying all the terms in B by these two, we accommodate the product terms in two parentheses D and D′, the contents of which look as follows

$$\begin{aligned} D &= e^{i\omega(l_1+l_2-1)} + e^{i\omega(l_1+l_2-2)} + ... + e^{i\omega o} + ... + e^{-i\omega(l_1+l_2-1)} \\ D' &= e^{i\omega(|l_1-l_2|-1)} + ... + e^{i\omega o} + ... + e^{-i\omega(|l_1-l_2|-1)} \end{aligned}$$

Thus

$$\begin{aligned} \chi^{(l_1)\times(l_2)} &= \chi(l_1).\chi(l_2) = C + D + E + ... + E' + D' + C' \\ &= \left(e^{i\omega(l_1+l_2)} + e^{i\omega(l_1+l_2-1)} + ... + e^{i\omega o} + ... + e^{-i\omega(l_1+l_2-1)} + e^{-i\omega(l_1+l_2)}\right) \\ &\quad + \left(e^{i\omega(l_1+l_2-1)} + e^{i\omega(l_1+l_2-2)} + .. + e^{i\omega o} + .. + e^{-i\omega(l_1+l_2-1)}\right) \\ &\quad + \cdots\cdots \\ &\quad + \cdots\cdots \\ &\quad + \left(e^{i\omega(|l_1-l_2|-1)} + e^{i\omega(|l_1-l_2|-2)} + .. + e^{i\omega o} + .. + e^{-i\omega(|l_1-l_2|-1)}\right) \\ &\quad + \left(e^{i\omega|l_1-l_2|} + ... + e^{i\omega o} + ... + e^{-i\omega|l_1-l_2|}\right) \end{aligned}$$

Actual evaluation of the summation of the terms may now be made by the separate G.P summations (cf eq. 11.27)

11.8 Group Theoretical Significance of Direct Product Representation with Angular Momentum Basis Functions, Addition of Angular Momenta:

The ultimate nature of the resultant $\chi^{(l_1)\times(l_2)}$, as exhibited in the foregoing example 11.2, allows us to draw the following conclusions.

[1] The direct product representation leads to the addition of the orbital moments in conformity with what we expect from the properties of orbital angular momentum of quantum mechanics. The resultant possible L values range from $(l_1 + l_2)$..... $\mid (l_1 - l_2) \mid$, an addition process. This, however, is not just confined to the total angular momentum, but also extends to cover the m's (the z-component angular momentum) indirectly

[M ranges from $(l_1 + l_2)$ to $(\mid l_1 - l_2 \mid)$ in of steps unity]

[2] The relevant part of Hilbert product space is block diagonalised into a direct sum of the possible resultant orbital angular momentum spaces (cf. Ex 11.2), viz.,

$$(\mathrm{L}) = (l_1)\times(l_2) = (l_1+l_2)\oplus(l_1+l_2-1)\oplus.....\oplus(l_1-l_2+1)\oplus(l_1-l_2) \tag{11.28}$$

The original product space is reduced into a number of irreducible subspaces. This type of direct sum relation is called Clebsch Gordan theorem.

[3] The basis functions of the direct product representation $\mid l_1 \mathrm{m}_1 >\mid l_2 \mathrm{m}_2 >$'s and the basis set of the possible total orbital momentum kets provide just two ways of spanning the same relevant part of the Hilbert space.

[4] Therefore, each member ket, such as $\mid l_1 l_2 \mathrm{LM} >$ of the block diagonally reduced space can be expressed as a linear combination of the product functions, viz., $\mid l_1 \mathrm{m}_1 >\mid l_2 \mathrm{m}_2 >$'s used as basis of the direct product representation or vice versa.

$$\mid l_1 l_2 \mathrm{LM} >= \sum_{\mathrm{M}=\mathrm{m}_1+\mathrm{m}_2} \mathrm{C}_{\mathrm{m}_1\mathrm{m}_2} \mid l_1 \mathrm{m}_1 >\mid l_2 \mathrm{m}_2 > \tag{11.29}$$

Or, conversely,

$$| l_1 m_1 >| l_2 m_2 >= \sum_{L,M} b_{L,M} | l_1 l_2 LM > \qquad (11.30)$$

The coefficients, $C_{m_1 m_2}$'s occurring in the linear combination are called Clebsch Gordan coefficients.

[5] Whatever has been described so far in relation to Clebsch Gordan theorem and coefficients in terms of l's and m's will be found to be valid when spin functions, spin angular momenta and total angular momenta are taken into account to describe the full symmetry of the atom, viz., the SU(2) group or the $R^*(3)$ double group which is isomorphic to SU(2)

11.9 The SU(2) group (Special Unitary Group- In Two Dimensions):

The set of all unitary unimodular transformations (det=+1), involving two fundamental variables, an allowable number of parameters and obeying group multiplication laws, constitutes an SU(2) group. This group is also a Lie group having an associated Lie algebra.

The example of SU(2) group are

[1] The set of all unitary matrices of the type $\begin{pmatrix} a & b \\ -b^* & a^* \end{pmatrix}$ with the det viz., $| a^2 + b^2 |= +1$. This is a group with the self representations by the same matrices containing at most 3- independent parameters. The basis set can be taken to be u and v represented by two row matrices
u=(1 0) and v=(0 1)

[2] All types of unitary transformations brought in by the above mentioned matrices on two basis vectors u and v or two functions.

[3] Rotations of complex vector by the diagonal matrices, a subgroup of the matrices referred to in (1).

As in R(3) we now form the infinitesimal operators and also the generators of the SU(2) group to arrive at the Lie algebra associated with it.

The infinitesimal operations upon the basis vectors u, v are

$$\begin{pmatrix} \mathrm{u}+\mathrm{du} \\ \mathrm{v}+\mathrm{dv} \end{pmatrix} = \begin{pmatrix} \mathrm{a}\ \mathrm{b} \\ -\mathrm{b}^*\ \mathrm{a}^* \end{pmatrix} \begin{pmatrix} \mathrm{u} \\ \mathrm{v} \end{pmatrix} = \begin{pmatrix} (1+\mathrm{a})\mathrm{u}+\mathrm{bv} \\ -\mathrm{b}^*\mathrm{u}+(1+\mathrm{a}^*)\mathrm{v} \end{pmatrix}.$$

Since 3 independent parameters can occur in SU(2), we put $a = i\zeta$, $b = \eta + i\xi$, $a^* = -i\zeta$ and $b^* = \eta - i\xi$, where ζ, η, ξ all very very small, are the parameters.

$$\left.\begin{aligned} \mathrm{du} &= i\zeta \mathrm{u} + (\eta + i\xi)\mathrm{v} \\ \mathrm{dv} &= (\eta + i\xi)\mathrm{u} - i\xi \mathrm{v} \end{aligned}\right\} \tag{11.31}$$

Now if F be a function of u and v

$$\mathrm{dF} = \frac{\partial \mathrm{F}}{\partial \mathrm{u}}\mathrm{du} + \frac{\partial \mathrm{F}}{\partial \mathrm{v}}\mathrm{dv}$$

Substituting du and dv by their equivalents from eq. (11.31) and arranging the coefficients of ζ, η and ξ, we have

$$\mathrm{dF} = \left\{\zeta i\left(\mathrm{u}\frac{\partial}{\partial \mathrm{u}} - \mathrm{v}\frac{\partial}{\partial \mathrm{v}}\right) + \eta\left(\mathrm{v}\frac{\partial}{\partial \mathrm{u}} - \mathrm{u}\frac{\partial}{\partial \mathrm{v}}\right) + \xi i\left(\mathrm{v}\frac{\partial}{\partial \mathrm{u}} + \mathrm{u}\frac{\partial}{\partial \mathrm{v}}\right)\right\}\mathrm{F}$$

Since ζ, η and ξ are arbitrarily chosen parameters, the infinitesimal operators, occurring above when equated are as follows

$$\left.\begin{aligned} -\tfrac{1}{2}i\left(\mathrm{u}\tfrac{\partial}{\partial \mathrm{u}} - \mathrm{v}\tfrac{\partial}{\partial \mathrm{v}}\right) &= X_3 \\ \tfrac{1}{2}\left(\mathrm{v}\tfrac{\partial}{\partial \mathrm{u}} - \mathrm{u}\tfrac{\partial}{\partial \mathrm{v}}\right) &= X_1 \\ \text{and}\quad \tfrac{1}{2}i\left(\mathrm{v}\tfrac{\partial}{\partial \mathrm{u}} + \mathrm{u}\tfrac{\partial}{\partial \mathrm{v}}\right) &= X_2 \end{aligned}\right\} \tag{11.32}$$

These lead to the commutation rules

$$[X_1,\ X_2] = X_3;\ [X_2,\ X_3] = X_1;\ [X_3,\ X_1] = X_2 \tag{11.33}$$

which the students can verify. These are the commutation relations of Lie algebra of SU(2) group. It is termed A_1 algebra. It is evident that Lie algebra A_1 of Su(2) and that of B_1 of R(3) are isomorphic compact algebras. We can naturally expect similar types of linear spaces of operators and algebraic relations and the commutation relations between the different field operators.

We shall now make a slight departure to review some items that have relevance to materials of our forthcoming discussion.

11.9.1 Diagonalization and Rotations, Isomorphism and Homomorphism, Higher Dimensional Representations:

Of all the (2 × 2) SU (2) matrices, there are many that are already diagonal. These belong to H subgroup of SU(2). Those which are not diagonal can be diagonalised readily to their equivalent diagonal eigenvalue matrices [see eq. (3.3) and (3.2)]. All the diagonal eigenvalue matrices, either existing in H- subgroup or resulting from diagonalisation, consist of pairs of complex conjugate numbers (eigenvalues), that can be written as $e^{i\omega/2}$ and $e^{-i\omega/2}$. Thus, in general, we can write an already diagonal or a diagonalised (2×2) SU (2) matrix as $\begin{pmatrix} e^{i\omega/2} & 0 \\ 0 & e^{-i\omega/2} \end{pmatrix}$ i.e., **U** (ω) briefly.

It can be shown that such a matrix indicates a rotation ω of the components of a vector, like an R(2) group element, about an arbitrarily chosen Z- axis (see Example 11.3).

Example 11.3

Show that the $\mathbf{U}(\omega)$ matrix causes a rotation of ω about any chosen Z-axis.

Let f and g form the basis set of components of $\mathbf{U}(\omega)$.

$$\left.\begin{aligned} f' &= fe^{i\omega/2} \text{ and } g' = ge^{-i\omega/2} \\ f'^2 &= f^2e^{i\omega} \text{ and } g'^2 = g^2e^{-i\omega} \\ &= f^2(\cos\omega + i\sin\omega) \text{ and } g'^2 = g^2(\cos\omega - i\sin\omega) \end{aligned}\right\} A$$

Using A, we can write down two relations

$$\left.\begin{aligned} \frac{f'^2-g'^2}{2} &= \frac{f^2-g^2}{2}\cos\omega - \frac{f^2+g^2}{2i}\sin\omega \\ \frac{f'^2+g'^2}{2i} &= \frac{f^2-g^2}{2}\sin\omega + \frac{f^2+g^2}{2i}\cos\omega \end{aligned}\right\} B$$

Substituting the symbols x and y for $\frac{f^2-g^2}{2}$ and $\frac{f^2+g^2}{2i}$ respectively we can write B as

$$\left.\begin{aligned} x' &= x\ \cos\omega - y\ \sin\omega \\ y' &= y\ \sin\omega + x\ \cos\omega \end{aligned}\right\} C$$

The relation C is suggestive of a rotation through an angle ω about Z-axis of two vector components x and y which may be complex.

It is thus found that a unitary unimodular matrix, $\begin{pmatrix} e^{i\omega/2}\ 0 \\ 0\ e^{-i\omega/2} \end{pmatrix}$ can be made to represent a two dimensional rotation R(ω) of R(2) group about the Z- axis under a suitable mapping of f^2-g^2 plain on to xy plain. Similarly, two other **U** matrices of the type

$$\pm\begin{pmatrix} \cos\omega/2 & i\sin\omega/2 \\ i\sin\omega/2 & \cos\omega/2 \end{pmatrix} \quad \text{and} \quad \pm\begin{pmatrix} \cos\omega/2 & \sin\omega/2 \\ -\sin\omega/2 & \cos\omega/2 \end{pmatrix}$$

can be moulded suitably to represent two dimensional rotation ω, viz., $R_X(\omega)$ and $R_Y(\omega)$ respectively of the R(2) group. The last two matrices are two specific cases of the general from $\begin{pmatrix} a\ b \\ -b^*\ a^* \end{pmatrix}$. It is thus found that all the R(2) matrices can be reciprocated by a couple of U matrices of the H- subgroup of SU(2) group.

This is illustrated by $R(0)=R(2\pi)=\begin{pmatrix} 1\ 0 \\ 0\ 1 \end{pmatrix}$ of R(2) to which two matrices of the H subgroup, viz., $U(0)=\begin{pmatrix} e^{i\omega/2}\ 0 \\ 0\ \ e^{-i\omega/2} \end{pmatrix} = \begin{pmatrix} 1\ 0 \\ 0\ 1 \end{pmatrix}$ and $U(2\pi)=-\begin{pmatrix} e^{i\omega/2}\ 0 \\ 0\ \ e^{-i\omega/2} \end{pmatrix} = -\begin{pmatrix} 1\ 0 \\ 0\ 1 \end{pmatrix} = \begin{pmatrix} -1\ 0 \\ 0\ -1 \end{pmatrix}$ with $\omega = 0$ and 2π respectively correspond. Similarly, it does happen for each matrix of R(2) group. We thus conclude that R(2) subgroup of R(3) and H sub group of SU(2) are not isomorphic, but are homomorphically related. But if we build R*(2) from R(2) group by introducing another generator group element $Q=\begin{pmatrix} -1\ 0 \\ 0\ -1 \end{pmatrix}$, the double group, R*(2), so formed will be isomorphic to the subgroup H of SU(2), It is apparent that the generator element Q, suggestive of a rotation of 2π being equal to -1, militates against reality. Thus R* double group is not totally a proper rotational group.

In an analogous manner, any rotation R(ωn) of R(3) group corresponds to two group elements of SU(2) group. The two groups are thus homomorphically connected. By enlarging the R(3) group to the double group R* with the introduction of Q, the R*(3) group becomes isomorphic to the SU(2) group. Thus R*(3) double group [or Su(2) group] plays the role of an atomic symmetry group (except H- atom). It is to be always remembered that a rotation of $R_n(\omega)$ is equivalent to (11.24 and 11.25) which involve rotations about R_z, R_x, R_y of a laboratory-fixed (with common origin of n- vector) XYZ axis system. Alternatively, $R_n(\omega)$ can be imitated by three proper Euler rotations in the XYZ axial system.

It is now time to turn our attention to higher dimensional IR's of SU(2) group. Starting with the basis (f, g) of the self representative SU(2) group of unitary unimodular (2×2) dimensional matrices of the type $U(a, b)=\begin{pmatrix} a & b \\ -b^* & a^* \end{pmatrix}$ we may build a 3- basis unit (f^2, fg, g^2). This can be conceived of as a basis set for a self direct product representation of a single IR of SU(2) group having the basis set (f, g). Alternatively, this 3- set basis may be obtained as self products and interproducts of fand g in the binomial expansion $(f+g)^2$, where the exponents of f and g together equal two. For details we turn to the next section.

11.9.2 Higher Dimensional IR's of SU(2) Group and Their character Values:

As just mentioned, let us take the basis set $\frac{f^2}{\sqrt{2}}$, fg and $\frac{g^2}{\sqrt{2}}$ and attempt to find the representation (3×3) matrix for the group element U(a, b) mentioned above. The introduction of the factor $\frac{1}{\sqrt{2}}$ with the assumed basis set is due to obtaining the final from of representation in a unitary matrix from,

---

Example 11.5:

Find the form of the transformation matrix for the operator

$$\begin{pmatrix} a & b \\ -b^* & a^* \end{pmatrix}$$

in the basis set $\left\{\frac{f^2}{\sqrt{2}}, \ fg, \ \frac{g^2}{\sqrt{2}}\right\}$ of a three- dimensional representation.

$$\begin{pmatrix} f'/2^{\frac{1}{4}} \\ g'/2^{\frac{1}{4}} \end{pmatrix} = \begin{pmatrix} a & b \\ -b^* & a^* \end{pmatrix}\begin{pmatrix} f/2^{\frac{1}{4}} \\ g/2^{\frac{1}{4}} \end{pmatrix} = \begin{pmatrix} af/2^{\frac{1}{4}} & +bg/2^{\frac{1}{4}} \\ -b^*f/2^{\frac{1}{4}} & +a^*g/2^{\frac{1}{4}} \end{pmatrix}$$
$$= \begin{pmatrix} af'' + bg'' \\ -b^*f'' + a^*g'' \end{pmatrix}$$

With this knowledge we can evaluate the following

$$\begin{aligned}
\frac{f'^2}{\sqrt{2}} &= (af'' + bg'')^2 = a^2f''^2 + 2abf''g'' + b^2g''^2 \\
f'g' = (fg)' &= (af'' + bg'')(-b^*f'' + a^*g'') = -ab^*f''^2 \\
&+ (aa^* - bb^*)f''g'' + a^*bg''^2 \\
\frac{g'^2}{\sqrt{2}} &= (-b^*f'' + a^*g'')^2 = b^{*2}f''^2 - 2a^*b^*f''g'' + a^{*2}g''^2
\end{aligned}$$

Using these three linear combinations, we write in matrix form

$$\begin{aligned}
\begin{pmatrix} \frac{f'^2}{\sqrt{2}} \\ (fg)' \\ \frac{g'^2}{\sqrt{2}} \end{pmatrix} &= \begin{pmatrix} a^2 & 2ab & b^2 \\ -ab^* & (aa^* - bb^*) & a^*b \\ -b^{*2} & -2a^*b^* & a^{*2} \end{pmatrix} \begin{pmatrix} f''^2 \\ f''g'' \\ g''^2 \end{pmatrix} \\
&= \begin{pmatrix} a^2 & \sqrt{2}ab & b^2 \\ -ab^* & \frac{1}{\sqrt{2}}(aa^* - bb^*) & a^*b \\ -b^2 & -\sqrt{2}a^*b^* & a^{*2} \end{pmatrix} \begin{pmatrix} \frac{f^2}{\sqrt{2}} \\ fg \\ \frac{g^2}{\sqrt{2}} \end{pmatrix}
\end{aligned} \tag{11.34}$$

on substituting f''^2, $f''g''$ and g''^2 by their equivalents.

We thus obtain representative matrix for $\begin{pmatrix} a\ b \\ -b^*\ a^* \end{pmatrix}$ group element in the 3- dimensional basis set.

It thus becomes feasible to have a polynomial expansion $(f+g)^{2j}$, where 2j is a positive integer with $j=\frac{1}{2}$ odd (+ve) integer or any positive integer. The (2j+1) number of homogeneous self- product and interproduct terms, thus given by the general form $f^{(j+m)}.g^{(j-m)}$ with m varying from o to j in steps of unity, provide the plank for the (2j+1) dimensional representation of the original (2×2) dimensional matrices of SU(2) group elements. Of course, like in 3- dimensional representation of the foregoing discussion, we associate, for the sake of ultimate turn up of a unitary matrix, a multiplying factor with the general form of the self and interproduct term. The ultimate form of general interproduct terms is

$$\frac{f^{(j+m)}.g^{(j-m)}}{\sqrt{(j+m)(j-m)}} \tag{11.35}$$

The general expression for the matrix element in the $(j)^{th}$ representation $D^{(j)}_{mm'}(a, b)$ can be painfully found out and does have a formidable look.

This is, as first given by Hamermesh (Eq. 11.36)

$$D^{(j)}_{mm''}(a, b) = \sum_{\mu} \frac{\{(j+m)!(j-m)!(j+m')!(j-m')!\}^{\frac{1}{2}}}{(j+m-\mu)!(j-m'-\mu)!\mu!(m-m'+\mu)!} \times a^{(j+m-\mu)} a^{*(j-m'-\mu)} b^{\mu}.(-b^{*})^{(m-m'+\mu)} \qquad (11.36)$$

Here m and m' vary from to j to -j and μ from 0 to m.

To utilize this for our purpose, it will be helpful to note the following

[1] In the (j) representation m and m′ vary from j to -j and the representation matrix is (2j+1)- dimensional.

[2] In the (j) representation, the diagonal matrix elements are $D^{(j)}_{mm}(a, b)$ where m=m′ and μ is specifically zero.

[3] The original SU (2) group elements U(a, b)'s, of which the higher dimensional representations are made, can initially be converted into their diagonal forms $U\left(e^{\frac{i\omega}{2}}, 0\right)$'s and their (2j+1) dimensional representations constructed starting with the original basis set {f, g}. The final matrix of the group element may trun out to be non- diagonal (eq. 11.34) but these will still be unitary. One should be careful in notating concretely the subscripts m, m′ of $D^{(j)}_{mm'}$ (a, b) and apportioning appropriate m and m′ values in the R H S of (11.36)

[4] Taking note of (2) we can write the diagonal matrix element $D^{(j)}_{mm}(e^{\frac{i\omega}{2}}, 0)$ from (11.36) remembering $\mu = 0$ and m=m′, a=$e^{\frac{i\omega}{2}}$, a*=$e^{\frac{-i\omega}{2}}$
b=(-b*)=0

$$\begin{aligned} D^{(j)}_{mm}(e^{\frac{i\omega}{2}}, 0) &= (e^{\frac{i\omega}{2}})^{2m} \\ &= e^{im\omega} \end{aligned} \qquad (11.37)$$

The character of the (j) representation

$$\begin{aligned} \chi^{(j)} &= \sum_{m=-j}^{j} e^{im\omega}, \text{similar to (11.27) a G.P sum of terms} \\ &= \sin \frac{\left(j+\frac{1}{2}\right)\omega}{\sin\frac{\omega}{2}} \end{aligned} \qquad (11.38)$$

It has been established that all these higher dimensional representations are irreducible.
For the R(3) rotation subgroup of SU(2)

$$\chi^{((j))}(\omega + 2\pi) = \chi^{((j))}(\omega) \tag{11.39}$$

Thus the role of l, orbital angular momentum is played by j when j=a positive integer. But when j=$\frac{1}{2}$, $\frac{3}{2}$, $\frac{5}{2}$ i.e., half of an odd (+ve) integer

$$\chi^{(j)}(\omega + 2\pi) = -\chi^{(j)}(\omega) \tag{11.40}$$

as can be concluded from (11.38)

It is thus seen that in SU(2) or R*(3) group, j, like l in R(3) group, performs the role of general angular momentum and like l_x, l_y, l_z the j_x, j_y, j_z components of general angular momentum can be related to the respective non commutation of the infinitesimal operators. (Relations 11.32, 11.33).

$$j_x = iX_1, \ j_y = iX_2 \text{ and } j_z = iX_3$$

$$[j_x, \ j_y] = [iX_1, \ iX_2] = iX_3 \text{ and so on}$$

As in orbital moments case (sec. 11.8), the kronecker product representation of $(j_1 \times j_2)$ leads to addition of general angular momenta and the corresponding product space is block diagonalised resulting into a direct sum of general angular momentum spaces [Example 11.2 and eq (11.28)]

$$J = j_1 \otimes j_2 = (j_1 + j_2) \oplus (j_1 + j_2 - 1) \oplus .... \oplus (j_1 - j_2) \tag{11.41}$$

All the algebraic formulations and results that follow from the Lie algebra B_1 of R(3) involving l_x, l_y, l_z, l, l_+, l_- are also mimicked by the corresponding parts of the j- operator. Since j value also permits the $(\frac{1}{2})$, $(\frac{3}{2})$ etc., representations, the properties of spin angular momenta find a natural explanation either as a component of the j operator or as the totality of the j operator itself when the orbital moment is absent or totally quenched.

From now on, we shall mostly us e the common symbol j to represent all types of angular momenta- spin, orbital or total. The nature of the angular momenta involved will either be clear from the background description or need no elaboration as the formulation under review will be valid for any of the three. In case all the three occur together, these will be differentiated as j_1 j_2 and J. For the Z- component angular momenta, m_1, m_2 and M symbols will be used.

In its use as the generator of group element, capital J will be used, viz., $e^{i\alpha J}$, $e^{i\alpha J_z}$ etc.

The treatments of the groups R(3), SU(2) suggest some related problems the results of which will serve to solve further problems of quantamechanical significance

[1] It can be shown that a unitary unimodular matrix $\begin{pmatrix} a & b \\ -b^* & a^* \end{pmatrix}$, where a=cos $\frac{\omega}{2}$ and b=isin $\frac{\omega}{2}$ induces a rotation ω about the X-axis and when a=cos $\frac{\omega}{2}$ and b=sin $\frac{\omega}{2}$, this causes an ω- rotation about the Y- axis

[2] The set of commutator relationships depicted just prior to the worked out example 11.1, come in handy for finding the general formulations of the matrix elements of the component angular momentum operators J_1, J_2, J_3 (i.e., J_x, J_y, J_z respectively) and also of J_+ and J_-.

[3] Once the general formulations are known, these can be utilised in forming the matrix representations of spin $J_\alpha^{\frac{1}{2}}$, spin J_α^1 or any such operator. Thus we can find the different Pauli matrices.

Besides, these formulations are helpful in the representations of the tensor product states involved in the block diagonal matrix from of the reduced product space.

We shall presently furnish some worked-out examples in respect of the items under (2) and (3).

Example 11.6: Find the general results of the following,

$$\text{(i)}\quad J_3 \mid j,\ m>; \quad \text{(ii)}\quad J'_+ \mid j,\ m>; \quad \text{(iii)}\quad J'_- \mid j\ m>$$

where $\mid j\ m>$ is a normalized basis eigenket of J_3

[i] $J_3 \mid j,\ m>= m \mid j,\ m>$ in units of $\hbar$

[ii]

$$\begin{aligned} J'_+ \mid j, m > &= \frac{J_+}{\sqrt{2}} \mid j,\ m > \\ &= \sqrt{j(j+1) - m(m+1)/2} \mid j,\ m+1 > \end{aligned}$$

[iii]

$$\begin{aligned} J'_- \mid j,\ m > &= \frac{J_-}{\sqrt{2}} \mid j,\ m > \\ &= \sqrt{j(j+1) - m(m-1)/2} \mid j,\ m-1 > \end{aligned}$$

From the results obtained in the foregoing example we can immediately write down the matrix element of J'_+ and J'_- between $\mid j,\ m' >$ and $\mid j,\ m >$

$$\begin{aligned} < j,\ m' \mid J'_+ \mid j,\ m > &= \sqrt{j(j+1) - m(m+1)/2}\delta m',\ m+1 \\ \text{and} \quad < j,\ m' \mid J'_- \mid j,\ m > &= \sqrt{j(j+1) - m(m-1)/2}\delta m',\ m-1 \end{aligned} \tag{11.42}$$

Since $J'_+ = \frac{1}{\sqrt{2}}(J_1 + iJ_2)$, Eq (11.42) is used to write down the matrix elements of J_1 and J_2 in the basis set of the eigenkets.

$$\begin{aligned} &\frac{1}{\sqrt{2}}\{< jm' \mid J_1 \mid jm > + < jm' \mid iJ_2 \mid jm >\} \\ &\qquad = \frac{1}{\sqrt{2}}\left[\sqrt{(j+m+1)(jm)}\delta_{m',\ (m+1)}\right] \end{aligned}$$

Similarly using J_-, one can write

$$\begin{aligned} &\frac{1}{\sqrt{2}}\{< jm' \mid J_1 \mid jm > - < jm' \mid iJ_2 \mid jm >\} \\ &\qquad = \frac{1}{\sqrt{2}}\left[\sqrt{(j-m+1)(j+m)}\delta_{m',\ m-1}\right] \end{aligned}$$

Adding up we have

$$\begin{aligned} < jm' \mid J_1 \mid jm > = &\frac{1}{\sqrt{2}}\left[\sqrt{\frac{(j+m+1)(j-m)}{2}}\delta_{m',\ (m+1)}\right] \\ &+ \frac{1}{\sqrt{2}}\left[\sqrt{(j-m+1)(j+m)/2}\delta_{m',\ (m-1)}\right] \end{aligned} \tag{11.43}$$

In a like manner, on subtraction,

$$< \mathrm{jm'} \mid \mathrm{J_2} \mid \mathrm{jm} >= \frac{1}{\sqrt{2}\mathrm{i}}\left[\sqrt{(\mathrm{j}+\mathrm{m}+1)(\mathrm{j}-\mathrm{m})/2}\,\delta_{\mathrm{m'},\ (\mathrm{m}+1)}\right] - \frac{1}{\sqrt{2}\mathrm{i}}\left[\sqrt{(\mathrm{j}+\mathrm{m})(\mathrm{j}-\mathrm{m}+1)/2}\,\delta_{\mathrm{m'},\ (\mathrm{m}-1)}\right] \tag{11.44}$$

For the matrix element of J_3, it is obvious from example (11.6) that

$$< \mathrm{jm'} \mid \mathrm{J_3} \mid \mathrm{jm} >= \mathrm{m} \mid \mathrm{jm} > \delta_{\mathrm{m'},\ \mathrm{m}} \tag{11.45}$$

All these relations, expressed in units of $\hbar$, are of importance in the matrix representations of J^j_1, J^j_2 and J^j_3 in the basis set of kets, $\mid jm >$'s.

The superscript j in the notation stands for the highest value of m occurring in the kets.

Before we turn to the next example (11.7), we need to elaborate a little on the practice of the usage of the indices of the matrix elements of the representative matrix. The number of basis kets in a representation is (2j+1). The ordinary conventional way in indicating the indices is for (kl) to have the row numbers k run from 1 to (2j+1) and also to allow column index l run the same range. Thus the matrix element

$$\left(\mathrm{J}^{\mathrm{j}}_{\alpha}\right)_{\mathrm{k}l} =< \mathrm{j},\ \mathrm{j}+1-\mathrm{k} \mid \mathrm{J}_{\alpha} \mid \mathrm{j},\ \mathrm{j}+1-l >$$

Another form of notation for the matrix element $< \mathrm{j},\ \mathrm{m'} \mid \mathrm{J}_\alpha \mid \mathrm{j},\ \mathrm{m} >$ is to indicate the rows and column directly by their m values in which m′ runs from j to -j in steps of unity along each row with simultaneous run of m in an identical manner down each column. The matrix element is indicated as $\left(\mathrm{J}^{\mathrm{j}}_{\alpha}\right)_{\mathrm{m'm}}$

where $\alpha = 1,\ 2$ or 3 and the superscript j is the highest possible value of m in the basis set of kets used.

This is called the highest weight construction (m=j highest) using the basis set $\mid \mathrm{j},\ \mathrm{m} >$. We shall now work out an illustrative case.

Example 11.7: Find the representation of total spin J=1 of a quantum system.

Here J=1, hence the maximum value of m=1 and the minimum m= −1. The basis set of kets to be considered for representations comprises

$|1,1>$, $|1,0>$ and $|1,-1>$. The m′ and m values run along the range 1, 0, -1 along the rows and columns respectively. The required representation is a collection of the representations $\mathbf{J}^1_1$, $\mathbf{J}^1_2$, $\mathbf{J}^1_3$, the three components of $\mathbf{J}^1_\alpha$. In order to feel at home in using (m′m) indices of the matrix element $[J^1_\alpha]_{m'm}$, we write the full complement of these in notations for $\mathbf{J}^1_1$ momentum component and then evaluate them.

$$\mathbf{J^1_1} = \begin{pmatrix} (J^1_1)_{11} & (J^1_1)_{10} & (J^1_1)_{1-1} \\ (J^1_1)_{01} & (J^1_1)_{00} & (J^1_1)_{0-1} \\ (J^1_1)_{-11} & (J^1_1)_{-10} & (J^1_1)_{-1-1} \end{pmatrix}$$

For the other components only J^1_1 is to be substituted by J^1_2 and J^1_3 respectively

Let us first evaluate $(J^1_1)_{10}$ and then $(J)_{-1-1}$ any two arbitrary selections. On applying (11.43)

$$\begin{aligned} (J^1_1)_{10} & \quad , \text{where } j = 1,\ m = 1\ m = 0 \\ &= \frac{1}{\sqrt{2}}\sqrt{(j+m)(j-m+1)/2}\delta_{m'},\ (m-1) \\ &+ \frac{1}{\sqrt{2}}\sqrt{(j+m+1)(j-m)/2}\delta_{m'},\ (m+1) \\ &= \frac{1}{\sqrt{2}} \end{aligned}$$

In exactly the same way, when j=1, m′=-1, m=-1, we find, using the same Eq (11.43) $(J^1_1)_{-1-1}=0$.

Similarly all the other matrix elements are evaluated
The final form of

$$\mathbf{J}^1_1 = \frac{1}{\sqrt{2}}\begin{pmatrix} 0 & 1 & 0 \\ 1 & 0 & 1 \\ 0 & 1 & 0 \end{pmatrix}$$

To find $\mathbf{J}^1_2$, we make a sample-test of the matrix element $(J^1_2)_{0-1}$ and use Eq.(11.44)

$$(J^1_2)_{0-1} = \frac{1}{\sqrt{2}i}[1] = -\frac{i}{\sqrt{2}} \qquad \text{where } j = 1,\ m' = 0,\ m = -1$$

Thus determining all the rest of the matrix elements $(J^1_2)_{m'm}$ using the same equation (11.44), we have

$$\mathbf{J}^1_2 = \frac{1}{\sqrt{2}}\begin{pmatrix} 0 & -i & 0 \\ i & 0 & -i \\ 0 & i & 0 \end{pmatrix}$$

Next we use Eq (11.45) and easily evaluate all the matrix elements $(J_3^1)_{m'm}$

$$\mathbf{J}_3^1 \begin{pmatrix} 1 & 0 & 0 \\ 0 & 0 & 0 \\ 0 & 0 & -1 \end{pmatrix}$$

The much talked Pauli spin matrices σ's are each 2 times the representations of total spin $J=\frac{1}{2}$.

$$\text{These are } \mathbf{J}_1^{1/2} = \frac{1}{2}\begin{pmatrix} 0 & 1 \\ 1 & 0 \end{pmatrix} = \frac{1}{2}\sigma_1$$

$$\mathbf{J}_2^{1/2} = \frac{1}{2}\begin{pmatrix} 0 & -i \\ i & 0 \end{pmatrix} = \frac{1}{2}\sigma_2$$

$$\text{and } \mathbf{J}_3^{1/2} = \frac{1}{2}\begin{pmatrix} 1 & 0 \\ 0 & -1 \end{pmatrix} = \frac{1}{2}\sigma_3$$

The readers may find it worthwhile to have a dig at their derivation.

The method (highest weight construction process) used in the construction of representations of J operators and also to be used in connection with some more, such as tensor product representation and block diagonalisation of a direct product space (see 12.4), stems from uses of operators of linear spaces defined by some Lie Algebra such as B_1 or A_1 etc., associated with a group. And since these operators also frequent the domain of quantum mechanics, infinite (Lie) groups and their algebras find wide applications in quantamechanical systems (sec 12.3, 12.4).

11.10 The Lie Group R(4)- Rotations in Four Dimensions:

We shall not delve into details, but only give a brief summary of its formulations.

[1] The system has six independent parameters (sec 11.3) and thus there can be formed six infinitesimal operators.

[2] These operators, viz, A_1, A_2, A_3 and B_1, B_2, B_3 lead to six commutator relationships, the starting point of the associated Lie Al-

gebra, called D_2 algebra. The commutator relations are

$$[A_1, A_2] = A_3, \ [A_2, A_3] = A_1, \ [A_3, A_1] = A_2$$

$$[B_1, B_2] = A_3, \ [B_2, B_3] = A_1, \ [B_3, B_1] = A_2$$

[3] There are other derivable commutation relations, such as,

$$[A_1, B_1] = [A_2, B_2] = [A_3, B_3] = 0 \text{ etc.}$$

[4] Two distinct sets of basis infinitesimal operators are linearly combined to give standard basic commutators. viz.,

$$J_i = \frac{1}{2}(A_i + B_i), \ K_i = \frac{1}{2}(A_i - B_i); \ i = 1, 2, 3$$

The commutation rules of the Lie Algebra, D_2, of $R^*(4)$ of which $R(4)$ is a subgroup, are

$$[J_1, J_2] = J_3, \ [J_2, J_3] = J_1, \ [J_3, J_1] = J_2$$

$$[K_1, K_2] = K_3, \ [K_2, K_3] = K_1, \ [K_3, K_1] = K_2$$

$$\text{and} \qquad [J_i, K_i] = 0$$

These two sub algebras are completely disjoint and each is compact. J_1, J_2 and J_3 may be regarded as components of general angular momentum J

[5] The D_2 algebra may be considered as a direct sum

$$D_2 = B_1 \oplus (cf.R(3))B_1 = A_1 \oplus (cf.SU(2))A_1$$

[6] Associated with the Lie Algebra, D_2, is the Lie group $R^*(4)$ isomorphic with $SU(2)\times SU(2)$. The group elements of $R^*(4)$ are given by the representations $(j_1 j_2)$ with

$$j_1 = 0, \ \frac{1}{2}, \ 1, \ \frac{3}{2}, \ 2......$$

$$j_2 = 0, \ \frac{1}{2}, \ 1, \ \frac{3}{2}, \ 2......$$

where $(j_1 j_2)$ may be any pair of combinations.

[7] The group elements of the subgroup $R(4)$, being real rotations in four dimensions, are obtained only when all the couples of the j values lying between $(j_1 + j_2)........|(j_1 - j_2)|$ add up to positive integers (including 0). This means the pair of values of j_1 and j_2 in (j_1, j_2) should either be both positive integers or both +ve half integers.

Chapter 12

Applications of Lie Groups In Quantamechanical Problems

12.1 General Remarks

A beginner in quantum mechanics encounters the so called simple problems.

(1) Particle in a box (2) Rigid rotator in a plain (3) Rigid rotator in three dimensional space. 4(a) simple harmonic oscillator in one-dimension 4(b)Isotropic simple harmonic oscillations in two dimensions 4(c) Isotropic oscillator in three dimensions (5) H-atom problem.

From group theory viewpoint, all these problems belong to the domains of different Lie groups (and Lie algebras) that are relatively harder to deal with than the finite molecular point groups (see chap 11). The foregoing quantum problems belong to the following continuous groups

(1) U(1)

(2) R(2)(3) R(3) 4(a) SU(1) 4(b) SU(2) 4(c) SU(3) (5) R(4) subgroup of R*(4)

A quantum theorist tries to solve the corresponding Schrodinger equation. Incidentally, by way of simplifying the solution, he proceeds generally along the following steps (i) forms the Hamiltonian (ii) finds the constants of motion (iii) Uses the operators associated with the components of the constant or constants of motion to form the corresponding number of basic commutator relations. Usually, the number of components of the constants of motion = number of basic quantamechanical operators formed = number of basic quantamechanical commutator relations (iv) finally solves the Schrodinger equation.

A group theorist approaches the problems in the following way (i) forms infinitesimal operators (ii)these operators span a linear operator space or spaces (iii) with the help of the operators basic commutator relations are formed. These constitute starting points of the corresponding Lie Algebra of the linear space. Usually the number of infinitesimal operators = number of basis operators spanning the operator space or spaces = number of basic commutator relations = number of independent parameters of the Lie group.

The point of confluence in the approaches of a quantum theorist and of a group theorist lies in

[1] The equalities of the various numbers mentioned in quantum approach are also similar to all the specific numbers of the group theoretical approach mentioned in the above paragraph.

[2] The nature of commutator relations drawn up by a quantum theorist are either (a)identical with (at most these may differ by a multiplicative +i or -i) or (b) proportional to those formed by the group theorist Moreover, these are isomorphic.

12.2 Total Angular Momentum, Casimir operator and the Hamiltonian operator

Casimir operator C is defined in a Lie group which contains all rotation operations or some rotation operations among its group elements. Such groups are R(2), R(3), SU(2), R(4) and SU(3) among the simpler ones. Since rotations are associated with angular momenta, Casimir operator is related to the square of the total angular momentum operator, viz. J^2 associated with each linear space. If more than one linear space is associated with the Lie group, more than one j^2 operator will be involved. In general the Casimir operator is simply related to J^2 as

$$2C = J^2$$

Again, since for rotational groups, J^2 is related to the hamiltonian operator H and also to the degeneracy of the quantum states n, Casimir operator can, therefore, be correlated with H or used in the depiction of the degeneracy of the quantum status.

The relations between the Casimir operator and the different infinitesimal operators that correspond with the component angular moment operators spanning the linear spaces involve a fair bit of complexity in their formulations. We shall skip these processes and simply write down the relations for the different groups we shall be concerned with.

[1] R(3) group. (B_1 algebra)

$$2C = L^2 = (l_x^2 + l_y^2 + l_z^2)$$

for R(2)group, l_x and l_y components are each zero, so that

$$2C = l^2 = l_z^2$$

[2] SU(2) group. A_1 algebra)
$2C = -(K^2 + L^2 + D^2)$, where L, K, D behave
like angular momentum components (see SHM in two dimensions)

[3] R(4) group (D_2 algebra)
$2C + 2C = -(A_1^2 + A_2^2 + A_3^2 + B_1^2 + B_2^2 + B_3^2)$

These are two disjoint linear spaces and hence have two separate angular momentum square operators J_1^2 and J_2^2 and when $J_1 = J_2 = J$, total is 2J and hence 4C casimir operators arise in the correlation above.

The IR notation of an IR is (j_1, j_2) in R(4) symmetry group. The basis set comprises $(2j_1 + 1)(2j_2 + 1)$ number product functions of the type $|j_1, m_1\rangle|j_2, m_2\rangle$. These are also eigenfunctions of j_1^2 and j_2^2 - the total angular momentum operators in an R(4) symmetry.

We can thus write

$$\begin{aligned} 2C|j_1, m_1\rangle|j_2, m_2\rangle &= (j_1^2 + j_2)|j_1, m_1\rangle|j_2, m_2\rangle \\ &= \{j_1(j_1 + 1) + j_2(j_2 + 1)\}|j_1, m_1\rangle|j_2, m_2\rangle \end{aligned}$$

If $j_1 = j_2 = j$, the total degeneracy of the eigenstate will be $\{(2_j + 1)^2 - 1\}$
Incidentally the operator C is also, as will be shown later, related to the hamiltonian of a system belonging to R(4) symmetry group.

12.3 Applications in some Quantamechanical Problems

1. Rigid rotation in a plain [R(2)] (fixed axis)

H is invariant under all rotations. H, L^2 and l_z commute. Since $l = r \times p$, p being linear momentum, analysis shows that l_x and l_y components are each zero. So l in identical with l_z. There is thus effectively one constant of motion, viz., the l_z component.

The basis set of l_z in R(2) are all nondegenerate $|l, m\rangle$, and each is thus represented by a one dimensional I.R. Each base function $|l, m\rangle$ is an eigenket of H.

Since $H = \frac{1}{2}l_z^2$ in proper units and $l_z|l, m\rangle = m|l, m\rangle$ (in units of $\hbar$) we have

$E = l_z^2 = m^2$ (in units of $\hbar^2/2\pi^2 I$).

Thus each quantum state of quantum mechanics refers to an IR of the corresponding symmetry group.

It is interesting to know that the results and the conclusions of the Lie group of "a particle in a one dimensional box" can be obtained as a special case of a two-dimensional rotation of a rigid rotator.

2. The Lie group R(3) - Rigid Rotator (Free - axis)

The hamiltonian is invariant under all forms of rigid rotation. $[H, L] = 0$,

l vector is a constant of motion and has three components associated with the motion. Group theoretically, therefore, it has three independent parameters, three infinitesimal operators, three basic commutator relations and a linear algebra B_1, with a linear operator space spanned by three operators of the aforesaid algebra. The irreducible representation (l) of the basic operator are each (2l+1)- dimensional and the spherical harmonics $Y_{L,m}$'s form the basis set of l^2, as well as the eigenfunctions of H. Therefore, the different quantum states of the hamiltonian have different degeneracies $(2l + 1)$ depending on the value of L. But $(2l + 1) = n$, an odd number denoting degeneracies.

$$\begin{aligned}
l^2|l, m\rangle &= l(l+1)|l, m\rangle \text{ in units of } \hbar^2 \\
C|l, m\rangle &= \frac{1}{2}l(l+1)|l, m\rangle \\
\hat{H} &= \frac{1}{2}l^2 \text{ (in proper units)} = C \\
E &= l(l+1), \left(\text{in appropriate energy units } \frac{\hbar^2}{2\pi^2 I}\right) = 2C
\end{aligned}$$

3. Simple Harmonic Motion - (in one dimension)

The H operator is invariant under all translations (say in the x-direction) positive or negative from the origin ranging from $+\propto$ to $-\propto$. The representations of such a group of translations subjected to a restoring force are given by an infinite set of unimodular unitary one dimensional matrices of SU(1) group. One dimensional matrices mean these are IR's. Each quantum state of a one-dimensional SHM - oscillator thus corresponds to one dimensional IR and hence is nondegenerate. The

nondegeneracy is also corroborated by the quantum mechanical energy expression
$E_n = (n + \frac{1}{2})$ in suitable units of energy where n = 0, 1, 2, 3, ...

4. Simple Harmonic Motion (isotropic) in Two-Dimensions
The result of two SHM's simultaneously imposed on a particle gives rise to a rotational motion (two-dimensional) about the origin. The apparent suggestion for the Lie group would be R(2). But it is more symmetric than this.

The hamiltonian H for a two-dimensional oscillator can be expressed (in suitable units) in a number of ways. We write two of these here.

$$H = \frac{1}{2}\left[p_1^2 + p_2^2 + q_1^2 + q_2^2\right] \text{..........} A$$

$$H = \frac{1}{2}(P^2 + Q^2) + (p_1p_2 + q_1q_2) \text{..........} B$$

where P and Q are $P = p_1 + p_2$ and $Q = q_1 + q_2$, with p's and q's denoting the linear momenta and coordinates of the particles.

Quantamechanical approach shows that there are three constants of motion: L, K and D where $L = \frac{1}{2i}(q_1p_2 - q_2p_1); K = \frac{1}{2i}(p_1p_2 + q_1q_2)$ and
$D = \frac{1}{4i}[(p_1^2 + q_1^2) - (p_2^2 + q_2^2)]$. These three behave like three components of a general angular momentum vector. There are thus three basic operators and three basic commutator relations:
$J^2 = L^2 + K^2 + D^2, [L, K] = -D, [K, D] = -L$ and $[L, D] = -K$
and

$$\begin{aligned} E_{n,m} &= (n + \frac{1}{2}) + (m + \frac{1}{2}) \text{in proper energy units} \\ &= (n + m + 1) = N + 1, \text{where } N = \text{ total degeneracy} \\ &\quad \text{of the quantum state } (n, m). \end{aligned}$$

Let us now look into group theoretical aspects of these observations.

1. There are three independent parameters (cf 3-constants of motion)
2. There are three infinitesimal operators spanning a linear operator space (cf. L,K,D operators)
3. There ought to be three basic commutator relations involving the infinitesimal operators similar to the commutator relations shown above.

These suggest that linear algebra is A_1 and the appropriate group is SU(2) and not R(2). The basis set is ψ_{jm}; of IR(j). The representation matrices are

(2j + 1) dimensional. Casimir operator $2C = J^2$ (sec. 12.2)

$2C\psi_{jm} = j^2\psi_{jm} = j(j+1)\psi_{jm}$ in units of $\hbar^2$, Since $[H, j^2] = 0 = [H, C]$, ψ_{jm}s are wave functions of S.H.M (two-dimensional) oscillator.

$$\text{Now } C = -\frac{1}{2}(K^2 + L^2 + D^2) \quad \text{[sec 12.2]}$$
$$= \frac{1}{8}(H^2 - 1) \quad \text{(C)}$$

The derivation of relation (C) is left as an exercise for the reader. The relation A and Heisenberg Uncertainty principle will be worthwhile here. Considering Eq (C) we can write

$C\psi_{jm} = \frac{1}{8}(H^2 - 1)\psi_{jm}$

$\frac{1}{2}j(j+1) = \frac{1}{8}(E^2 - 1)$

$E^2 = 4j^2 + 4j + 1 = (2j+1)^2$

$E = (2j+1) =$ Degeneracy number (in suitable units of energy)

This group theoretical conclusion supports the independent quantamechanical derivation, viz.

$$E_{n,m} = (n + \frac{1}{2})h\nu + (m + \frac{1}{2})h\nu$$
$$= (n + m + 1)h\nu$$
$$= (n + m + 1) = \text{Degeneracy number in units of } h\nu$$

Each quantum state is $(2j+1)$ i.e. $(n+m+1)$ fold degenerate.

The isotropic SHM of an oscillator in three dimensions belongs to SU(3)symmetry. Since we have not discussed the general features of SU(3) formulations, its description, from the viewpoint of symmetry, is skipped.

5. Hydrogen and hydrogen - like atoms

For theoretical treatment of atoms, "a single-electron spin orbital" model unit of architecture has been used for their $\psi-$ construction. On this basis, coupled with no magnetic interaction, all the atoms, including H atom, should belong to SU(2) or its isomorphic $R^*(3)$ symmetry group. But the singular so called "quantum property of accidental degeneracy" was, from group theory viewpoint, extremely baffling. All the atoms, in common with H-atom, have the commutator $[H, L] = 0$, suggesting one common constant of motion. In 1966, Hermann established that the Runge Lenz vector of H-atom commutes with the hamiltonian and forms a second constant of motion. This led Fock to suggest that H-atom might belong to a higher symmetry group R(4) rather than to SU(2)/R*(3) group.

There are two vectors, orbital angular momenta **L** and the Runge Lenz vector $\mathbf{R} = \frac{1}{2}(\mathbf{l}\times\mathbf{p} - \mathbf{p}\times\mathbf{l}) + Z\frac{\overrightarrow{\mathbf{r}}}{\mathbf{r}}$ giving (1) altogether six components of the constants of motions. (2) The six components provide six quantamechanical operators and six commutator relations - 3 of those in between the components of l and the other 3 CR's among the components of R

(3) The commutator relations are

$$
\begin{aligned}
&(a)\ [l_x, l_y] = il_z,\ \ [l_y, l_z] = il_x,\ \ [l_z, l_x] = il_y\\
&(b)\ [R_x, R_y] = -2iHl_z, [R_y, R_z] = -2iHl_x, [R_z, R_x] = -2iHl_y
\end{aligned}
\tag{12.1}
$$

H being the hamiltonian operator. These three aspects of the quantamechanical approach find their echo in group theoretical analysis of H-atom in the light of R(4) Lie group. (Sec 11.10)

1. There are six infinitesimal operators and six independent parameters (cf 1 and 2 above)

2. These six operators provide six commutation rules among the field operators A_1, A_2, A_3 and B_1, B_2, B_3 (cf. 2 above). The field operators span two (three-basis spanned) linear fields.

3. The six-commutator relations divide themselves into two sets, each comprising 3 CR's (cf. 3a and 3b above). These are, with $A_3 = -il_z$ etc.

$$
\left.
\begin{aligned}
&(a)\ [A_1, A_2] = A_3, [A_2, A_3] = A_1, [A_3, A_1] = A_2\\
&(b)\ [B_1, B_2] = A_3, [B_2, B_3] = A_1, [B_3, B_1] = A_2
\end{aligned}
\right\}
\tag{12.2}
$$

4. The two distinct sets of infinitesimal operators are linearly combined to give standard basic commutators, viz.,

$$
J_j = \frac{1}{2}(A_j + B_j) \text{and} K_j = \frac{1}{2i}(A_j - B_j)
\tag{12.3}
$$

It may be observed that while the l_i's and A_i's are almost identical except for (sign-involving) i factor the second set of commutator relations are merely proportional.

$$\left.\begin{aligned}[B_1, B_2] &= \left[\frac{iR_z}{\sqrt{-2H}}, \frac{iR_y}{\sqrt{-2H}}\right] = -il_z = A_3 \\ [B_2, B_3] &= \left[\frac{iR_y}{\sqrt{-2H}}, \frac{iR_z}{\sqrt{-2H}}\right] = -il_x = A_1 \\ \text{and } [B_3, B_1] &= \left[\frac{iR_z}{\sqrt{-2H}}, \frac{iR_x}{\sqrt{-2H}}\right] = -il_y = A_2\end{aligned}\right\} \tag{12.4}$$

where, evidently, $B_1 = iR_x/\sqrt{-2H}$, $B_2 = iR_y/\sqrt{-2H}$ and $B_3 = iR_z/\sqrt{-2H}$. From sec 12.2, we can write for the Casimir operator C.

$$C = -\frac{1}{4}[A_1^2 + A_2^2 + A_3^2 + B_1^2 + B_2^2 + B_3^2] \tag{12.5}$$

From (12.2) $A_1^2 = -l_x^2, A_2^2 = -l_y^2, A_3^2 = -l_z^2$

and from (12.4) $B_1^2 = R_x^2/2H, B_2^2 = R_y^2/2H, B_3^2 = B_3^2 = R_z^2/2H$

With these in view

$$C = \frac{1}{4}[l^2 - R^2/2H] \tag{12.6}$$

It can be shown that $R^2 = 2Hl^2 + 2H + Z^2$
(see worked out example)

With this substitution, (12.6) changes to

$$C = -\left(\frac{Z^2}{8H} + \frac{1}{4}\right) \tag{12.7}$$

From previous treatment (sec 12.2), it was shown group theoretically that the eigenvalue of C in R(4) symmetry is given by

$\frac{1}{4}(n^2 - 1)$, where $n^2 = (2j + 1)^2$ = degeneracy of the quantum eigenstate, Relation(12.7) thus demands that the operator $-\left(Z^2/8H + \frac{1}{4}\right)$ have an eigenvalue $\frac{1}{4}(n^2 - 1)$
i.e., the eigenvalue

$-Z^2/8H$ is $\frac{1}{4}n^2$

which is equivalent to H having an eigenvalue $-2Z^2/n^2$ for the energy of the H atom of R(4) symmetry in the quantum state n with n^2 fold degeneracy.

From R(3) symmetry viewpoint of H-atom, the 2s, 2p electronic states have the same energy and the quantum states are four-fold degenerate. Such a degeneracy of an s and p states of electron was "unexpected" from symmetry angle and was attributed to causes lying beyond the province of symmetry. The degeneracy was described as "accidental". Instead now it is realized that a H-atom belongs to R(4) and not R(3) or R*(3) and the 4-fold degeneracy is normal for a state in $(\frac{1}{2}\frac{1}{2})$ representation of R(4).

Example 12.1

The Range-Lenz vector in H-atom problem is given by
$R = \frac{1}{2}(\vec{l} \times \vec{p} - \vec{p} \times \vec{l}) + Z\frac{\vec{r}}{r}$, Show that it can also be written in equivalent forms, viz.,

$$\underline{R} = \frac{Z\vec{r}}{r} - \vec{r}p^2 + \vec{p}(\vec{r}.\vec{p});\ \underline{R} = \vec{p}(\vec{r}.\vec{p}) - 2\vec{r}H - \frac{Z\vec{r}}{r}$$

$R = \frac{1}{2}(r \times p \times p - p \times r \times p) + \frac{Z\vec{r}}{r}$

On applying the cross-product rule of triple vectors,

$$\begin{aligned} R &= \frac{1}{2}\left[-2rp^2 + 2p(r.p)\right] + \frac{Zr}{r} \\ &= Z\frac{r}{r} - rp^2 + p(r.p) \end{aligned}$$

Now the Hamiltonian in operator H for hydrogen like atom in atomic units

$$H = \frac{1}{2}p^2 - Z/r \text{ (On multiplying by the vector } \vec{r}\text{)}$$

On substituting this in the foregoing relation, we have

$$\vec{R} = \vec{p}(\vec{r}.\vec{p}) - 2\vec{r}H - \frac{Z\vec{r}}{r}$$

Example 12.2

Show that $R^2 = 2Hl^2 + 2H + Z^2$

Hints: We have already utilized an expression for R^2 involving the B operator in arriving at relation (12.6). Now try to extract a relation for R^2 from the Runge Lenz vector R^2 which appears in three alternative forms. Then on utilizing the expression for the hamitonian operator (in atomic units), you will achieve the desired result.

Example 12.3

Show that in a two-dimensional (isotropic) simple harmonic oscillator problem (sec 12.3 item 4), the following relations are valid:

[a] Hamiltonian operator $H = \frac{1}{2}(P^2 + Q^2) - (p_1p_2 + q_1q_2)$ where $P = p_1 + p_2$ and $Q = q_1 + q_2$

[b] Casimir operator $C = \frac{1}{8}(H^2 - 1)$, where C is defined as $C = -\frac{1}{2}(K^2 + L^2 + D^2)$, the expressions for K, L and D are given in item 4 (sec 12.3)

[a] $P.P = (p_1 + p_2).(p_1 + p_2) = p_1^2 + p_2^2 + (p_1p_2 + p_2p_1)$

$P^2 = p_1^2 + p_2^2 + 2p_1p_2 = (p_1 + p_2)^2$

Similarly $Q^2 = q_1^2 + q_2^2 + 2q_1q_2$

$\frac{1}{2}(P^2 + Q^2) = \frac{1}{2}(p_1^2 + p_2^2 + q_1^2 + q_2^2) + (p_1p_2 + q_1q_2)$

$$\begin{aligned} \text{i.e.}\frac{1}{2}(P^2 + Q^2) - (p_1p_2 + q_1q_2) &= \frac{1}{2}(p_1^2 + p_2^2 + q_1^2 + q_2^2) \\ &= H, \text{(in atomic units)} \end{aligned}$$

which is the form of the hamiltonian (in atomic units) for a two dimensional isotropic oscillator.

[b] Using the expressions for K, L and D we find, after a bit of labour and with due care for Heisenberg commutation principle, viz. $[p_i, q_i] = -i$ (in units of $\hbar$)

$$\begin{aligned} L^2 + K^2 + D^2 &= -\frac{1}{16}[\{p_1^4 + p_2^4 + q_1^4 + q_2^4 + 2p_1^2p_2^2 + 2q_1^2q_2^2 + 2p_1^2q_2^2 + \\ &\quad 2q_1^2p_2^2 + p_1^2q_1^2 + q_1^2p_1^2 + p_2^2q_2^2 + q_2^2p_2^2\} + \\ &\quad (4p_1p_2q_1q_2 + 4q_1q_2p_1p_2 - 4q_1p_2q_2p_1 - 4q_2p_1q_1p_2)] \\ &= -\frac{1}{16}\left[\{p_1^2 + p_2^2 + q_1^2 + q_1^2\}^2 + \{(4p_1p_2q_1q_2 - 4q_1p_2q_2p_1) + \right. \\ &\quad (4q_1q_2p_1p_2 - 4q_2p_1q_1p_2)\}] \end{aligned}$$

$$= -\frac{1}{16}\left[4\mathrm{H}^2 + 4(\mathrm{p}_1\mathrm{q}_1 - \mathrm{q}_1\mathrm{p}_1)\mathrm{p}_2\mathrm{q}_2 + 4(\mathrm{q}_1\mathrm{p}_1 - \mathrm{p}_1\mathrm{q}_1)\mathrm{q}_2\mathrm{p}_2\right]$$
$$= -\frac{1}{4}\left[\mathrm{H}^2 - i\mathrm{p}_2\mathrm{q}_2 + i\mathrm{q}_2\mathrm{p}_2\right] = -\frac{1}{4}[\mathrm{H}^2 - 1],$$
using commutation principle (12.8)

Casimir operator
$\mathrm{C} = -\frac{1}{2}(\mathrm{L}^2 + \mathrm{K}^2 + \mathrm{D}^2) = \frac{1}{8}(\mathrm{H}^2 - 1)$

12.4 Atomic Symmetry Group $SU(2)/R^*(3)$– Applications in Angular Momenta Aspects

The main results of these applications have already been given in the shape of "Clebsch Gordan Theorem" (see example 11.2 Eqs. (11.28, 11.41) and especially sec 11.8). The salient features of a tensor product representation are

[1] Formation of a linear product space spanned by binary product of basis kets.

[2] Decomposability of reducible direct product representation into a direct sum of IR's, leading to block diagonalisation of the relevant part of the Hilbert space.

[3] The dimensionalities of the IR's confirm the J values, the net possible angular momenta that may result from the addition of j_1, j_2 in direct product $(j \times j)$ i.e., the addition theorem of component angular momenta vectors.

We shall demonstrate here the technique of decomposing the product space (cf. salient feature 2) into linear combination of the basis kets of the IR with the help of very elementary examples

Secondly we shall show how to obtain the coefficients (called coupling coefficients) in the addition of basis set of product space (cf. salient feature no. 3) into another basis unit with larger value of angular momentum.

(a)Decomposition: of Product space

Consider the product space spanned by two spin kets of total spin 1 (S = J = 1) and $\frac{1}{2}$(S = J = $\frac{1}{2}$). The basis set spin kets involved are $|\,1, m_s>$ and $|\,\frac{1}{2}, m_s>$.

The highest weight state

$$|\frac{3}{2},\frac{3}{2}>=|\frac{1}{2},\frac{1}{2}>|1,1>\cdots\cdots(A1)$$

The corresponds to J value $\frac{3}{2}$. Since the J values can range from (j_1+j_2) to $|(j_1-j_2)|$, the possible J values are $\frac{3}{2}$ and $\frac{1}{2}$. The state J = $\frac{3}{2}$ having been furnished by A_1 the other possible highest weight state distinct from and orthogonal to A_1 is

$$|\frac{1}{2},\frac{1}{2}>=|\frac{1}{2},-\frac{1}{2}>|1,1>\cdots\cdots(B1)$$

From A, we shall obtain the other $\frac{3}{2}$ states by applying the lowering operator J_-.

$J_-\,|\,\frac{3}{2},\frac{3}{2}>=J_-\left(|\,\frac{1}{2},\frac{1}{2}>|\,1,1>\right)$

[Note: J_- on the RHS is equivalent to $(J_{1-}+J_{2-})$]

$$\sqrt{\tfrac{3}{2}\left(\tfrac{3}{2}+1\right)-\tfrac{3}{2}\left(\tfrac{3}{2}-1\right)/2}\;|\tfrac{3}{2},\tfrac{1}{2}> = \sqrt{\tfrac{1}{2}\left(\tfrac{1}{2}+1\right)-\tfrac{1}{2}\left(\tfrac{1}{2}-1\right)/2}\times$$

$$|\frac{1}{2},-\frac{1}{2}>|1,1>+\sqrt{2-1(1-1)/2}|\frac{1}{2},\frac{1}{2}>|1,0>$$

$$|\tfrac{3}{2},\tfrac{1}{2}>=\sqrt{\tfrac{2}{3}}|\tfrac{1}{2},-\tfrac{1}{2}>|1,1>+\sqrt{\tfrac{1}{3}}|\tfrac{1}{2},\tfrac{1}{2}>|1,0>\ \ldots \qquad (A_2)$$

Continuing with J_- operation

$$\sqrt{\frac{3}{2}\left(\frac{3}{2}+1\right)-\frac{1}{2}\left(\frac{1}{2}-1\right)/2}|\frac{3}{2},-\frac{1}{2}>$$

$$=\sqrt{\frac{4}{3}}|\frac{1}{2},\ -\frac{1}{2}>|1,0>+\sqrt{\frac{2}{3}}|\frac{1}{2},\frac{1}{2}>|1,-1>$$

i.e., $$|\tfrac{3}{2},-\tfrac{1}{2}>=\sqrt{\tfrac{2}{3}}|\tfrac{1}{2},-\tfrac{1}{2}>|1,0>+\sqrt{\tfrac{1}{3}}|\tfrac{1}{2},\tfrac{1}{2}>|1,-1>\ \ldots \qquad (A_3)$$

A J_- operation once more, brings us to the final stage

$$|\frac{3}{2},-\frac{3}{2}>=|\frac{1}{2},-\frac{1}{2}>|1,-1>\ \ldots \qquad (A_4)$$

So, A_1, A_2, A_3 and A_4 corresponding basis sets occur in the reduced product space

Before we apply J_- on to next highest weight state $|\frac{1}{2}, \frac{1}{2}>$ we have to write B_1 in a form orthogonal to the A_i states

$$|\tfrac{1}{2}, \tfrac{1}{2}> = \sqrt{\tfrac{2}{3}}|\tfrac{1}{2}, -\tfrac{1}{2}>|1,1> - \sqrt{\tfrac{1}{3}}|\tfrac{1}{2}, \tfrac{1}{2}>|1,0> \ldots \qquad (B_2)$$

Applying J_- on B_2

$$|\tfrac{1}{2}, -\tfrac{1}{2}> = \sqrt{\tfrac{1}{3}}|\tfrac{1}{2}, -\tfrac{1}{2}>|1,0> - \sqrt{\tfrac{2}{3}}|\tfrac{1}{2}, \tfrac{1}{2}>|1,-1> \ldots \qquad (B_3)$$

Thus the basis set of product kets of the block-diagonally reduced overall J states of the relevant Hilbert product space are obtained.

(b) **Coupling of Two Angular Momenta & Coupling Coefficients, Change of Basis set**

Attention need be paid to sec. 11.8 and especially to items noted under (3) and (4) of that section.

We are to find each member that $|j_1 j_2 JM>$ of the block diagonally reduced J state expressed as a linear combination of the component product kets $|j_1 m_1>|j_2 m_2>$. This reduced J state is one of the coupled states and, hence, the coefficients appearing in the linear combination are called coupling coefficients. This is just a change of basis set and not fundamentally different from decomposition described under (a). Hence is the mathematical content essentially the same as in (a)

We write

$$|j_1 j_2 JM> = \sum_{M=m_1+m_2} C_{m_1 m_2}|j_1 m_1>|j_2 m_2>$$

Here $|j_1 j_2 JM>$'s and $|j_1 m_1>|j_2 m_2>$'s are two different basis sets for the same reduced state $|J|$ of A_1 algebra of SU(2) group. J is one of the possible values in the range $(j_1 + j_2) \ldots |(j_1 - j_2)|$ and $M = m_1 + m_2$ and $J_3|JM> = M|JM>$ $(J_3 \equiv J_z)$

$< j_1 m_1 j_2 m_2|j_1 j_2 JM> = C_{m_1 m_2}$, since $< j_1 m_1 j_2 m_2|j_1 m_1>|j_2 m_2> = 1$ and the rest of the integrals on the RHS vanish due to orthogonality. It should be noted that $|j_1 m_1 j_2 m_2>$ is a second optional way of expressing $|j_1 m_1>|j_2 m_2>$. This option is also valid for bra notation.

We can thus write

$$|j_1 j_2 JM> = \sum_{M=m_1+m_2} < j_1 m_1 j_2 m_2|j_1 j_2 JM>|j_1 m_1>|j_2 m_2>$$

We now try to concretise the problem $J = 2$ with $j_1 = 1$ and $j_2 = 2$ (fixed values) M=2, highest weight when $J = 2$. m_1 and m_2 take variable values in the bra $< j_1 m_1 j_2 m_2|$ and the kets $|j_1 m_1 > |j_2 m_2 >$ in conformity with the fixed M value = 2. $J_{\pm} = (J_1 \pm iJ_2)/\sqrt{2}$ and also $J_{\pm} = (J_{1\pm} + J_{2\pm})$. This specific problem is worked out in the following way

$$J_+ = \sqrt{\tfrac{1}{2}}\sqrt{J(J+1) - M(M+1)} \mid J + M + 1 >$$

$$J_{1+} = \sqrt{\tfrac{1}{2}}\sqrt{j_1(j_1+1) - m_1(m_1+1)} \mid j_1 + m_1 + 1 >, \text{ m= suitable value}$$

$$J_{2+} = \sqrt{\tfrac{1}{2}}\sqrt{j_2(j_2+1) - m_2(m_2+1)}|j_2 + m_2 + 1 >, m_2 = \text{suitable value}$$

$$\left.\begin{array}{l} J_+|1,2,22 >= 0 =< 1022|1222 > \{J_{1+}|1,0 > |22 > +|1,0 > J_{2+}|22 >\} \\ \qquad + < 1121|1222 > \{J_{1+}|11 > |21 > +|11 > J_{2+}|21 >\} \end{array}\right\} \quad ...(A)$$

Of the four operations with J_{1+} and J_{2+} in (A) the second and the third vanish, since

$$J_{2+}|22 >= 0 = J_{1+}|11 >$$

The surviving non zero operations are worked out

$$\begin{aligned} J_{1+}|10 > |22 > &= \frac{1}{\sqrt{2}}\sqrt{1(1+1) - 0(0+1)}|11 > |22 > \\ &= |11 > |22 > \\ |11 > J_{2+}|21 > &= \frac{1}{\sqrt{2}}\sqrt{2(2+1) - 1(1+1)}|11 > |22 > \\ &= \sqrt{2}|11 > |22 > \end{aligned}$$

Inserting these in A, we get

$$\left[< 1022|1222 > +\sqrt{2} < 1121|1222 >\right] |11 > |22 >= 0... \quad (B)$$

Since the kets $|11 > |22 >$ are each nonzero,

it follows

$$< 1022/1222 >= -\sqrt{2} < 1121|1222 > ... \quad (C)$$

The normalization condition requires

$$< 1022|1222 >^2 + < 1121|1222 >^2 = 1$$

i.e. from (C)

$$2 < 1121|1222 >^2 + < 1121|1222 >^2 = 1$$

$$\text{i.e.} < 1121|1222 >^2 = \frac{1}{3}$$

So the normalising coefficients C_{11} and C_{02} are respectively,

$$< 1121|1222 > = \frac{1}{\sqrt{3}} \text{ and } < 1022|1222 > = \sqrt{2}/\sqrt{3}$$

For

$$|j_1 j_2 JM > = \sum_{M=m_1+m_2} C_{m_1 m_2} |j_1 m_1 > |j_2 m_2 >$$

we can write now

$$\begin{aligned} |1222 > &= C_{02}|11 > |22 > + C_{11}|11 > |22 > \\ &= \sqrt{\frac{2}{3}}|11 > |22 > + \frac{1}{\sqrt{3}}|11 > |22 > \\ &= \frac{1}{\sqrt{3}}(\sqrt{2} + 1)|11 > |22 > \end{aligned}$$

The converse process of expressing
$|j_1 m_1 > |j_2 m_2 >$ i.e., $|j_1 m_1 j_2 m_2 >$ in terms of $|j_1 j_2 JM >$'s can similarly be accomplished

In molecules, all of which sans $C_{\infty v}$ and $D_{\infty h}$, belong to finite point groups, orbital angular momenta are more or less totally quenched. However, in their cases, coupling of representations, instead of coupling of momenta, may be effected with accompanying coupling coefficients.

In quantum chemistry we are usually concerned with linear transformations which convert one orthonormal basis set into another orthonormal basis set. Let us try to understand concretely how far is this maxim obeyed in the two applications (sec 12.4 a,b) of atomic symmetry.

Let us consider a tensor product representation $(j_1 \times j_2)$ involving 2 electrons with $j_1 = 3$ and $j_2 = 2$. The full product space will be (2.3+1) $(2.2 + 1) = 35$ - dimensional spanned by 35 binary product basis functions like $|j_1 m_1 > |j_2, m_2 >$.

Clebsch Gordan theorem tells us that this 35 dimensional binary - product basis space can be diagonalised into block-diagonal forms characterized by total J values $(3 + 2)...|(3 - 2)|$ i.e. J values 5,4,3,2,1. The matrices of these diagonalised blocks will be 11-, 9-, 7-, 5- and 3- dimensional (total 35) spanned by 11, (9, 7, 5, 3) (new) binary product basis functions. (new in the sense that each is either the same or some linear combination), determined by the highest weight consideration (m =j). For example, the block diagonal (J = 3), which is 7-dimensional, has each base function a binary product function or some linear combination of two binary product functions. All these 7-basis set functions comprise product components that originally are present in the pack of 35 binary product basis.

Application of atomic symmetry 12.4(a) is made at this stage to find the resultant degenerate set of 7 functions of the reduced (J = 3) in terms of the linear combination of binary product functions of the basis set. Application of sec 12.4(b) has its first part substantially the same as in 12.4(a). Additionally, however, it suggests the feasibility of the change of basis set for the reduced space J=3. One can express each binary product of the 7-basis set as a linear combination of the seven degenerate resultant functions of the (J = 3) reduced space. In both 12.4(a) and 12.4(b) addition of angular momenta is involved intrinsically.

There are other applications of atomic symmetry, viz., passage to the IR's of molecular symmetry group from the IR's of atomic symmetry group via a process called branching of IR's, The principle is the same as discussed under molecular symmetry groups. Moreover, atomic symmetry concepts are helpful in the classification of polyelectronic atomic states. These are left out here.

Chapter 13

Symmetry And Stereochemistry Of Reactions

Reactions which occur in concert e.g., dimerisations of olefinic compounds, cyclo additions, many Diels Alder reactions, electro-cyclic reactions, sigmatropic reactions etc., have been explained on the basis of stereospecitic mechanisms in which symmetry conservation plays a vital role. We shall just confine ourselves to a certain typical example of an electrocyclic reaction first to exhibit the role of symmetry in deciding the mechanism. Electrocyclic reactions may be defined as those in which bonding of the two terminal carbon atoms of a polyene occurs to form a ring or the reverse of it to form a polyene. Concrete illustrations typifying electro-cydic reactions are the conversions of cis-butadiene into cyclobutene and of hexatriene into cyclohexadiene.

Sometimes this cyclisation occurs in parts of a bigger molecule such as ring closure of previtamin D to ergosterol. Often again substituents may be present in conjugated polyenes which undergo electrocyclic reactions. We shall consider the reaction

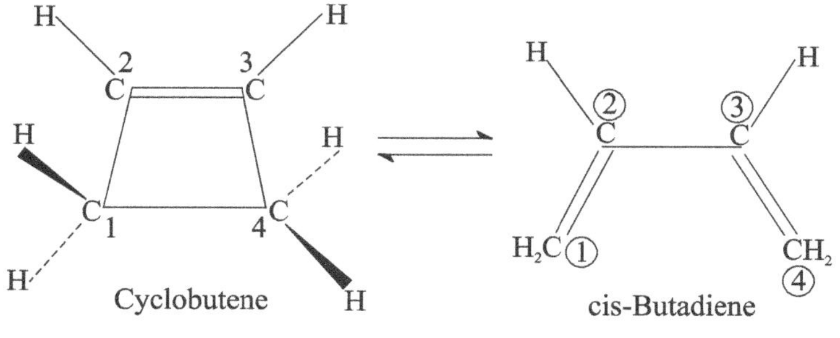

13.1 Molecular Orbital Background.

Before one delves into investigating the role of symmetry in unravelling the mechanism of conversion of cyclobutene into butadiene, it is necessary to be equipped with relevant detail of the mo's of the reactant and the product.

cis-Butadiene: It has the C_{2v} symmetry and its $\pi-$mo's are found (on reduction of the reducible representation) to conform to the symmetries of A_2 and B_1. The forms of $\pi-$symmetry orbitals (cf. trans-butadiene Chap. 9) are

$$\psi_1 = \frac{1}{2}(\phi_1 + \phi_2 + \phi_3 + \phi_4) \text{ symmetry } B_1$$
$$\psi_2 = \frac{1}{2}(\phi_1 + \phi_2 - \phi_3 - \phi_4) \text{ symmetry } A_2$$
$$\psi_3 = \frac{1}{2}(\phi_1 - \phi_2 - \phi_3 + \phi_4) \text{ symmetry } B_1$$
$$\psi_4 = \frac{1}{2}(\phi_1 - \phi_2 + \phi_3 - \phi_4) \text{ symmetry } A_2$$

We can thus compile the following table for cis-butadiene.

Orbital symmetry (IR)	$\pi-$symmetry^2 orbital	$\pi-$electronic configuration	Overall state symmetry (from Direct product)
B_1	$\psi_1(b_1)$	(i) Ground state $= (b_1)^2(a_2)^2 = \psi_1^2\psi_2^2$	A_1
A_2	$\psi_2(a_2)$	(ii) First excited state $b_1^2 a_2 b_1^* = \psi_1^2\psi_2\psi_3$	B_2
B_1	$\psi_3(b_1)$	(iii) Another excited state $b_1^2(b_1^*)^2 = \psi_1^2\psi_3^2$	A_1
A_2	$\psi_4(a_2)$	(iv) Excited state $= (b_1)(a_2)^2 a_2^* = \psi_1\psi_2^2\psi_4$	B_2

Cyclobutene: In the valence shell basis, cyclobutene has altogether 22 atomic orbitals comprising six-1s H-atom orbitals and four atomic orbitals (one $2s$ and $2px$, $2py$, $2pz$) from each of the four C-atoms. The molecule belongs to C_{2v} point group. Using the 22 atomic orbitals as the basis, the following reducible representation is obtained.

C_{2v}	E	C_2	$\sigma_v^{(xy)}$	$\sigma_v^{(yz)}$
χ^{red}	22	0	0	14

On employing Theo. 5.8 and using the character table of C_{2v} it is found

$$\Gamma^{red} = 9A_1 \oplus 2A_2 \oplus 2B_1 \oplus 9B_2$$

These 22 mo's in principle extend over the whole nuclear framework, but in practice these orbitals are (or are rendered into, if necessary, with the help of a suitable unitary transformation of the basis set) localised mo's fitting with our chemical concepts of practically two-centre σ and π orbitals. In this context, the lower $\sigma(a_1)$ orbital is σ-bonding, the upper a_2 is antibonding π^*, The lower b_1 is π-bonding, the upper b_2 is antibonding σ^*. One may thus have the following table for cyclobutene in respect of the π and their nearest σ-orbitals.

Symmetry of mo's	Relevant orbitals	Electronic configuration of the two upper occupied orbitals (excluding the low lying nine orbitals)	Overall state symmetry (From Direct Product)
A_1	$\sigma(a_1)$	Ground state=	A_1
B_1	$\pi(b_1)$	$(a_1)^2(b_1)^2 = \sigma^2\pi^2$	
A_2	$\pi^*(a_2)$	First excited state=	B_2
B_2	$\sigma^*(b_2)$	$(a_1)^2 b_1 a_2 = \sigma^2\pi\pi^*$	

13.2 Symmetry Control Of Electrocyclic Reactions

The reactions of the above category preserve one or other symmetry element during their course. Consider the reaction Cis-Butadiene. (see page 273)

$$\text{Cyclobutene} \longrightarrow \text{Cis} - \text{Butadiene}.$$

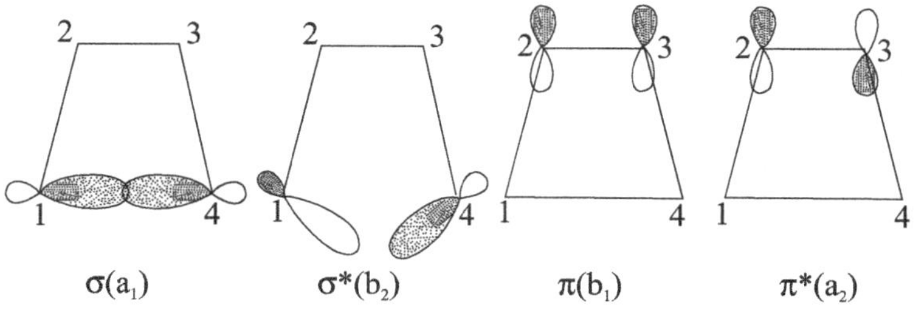

Fig. 13.1: Boinding and antibonding mo's in cyclobutene. The dotted portions represent the positive lobes of the constituent atomic orbitals

This simple conversion involves in the forward direction the opening of the ring. The vital bond between 1C–4C must break, i.e., must disgorge the constituted atomic orbitals which ought to rotate in suitable direction to form the π-orbitals in butadiene. Fig. 13.1 depicts the bonds in the four mo's of cyclobutene.

It is to be remembered that the planar molecule, belonging to C_{2v}, has the C_2 axis (and hence the z-axis) in the plane of the paper. The π-bond is due to the overlap of two p_x orbitals normal to the paper plane.

Fig. 13.2 shows the $\pi-$orbitals in the four symmetry orbitals of cis-butadiene.

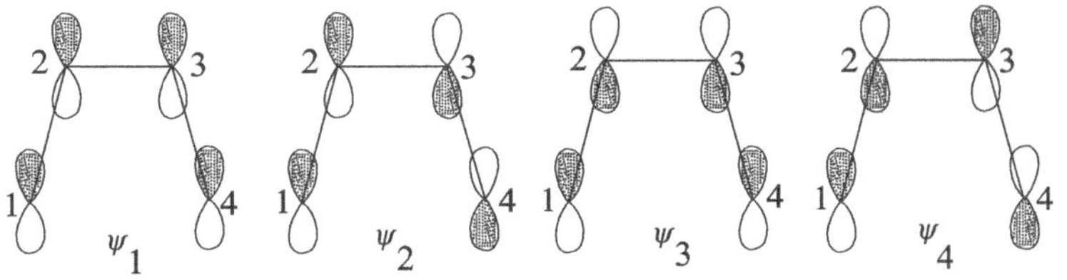

Fig. 13.2: The π-symmetry orbitals in butadiene. The p_s orbital forming the π bonds are normal to the paper plane. Overlap are there, but not shown

Now the imaginable ways of opening of the cyclobutene ring are either by

(i) Conrotation (i.e. rotations in the same sense. Fig. 13.3) of the newly disengaged atomic orbital components of the erstwhile 1C-4C σ-bond and simultaneous readjustment of the π-bond. or by

(ii) disrotation (i.e., rotations in opposite sense, Fig. 13.3) of the newly disengaged atomic orbital components of the erstwhile 1C-4C σ-bond.

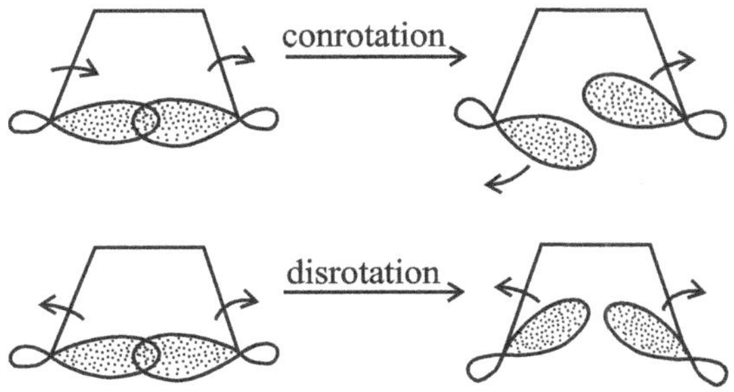

Fig. 13.3: Conrotation and disrotation of the 1C-4C bond. The separated orbitals tend to assume normal orientation finnaly with respect to paper plane.

It should be noted that cyclobutene (reactant) and butadiene (product) both belong to C_{2v} point group and possess all the symmetry elements of the group. The point groups of the arrangements of the nuclei, the relevant disengaged atomic orbitals and the incipient molecular orbitals in the series of intermediate conformations including the transition state do not retain all the symmetry elements of the original point group C_{2v}. However, one or the other symmetry element is preserved throughout the entire reaction course depending on the nature of the process, viz., conrotation or distrotation.

The symmetry rules, given by Woodworth and Hoffmann, for the actual mechanism are summed up as

1. During the entire process of transformation, there should be a conservation of one symmetry element.
2. The orbital symmetries with respect to this conserved symmetry element remain unchanged (conserved) in the transformation. An orbital which is symmetric (S) with respect to the conserved symmetry element remains so all through until it merges (correlates) with a symmetric orbital of the product. Similarly an antisymmetric one (A) merges with an antisymmetric orbital.
3. An orbital symmetry correlation diagram can thus be drawn up in which the noncrossing rule will apply.

These three symmetry rules are supplemented by a couple of energy rules:

(a) In thermal reaction, a ground state configuration of the reactant must correlate with ground state configuration of the product in the orbital symmetry correlation diagram.

(b) In photochemical reaction, the first excited state configuration of the reactant should merge into the first excited state of the product.

In controtation, it is the C_2 symmetry axis which is conserved in all the conformations starting from the initial cyclobutene to the end material butadiene. The behaviours of the orbitals
(σ, π, π^*, σ^*, ψ_1, ψ_2, ψ_3, ψ_4) and also of the newly dissengaged atomic

orbitals as symmetric or antisymmetric with respect to C_2 operations are only considered.

On the other hand in disrotation it is the σ_v^{xz} which is conserved in all the stages from the starting material to the end product. The behaviours of the orbitals and also of their successive evolutions during transition with respect to σ_v^{xz} are only to be judged.

The reactant-product orbital symmetry correlation diagram from the initial ground state configuration of cyclobutene is shown in Fig. 13.4 both for conrotation and disrotation mechanisms from which the final selection will be made.

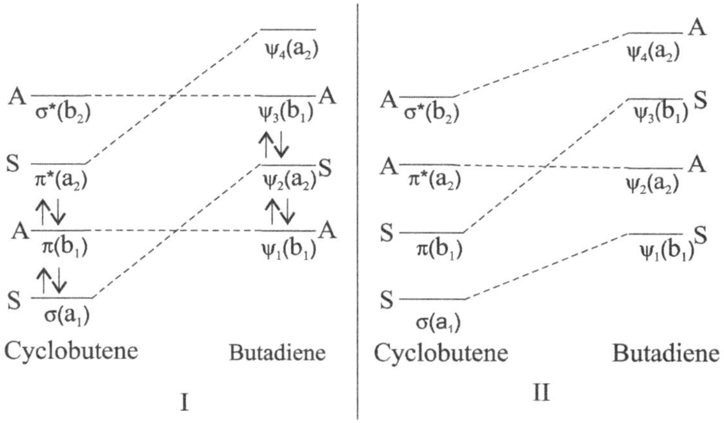

Fig. 13.4: Orbital symmetry correlation diagram of cyclobutene-butadiene. I Conroration. II Disrotation. Initial state is the ground configuration of cyclobutene.

The following observations may be made from the two diagrams.

1. Since in disrotation mechanism, the end product is energetically much higher, this cannot be the true mechanism for thermal conversion. The conrotation, in which the initial ground state converts into the final ground stale, represents *the true mechanism.*

2. The overall state correlations both in conrotation and in disorotations are $A_1 \longrightarrow A_1$, (cf. the tables of Sec. 13.1).

3. In the diagrams, an S (symmetric) orbital correlates with an S orbital and an A with A. One should not feel confused when he finds in I (Conrotation) that an a_1, orbital merges, into a_2. It is to be remembered that only the behaviour with respect to C_2 is all

that needs be considered. This is denoted by 'a', the subscripts 1 and 2 in a_1 and a_2 refer to the behaviour with respect to σ_v^{xz} which is a redundant and an inappropriate consideration in conrotation.

4. Exactly similar arguments apply to disrotation where the subscripts matter and not the alphabets a, b.

   We now consider the orbital symmetry correlation diagrams for conrotation and disrotation starting from the first excited state of cyclobutene to ascertain the mechanism of photochemical conversion (Fig. 13.5).

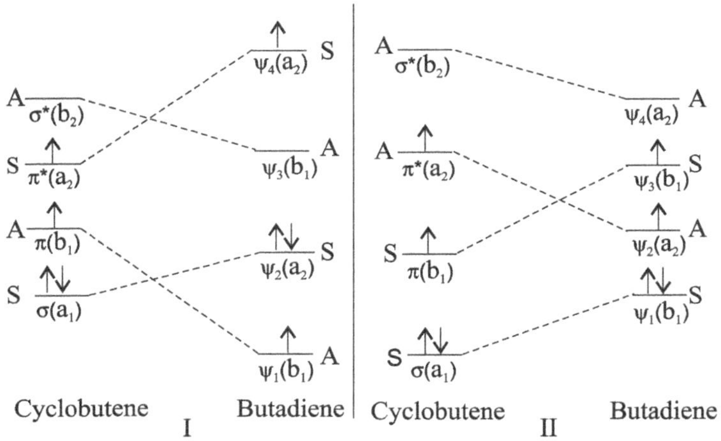

Fig. 13.5: Orbital symmetry correlation diagram. Initial state is the first state of cyclobutene. I-Conrotation II-Distrotation

It is seen that conrotation leads to highly excited state which for energy reasons is apprarently unfavourable. The disrotation technique leads from the first excited state of cyclobutene to the first exited state of butadiene and is manifestly the true path in photochemical conversion.

The overall state correlations, both in conrolation and disrotation are the same, viz., $B_2 \longrightarrow B_2$. The reverse process of ring closure of butadiene will similarly be found to be one of conrotation for the thermal conversion and of disrotation in photochemical transformation.

Orbital symmetry control of the stereochemical course of electrocyclic reactions at once enables us to predict the stereospecific products when substituents are present. Thus the thermal isomerisation of cis 1,4-dimethyl cyclobutene gives cis, trans 2,4-hexadiene. This is what is to be expected if the thermal opening of the ring is accompanied by a conrotation which we know to be the one operative for cyclobutene (and its derivative), [see fig. p-280]

The incorporation of the substituted derivatives within the range of discussion of symmetry control often presents an apparent difficulty, viz., the symmetry is often lost. However, in such case the symmetry is considered without reckoning the presence of the substituents. This is quite a permissible procedure if the substituents do not exert a large perturbation to alter the stereochemical course of the reaction.

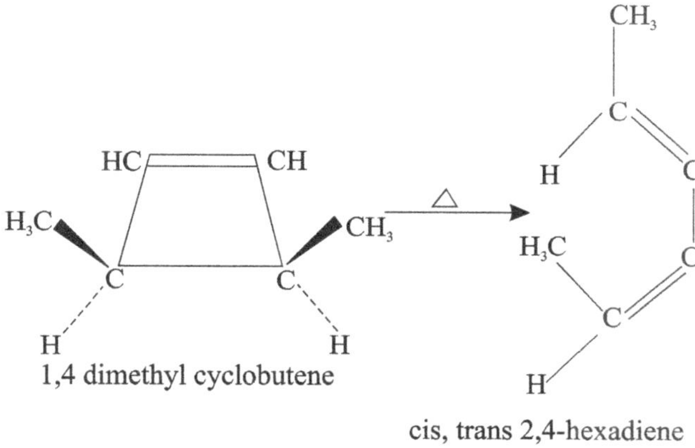

The construction of orbital symmetry correlation diagram is somewhat time consuming. There is, however, a thumb rule to predict quickly the stereochemical course of symmetry-allowed *electrocyclic reactions.* The rule is

"The symmetry of HOMO (highest occupied molecular orbital) in the open chain partner determines the stereochemical course". If the HOMO lacks a σ_v (symmetry A_2, B_2), but possesses a C_2, the process will be conrotatory. If it is the other way round, a disrotation will ensue.

Consider the reaction

$$\text{Cyclohexadiene} \rightleftharpoons \text{Hexatriene.}$$

The π-symmetry orbitals in the open chain hexatriene are ψ_1, ψ_2, ψ_3, ψ_4, ψ_5 and ψ_6. In the ground state the HOMO is ψ_3, which has two nodes. This indicates the HOMO has the same sign at the terminal

C-atoms and necessarily also at the other pairs [viz, 2 & 5 and 3 & 4] of C-atoms [cf. Fig. 13.2]. The presence of σ_v is thus indicated.

The conversion, which is thermal (since the ground state is involved), is thus disrotatory in contrast to cyclisation of butadiene. The first excited state of hexatriene involves ψ_4 as the HOMO for the photochemical conversion. This symmetry orbital has three nodes, and, therefore, lacks a σ_v but possesses a C_2. The photochemical conversion of hexatriene is thus predicted to be con-rotarory in nature, again in contrast to what happens in the butadiene case. We thus detect an alternation of mechanisms as the polyene chain is ascended.

In addition to the electrocyclic reactions, symmetry plays a controlling role in cycloaddilions, Diels Alder and sigmatropic processes.

13.3 Symmetry and Cycloaddition Reactions

Symmetry plays its role in the prediction of the streochemical course of Cycloaddition reactions including Diels Alder reactions. In a Cycloaddition, there is produced a cyclic compound from two acyclic components which may be of diverse types including dienes and dienophiles. The Cycloaddition reactions are classified from the standpoint of three characteristics considered collectively.

(i) The number of electrons which participate in the new bondings. It is denoted by an integer.

(ii) The nature of the bonds (π or σ) which are reorganised or converted during the new bond formations. It is represented by a prefix π or σ.

(iii) The nature of stereodispositions of the overlapping orbital lobes during Cycloaddition. This suprafacial or antara-facial approach for bond formation is symbolised by a suffix "*s*" or "*a*" respectively.

The reaction between two ethylene units to form cyclobutane is represented as $\pi^2S + \pi^2S$. The arrows in the figure below indicate suprafacial approaches in both the units for bond formation

Another example is the cycloaddition of ethylene to butadiene for which $\pi^2S + \pi^4a$, $\pi^2a + \pi^4S$, $\pi^2S + \pi^4S$ represent some processes among the different possibilities.

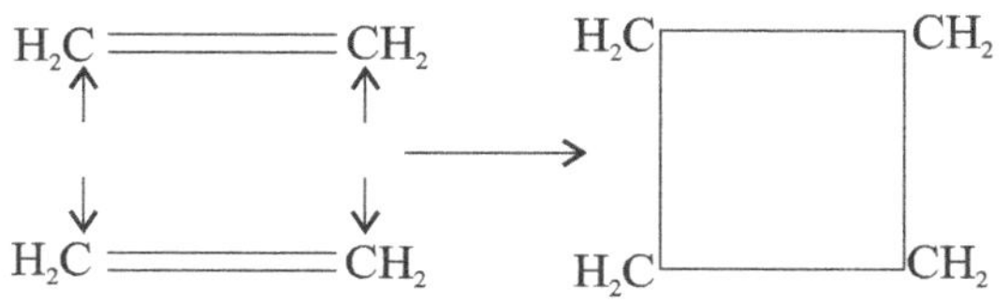

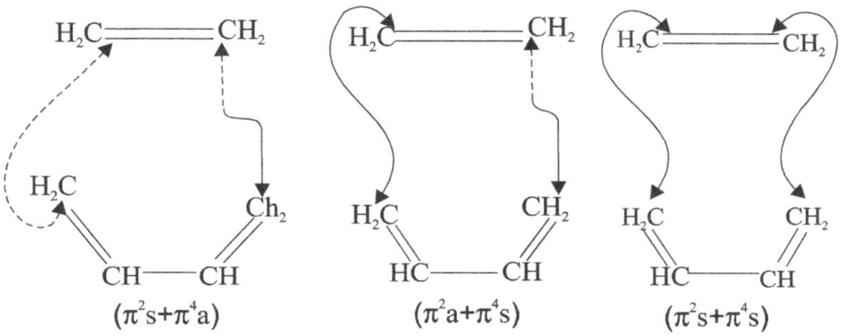

Without indulging into further examples, let us focus our attention on the symmetry aspects only in the cycloaddition of two ethylene units. Let one ethylene molecule approach a second one from a vertical direction. The relative approaches may be of the following types for a supra-supra addition

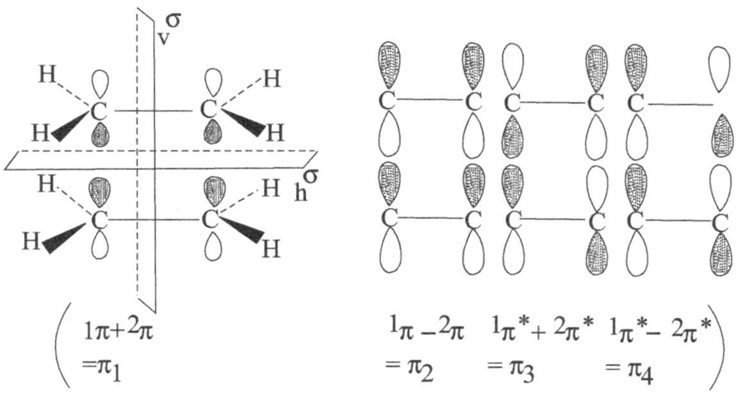

All these reaction schemes pertain to $(\pi^2S + \pi^2S)$ cycloadditions. when the $\pi-$orbitals are fully converted into σ-bonds, the resultant cyclobutanes will have the following natures of overlaps.

Now, the symmetry principles are

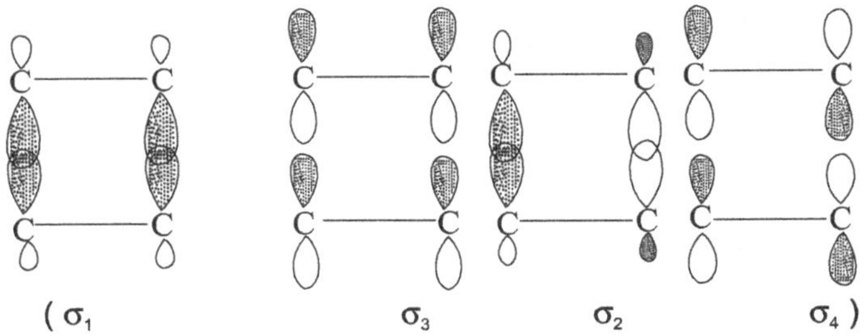

(i) The cycloaddition reactions are characterised by the conservation of one or more symmetry planes of the interacting units and the product.

(ii) The symmetry of the orbitals with respect to the conserved symmetry plane (or planes) remains unchanged throughout the reaction. In other words an antisymmetric (A) orbital merges into an antisymmetric (A) orbital of the product. Similarly a symmetric (S) orbital correlates with the product orbital of symmetric nature.

(iii) One can tnus draw an orbital symmeiry correlation diagram keeping in view the non-crossing rule. This correlation diagram tells which reaction is feasible and which is not.

(iv) The energy rules similar to those stated for electrocyclic reactions are valid.

In the approach of two ethylene units, the configuration has two reflecting planes σ_v and σ_h which remain conserved throughout. The symmetry of an incipient orbital, say, π_4, is antisymmetric (A) with respect to both σ_v and σ_h. The final product orbital σ_4 of cycfobutane that is obtained from it is also antisymmetric both with respect to σ_v and σ_h. In the case of π_2, the orbital is S with respect to σ_v and A with respect to σ_h. Conservation of orbital symmetry demands that the product orbital generated from it be similarly S and A under σ_v and σ_h operations. It is found from the figure to be the σ_3 orbital of cyclobutane. We shall now draw two orbital symmetry correlation diagrams one for the ground state and the other for the first excited state of two reacting ethylene units with incipient bonds (Fig. 13.6 and 13.7).

Since the ground slate configuration $\pi_1^2\pi_2^2$ leads to the excited state $\sigma_1^2\ \sigma_3^2$, this thermal supra-supra cycloaddition is ruled out on energy

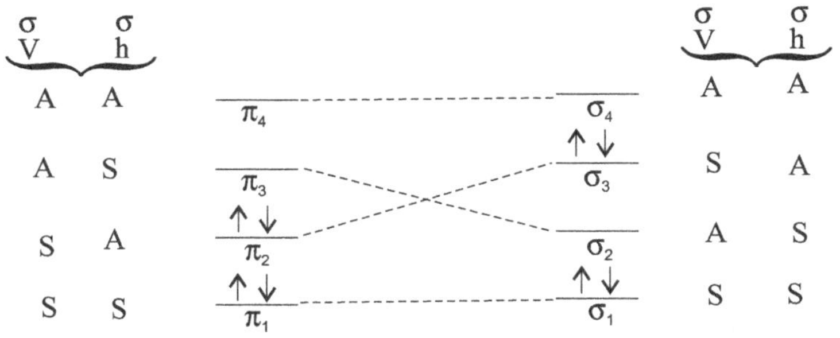

Fig. 13.6: Correlation diagram for cycloaddition (supra-supra) of two ethylene units. Ground states-Excited state.

grounds. The same correlation diagram (Fig. 13.7) drawn for the excited state is

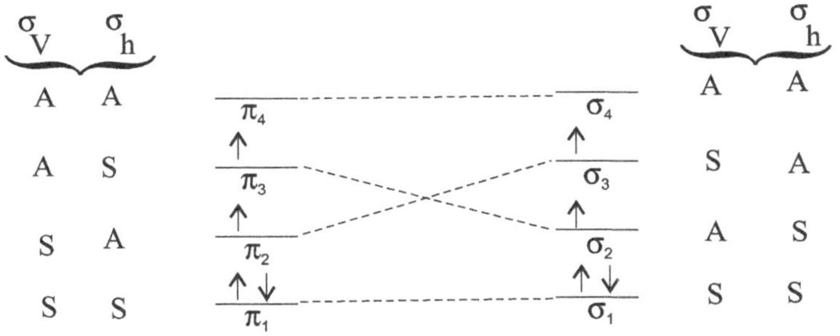

Fig. 13.7: Correlation diagram for cycloaddition (supra-supra) of two ethylene moecules. Initial configuration- First excited state.

It is seen that the first excited state configuration leads to the first excited state of cyclobutane. Hence, the reaction is feasible via a photochemical path with a suprafacial-suprafacial addition.

This symmetry-based approach for predictions of thermal and photochemical changes with corresponding stereochemical course covers a wide variety of cycloaddition reactions. In many cases, however, there may be conservation of one symmetry plane instead of two. A series of generalisations have been made correlating the number of electrons and the number of nodes vis a vis the possibilities of thermal and photochemical ructions with supra-supra, supra-antara and antara-antara approaches.

13.4 Symmetry And Sigmatropic Processes

Before concluding, a brief reference is made to the role of symmetry in sigmatropic reactions. These reactions are characterised by the shift of a sigma bond from one part of a molecule to another part of the same molecule in the absence of catalysis. This actually happens due to migration of a H-atom or a group from one carbon atom to another. A simple example is the conversion of 1 methyl pentadiene into 5 methyl pentadiene.

During the migration, the migrating atom or group may remain on one side of the molecular plane of the C-nuclear frame or it may burrow through the plane from one side to the other (Fig. 13.8 a & b). The former stereochernical course is called suprafacial and the latter antarafacial. Unlike the electrocyclic reactions there is no conserved symmetry element common to the starting material, the intermediate conformations and the final product in sigmatropic processes.

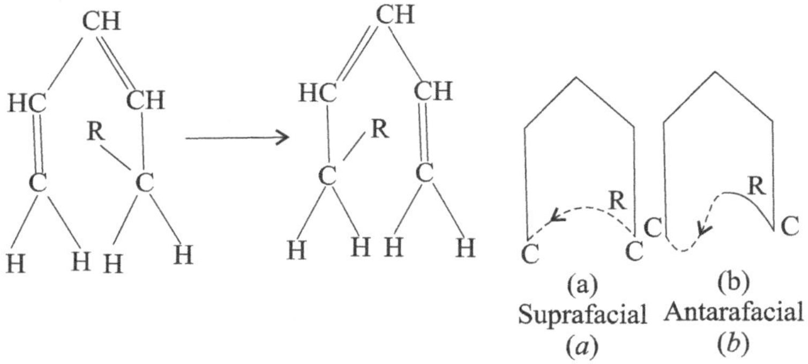

Fig. 13.8: Sigmatropic shift. (a) Suprafacial (b) Antarafacial

Thus there cannot be drawn any orbital symmetry correlation diagram in such reactions. But the transition state (mainly planar with a polyene framework) acquires a C_2 symmetry element or a σ_v, symmetry element depending on whether the course is antarafacial or suprafacial. The antarafacia or the suprafacial transfers are connected in their turn with the symmetry of the π-HOMO. The π-HOMO decides whether the process is to be a thermal or a photochemical one. If it is a ground state π-HOMO, it becomes a thermal shift; if, however, the π-HOMO be one representing the first excited state, the process is a photochemical one. The whole chain of links is depicted below.

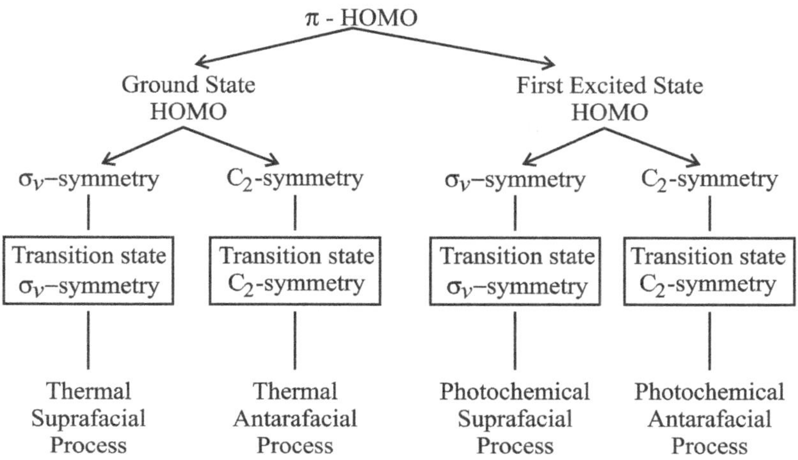

The reader must now be wondering at the power and beauty of Group Theory and the subtle ways in which it serves chemistry. The principles of symmetry and their wide ranging intelligent applications make the study of chemistry more complete and hence more delightful.

Problems & References

(Chapter 1, 2, 3, 4, 5 and 6)

1. List the symmetry elements in and the symmetry operations performable on the following:
(i) a burette without graduation marks (ii) the letter "Z" (iii) phenanthrene molecule (iV) B-hydroxyquinoline (v) two similar cylinders of equal length placed crosswise one above the other and rivetted at the midpoints (vi) unstable forms of fully and partially staggered C_2H_4 molecule (vii) monoclinic crystal lattice ($a \neq b \neq c$; $\alpha = \gamma = 90° \neq \beta$) (viii) a simple cube (ix) boat form of cyclohexane (x) H_3BO_3.

2. Write down, giving justification, the point groups of (ii), (iii), (v), (viii) and (ix) of the foregoing examples. Construct the group multiplication tables of (iii) and (v).

3. What are the generators in (ii), (v) and (x)? Utilise the generator-coset method of writing down the group elements. Arrange the latter class wise.

4. Write down the metrical matrix of (i) the normalised (d-orbital basis set (ii) unit base vectors (l_1, l_2, l_3) of Example 4.1.

5. When is a similarity transformation termed (i) orthogonal transformation (ii) unitary transformation? Show that any two rows (and also columns) of a unitary matrix behave like two mutually orthogonal vectors.

6. Using the translational vectors (T_x, T_y, T_z) as the basis set find the representations of the point groups of H_2O and NH_3.

7. Use the three rotational vectors to find the representation of C_{4h}. Can you write down the character values without forming the full transformation matrices?

8. Name some point groups wherein the vector z and rotational vector R_z belong to (i) the same IR (ii) different IR's. Do you detect the existence of any distinctive group element or elements causing their classification under different IR's?

9. Use the orthogonal unit base vector (d_1, d_2, d_3) of Example 4.2 to find the representation of the point group C_{3v}.

10. Find the rotational matrix for a rotation θ about C_3 in the basis set of unit vectors (e_1, e_2, e_3) where e_1 and e_2 are mutually inclined at an angle α and e_3 is perpendicular to both e_1 and e_2.

11. The characters of an IR of a certain point group are

	E	$2C_4$	C_2	$2C_2'$	$2C_2''$	i	$2S_4$	σ_h	$2\sigma_v$	$2\sigma_d$
χ	1	−1	1	−1	1	−1	1	−1	1	−1

What is the IR in Mulliken notation? Name also the point group.

12. What are the specific IR's of D_{4d} point group, the characters of which are

	E	$2C_4$	$2S_8$	$2S_8^3$	C_2	$4C_2'$	$4\sigma_d$
χ_1	1	1	−1	−1	1	−1	1
χ_2	1	1	−1	−1	1	1	−1

13. Construct the character table of the point group to which partially staggered ethane molecule belongs. Do (d_{xz}, d_{yz}) constitute a doubly degenerate basis set for an IR of this point group?

14. Form SALC's from d-orbital space that can form basis of IR's of the C_{4v}, point group.

15. Can you form a SALC of A_{2u} symmetry using the p_z-orbitals of the nitrogen and oxygen atoms in NO_2 molecule?

16. The transformations of the basis functions $(\phi_1\ \phi_2\ \phi_3\ \phi_4)$ under symmetry operations of D_4 are given in the following table:

	E	C_4	C_4^3	C_2	$C_{2a'}$	$C_{2b'}$	$C_{2a''}$	$C_{2b''}$
ϕ_1	ϕ_1	ϕ_2	ϕ_4	ϕ_3	$-\phi_4$	$-\phi_2$	$-\phi_1$	$-\phi_3$
ϕ_2	ϕ_2	ϕ_3	ϕ_1	ϕ_4	$-\phi_3$	$-\phi_1$	$-\phi_4$	$-\phi_2$
ϕ_3	ϕ_3	ϕ_4	ϕ_2	ϕ_1	$-\phi_2$	$-\phi_4$	$-\phi_3$	$-\phi_1$
ϕ_4	ϕ_4	ϕ_1	ϕ_3	ϕ_2	$-\phi_1$	$-\phi_3$	$-\phi_2$	$-\phi_4$

From a normalized function belonging to B_1 symmetry.

17. An integral occurring in the quantum mechanical treatment of benzene molecule involves an integrand which is a product of three functions belonging to the IR's E_{1u}, E_{2g} and B_{2u} respectively. Will the integral survive?

18. A d^2-transition metal ion in infinitely strong octahedral crystal field has a configuration $t_{2g}^1\ e_g^1$ and is characterized by the the direct product representation $\Gamma\,(T_{2g} \times E_g)$. What IR's are involved?

19. What IR's are spanned by the six 1s orbitals of the H-atoms in ethane molecule [(i) staggered (ii) eclipsed]?

20. What can you say about the descent in symmetry of the atomic d-orbitals when the atom is placed in a crystal field environment of tetragonal symmetry? (compression along z-axis).

21. How many nonequivalent IR's are there in the point group C_{4h} and what are their dimensionalities? Reconcile your findings with the fact that the character table shows six IR's.

22. The point group D_{4h} is a direct product group, viz., D_{4h}=D_4×S_2. Construct the character table of D_{4h} by using those of D_4 and S_2.

23. Work out all the direct products $\Gamma_i \times \Gamma_j^1 = \Sigma n_k\ \Gamma_k$ of the IR's of the group C_{3v}.

24. Draw a flow sheet diagram depicting the process of reduction of a reducible representation.

25. Show that the off diagonal matrix elements, $H_{\mu v}$ of the hamiltonian operator H vanishes in the basis set where ϕ_μ, ϕ_v's etc., are s, p, d-atomic orbitals.

(Chapters 7, 8, 9, 10, 11, 12 and 13)

26. Examine the occurrence or the lack of alternating axis in the following molecules and label these as optically active or inactive. Indicate the location of the S_n axis wherever it exists. (i) H_2O (ii) cyclohexane (chair form) (iii) CHFClBr (iv) Monochlorocyclohexane (equatorial and axial) (v) cis 1(ax) 2(eq) dichlorocyclohexane (vi) chloroformaldehyde.

27. Argue from symmetry viewpoint whether the following molecules are likely to possess dipole moment or not. (i) formaldehyde (ii) acetaldehyde (iii) ethyiene (iv) hydrogen sulphide (v) methylchloride (vi) nitrogen trifluoride (vii) pyridine (viii) anthracene.

28. Find the symmetry species of the second vibrationally excited state wave function of CO molecule. What are the Raman active and *ir* active fundamentals of the molecule?

29. To which IR's do the following vibrations belong in trans N_2F_2? (i) N-N stretch (ii) symmetric N-F stretch (iii) antisymmetric N-F stretch. Can the N-N stretch and N-F stretch couple?

30. Classify the in-plane and out-of-plane normal vibrations, if any, in the following species: (i) NO_2 (ii) CO_2 (iii) C_2H_4 (iv) $[PtCl_4]^{2-}$

31. What symmetry species (i.e., IR's) characterise the stretching and the angle-bending modes of vibrations in phosphine molecule? Is there any mixing of the two modes disallowing pure stretching type?

32. One of the LCAO SCF MO's of the formaldehyde molecule in the valence shell basis set of atomic orbitals $(\phi_1\ \phi_2 \cdots \phi_{10})$ is given by

$$= 0.544\phi_1 - 0.406\phi_4 - 0.391\phi_5 - 0.136\phi_8 + 0.429\phi_9 + 0.429\phi_{10}$$

Here $\phi_1 \cdots\cdots \phi_4$ represent C$-2s$, $2p_x$, $2p_y$, $2p_z$, ao's $\phi_5 \cdots\cdots \phi_8$ are the O-orbitals in the similar sequence and $\phi_9\ \phi_{10}$ signify the two H-orbitals. What is the symmetry of this MO?

33. A homonuclear diatomic vibrating molecule behaves as one-dimensional harmonic oscillator. Quantum mechanics lays down the selection rule for vibrational transition as $\Delta n = \pm 1$. Is this strictly in quantitative agreement with what follows from symmetry principles?
Rope in hetero nuclear diatomic molecules for further discussion.

34. The species $[VO(H_2O)_5]^{2+}$ is a tetragonally distorted octahedron with V=O bond pointing along the z-axis and metal atom at the centre. The HOMO and the LUMO of the species are of B_2 and E symmetries respectively. Justify or contradict the feasibility of $e \leftarrow b_2$ excitation. Indicate the nature of polarization, if any.

35. Find the nature and the quantitative forms of the hybrids of the P-atom necessary for formation of the molecule PCl_5. Neglect π-hybridization.

36. Construct the π-mo's of pyridine molecule. Are the $\pi^* \leftarrow n$, $\pi^* \leftarrow \pi$ transitions permissible? In what way does the process of energy evaluation of the π-orbitals of pyridine differ from that of benzene?

37. How do the d-orbitals of a transition metal split if the latter be located in rhombic environment?

38. What vibrational modes in $(PtCl_4)^{2-}$ can couple with the electronic wave functions? Use table 10.5 as an aid.

39. A metal ion (d^1) in an O_h symmetry is subjected to a rhombic perturbation $\left[\left(P_{x^2} + Q_{y^2}\right) - (P + Q)_{z^2}\right]$. What will be the general natures of the elements in the perturbation matrix? [Hint: First find the IR of the perturbation]

40. To what IR of O_h does $z(x^2 - y^2)$ belong? What are its partners?

41. Discuss the nature of the electrocyclic ring opening of cis 7, 8-dimethyl octa −1, 3, 5-triene in thermal and photochemical conversions.

42. What sigmatropic shift takes place when 7-methoxycyclohep-tatriene is irradiated?

REFERENCES

Recommended Books (General-Chapters 1 to 12)

1. Chemical Applications of Group Theory, F. A. Cotton, Inter-science, New York, 1967.

2. Symmetry-R. McWeeny, Pergamon Press, Oxford, 1963.

3. Group Theory and Quantum Mechanics-M. Tinkham, Tata Mc-Graw Hill, New Delhi, 1974.

4. Group Theoretical Techniques in Quantum Chemistry, C. D, H, Chisholm, Academic Press, London, 1976.

5. Molecular Symmetry-D. S. Schonland, D. Van Nostrand Co. London, 1965.

6. Group Theory and Chemistry-D. M. Bishop, Clarendon Press, Oxford, 1973.

7. Symmetry in Chemistry-H. H. Jaffe and M. Orchin, John Wiley and Sons, London, 1965.

8. Group Theory and its Applications to Physical Problems;, M. Hammermesh, Pergamon Press, London, 1962.

9. The Classical Groups-H. Weyl, Princeton University Press, 1946.

10. Group Theory and Symmetry in Chemistry-L. M. Hall, McGraw Hill, New York, 1969.

11. Elements of Group Theory for Physicists
−A.W.Joshi, Wiley Eastern Limited, 1985 Kolkata.

12. Lie Algebras in Particle Physics – Howard Georgi Levant Books, 2008.

Recommended Books (other than Group Theory)

(For Chapters 1 to 6)

1. Elementary Quantum Chemistry−F. Pilar, McGraw Hill, New York, 1968. (Chap. 3)

2. A Text Book of Matrices-Shanti Narayan, S. Chand and Co. New Delhi, 1972 (Chap. 3).

3. Mathematics of Quantum Chemistry-J. M. Anderson, Ben-amin, New York 1966 (Chap. 3).

(For Chapters 7 to 12)

4. Molecular Quantum Mechanics Vol. I-P. W. Atkins, Clarendon Press, Oxford, 1980 (Chap. 7).

5. Methods in Molecular Orbital Theory-A. G. Turner, Prentice-Hall, London, 1974 (Chap. 7).

6. Topics in Current Chemistry $-\sigma$ and π Electrons in Organic Compounds W. Kutzclnigg, G. Del Re and G. Berthier, Springer Veriag, Berlin, 1971 (Chap. 9).

7. Introduction to Ligand Field Theory-C. J, Ballhausen, McGraw Hill, New York, 1962 (Chap. 10).

8. Inorganic Electronic Spectroscopy-A. B. P. Lever, Elsevier,. Amsterdam, 1968 (Chap, 10).

9. Introduction to Ligand Fields-B. N. Figgis, Wiley Eastern. New Delhi, 1976 (Chap. 10).

9A. Atoms And d^n Ions In Perturbation Fields– S.C.Rakshit. Levant Books, Kolkata 2010

(For Chapter 13)

10. The Conservation of Orbital Symmetry-R. 8. Woodward and R. Hoffman, Academic Press, London, 1970.

11. Molecular Reactions and Photochemistry-C. H, De Puy and O, L. Chapman, Prentice Hall, New Delhi, 1975.

12. Fundamentals of Photochemistry-K. K. Rolmtgi Mukherjee, Wiley Eastern, New Delhi, 1978.

Literature References:

1. H. C. Longuet Higgins-Molecular Orbitals In Chemistry, Physics and Biology, Ed. by P. Lowdin and B. Pullmann. Academic Press, New York, 1964, pp. 113.

2. M. Tinklmm-Group Theory And Quantum Mechanics, Tata McGraw Hill, New Delhi, 1974, pp. 23.

3. D.M.Bishop-Group Theory And Chemistry, Clarendon Press, 1973, pp. 143.

4. D. M. Bishop-Opp. cit., pp. 145.

5. M. Tinkham-Opp. cit., pp. 28.

6. J. D. Dioxon-Numerische Mathematik, 1967, 10, 446.

7. E. Blokker-Int. J. Quant. Chem., 1972, 6, 925.

8. H. Goldstein-Classical Mechanics, Addison Wesley, London, 1964, pp. 23.

9. H. Margenau and G. M. Murphy-Mathematics of Physics and Chemistry, D. VanNostrand, New York, 1952, pp. 117, 342.

10. A. B. Sannigrahi-J. Chem. Edn., 1975, 37.

11. C. C. J. Roothaan-Rev. Mod. Phys., 1951, 23, 69.

12. P. G. Lykos-Pi-Electron Approximation. Advances in Quantum Chemistry, Ed. P. Lowdin. Vol. I, 1964, 1971.

13. H. Eyring, J, Walter and G. E. Kimbail-Quantum Chemistry, John Willey and Sons, New York, 1944, pp. 128.

14. F. A. Cotton and G. Wilkinson-Advanced Inorganic Chemistry Wiley Eastern, New Delhi 1969, pp. 600.

15. H. E. White-Atomic Spectra, McGraw Hill, New York, 1934, pp. 296.

16. E. U. Condon and G. H. Shortley-Theory of Atomic Spectra, Cambridge University Press, London/1967, pp. 176.

17. L. Harris and L. Loeb-Wave Mechanics, McGraw Hill, New Work. 1963, pp. 175.

18. Ref. 13, pp. 370.

19. Tanabe and Sugano-J. Phys. Soc. Japan, 1954, 9, 753.

APPENDIX I

Commutation of the hamiltonian and the symmetry operators

In the present volume, molecular orbitals, atomic orbitals and vibrational properties amongst others have been brought within the domain of applications of symmetry. Consequently one is concerned with the electronic hamiltonian and vibrational hamiltonian. Since electronic motions are much faster than vibratory motions of the nuclei, Born-Cppenheimer approximation allows separate treatments of the two.

1. Electronic Motions

The hamiltonian operator, without the spin-orbit interaction. for any given nuclear configuration is

$$H = -\frac{\hbar^2}{2m}\sum_i \nabla i^2 - \sum_i \sum_a \frac{Z_a e^2}{r_{ai}} + \sum_{i<j} \frac{e^2}{r_{ij}}$$

where Z_a is the nuclear charge on the ath nucleus.

(i) Potential energy operator—This operator V(X), representing the last two terms of the hamiltonian, depends on r_{ij}'s and r_{ai}'s. Here X denotes the set of electronic coordinates, $(x_1^1,\ x_2^1,\ x_3^1)$, $(x_1^2,\ x_2^2,\ x_3^2)\cdots$ $(x_1^n,\ x_2^n,\ x_3^n)$ in the configuration space with a cartesian reference frame having the origin at some point which remains unmoved during symmetry operations.

A symmetry operation R causes equivalent nuclei to trade places, but the electrons need not necessarily exchange coordinates. The latter simply take up new coordinates symbolised by X′. Since during a symmetry operation, R, all the nuclei and electrons move in the same manner, there occurs no change in the relative inter-electron and nuclear-electron distances. The potential energy thus remains, invariant under R., i.e., V(X)=V(X′).

$$RV(X)\psi = V(X)R\psi$$

(ii) Kinetic energy operator- The commutability of this operator with W requires a skilful handling[2]. Consider first a function $\phi(x_1,\ x_2,\ x_3)$. A symmetry operation R applied on it will ensure [cf. Sec. 3.5]

$$\phi'(x_1'x_2'x_3') = \{\mathbf{R}\ \phi\}\,(x_1'x_2'x_3') = \phi(x_1x_2x_3) \tag{I.1}$$

and also

$$x_i' = \sum_{k=1}^{3} C_{ik}x_k \tag{I.2}$$

Here i runs through 1 to 3 and the row matrices $\mathbf{C}_1$, $\mathbf{C}_j$ etc., are mutually orthogonal, i.e.,

$$\sum_{k=1}^{3} \mathrm{C}_{ik}\mathrm{C}_{ij} = \delta_{ij} \tag{I.3}$$

$$\text{Additionally from (I} - 2),\ \mathrm{C}_{ik} = \left(\frac{\partial x_i'}{\partial x_k}\right) \tag{I.4}$$

Now starting from the original function we can construct a new function $\nabla^2\phi(x_1x_2x_3)$ and follow it up by an R operation. Remembering (I-1),

$$\begin{aligned} \{\mathrm{R}\nabla^2\phi\}\,(x_1'x_2'x_3') &= \nabla^2\phi(x_1x_2x_3) = \nabla^2\,\{\mathbf{R}\phi\}\,(x_1'x_2'x_3') \\ &= \nabla^2\psi'(x_1'x_2'x_3') \end{aligned} \tag{I.5}$$

In (I-5), ψ' is a function of $x_1',\ x_2',\ x_3'$, but ∇^2 represents two successive differentiations with respect to $x_1,\ x_2,\ x_3$. To evaluate $\nabla^2\psi'$ we proceed as follows.

$$\frac{\partial\psi'}{\partial x_k} = \frac{\partial\psi'}{\partial x_1'}\left(\frac{\partial x_1'}{\partial x_k}\right) + \frac{\partial\psi'}{\partial x_2'}\left(\frac{\partial x_2'}{\partial x_k}\right) + \frac{\partial\psi'}{\partial x_3'}\left(\frac{\partial x_3'}{\partial x_k}\right)$$

i.e., $\frac{\partial\psi'}{\partial x_k} = \sum\limits_{i=1}^{3} \mathrm{C}_{ik}\left(\frac{\partial\psi'}{\partial x_i'}\right)$, where relation (1-4) has been used. Differentiating once again, summing over k and using (I-3),

$$\begin{aligned} \sum_k \frac{\partial^2\psi'}{\partial xk^2} &= \sum_i\sum_j\sum_k \mathrm{C}_{ik}\mathrm{C}_{jk}\frac{\partial}{\partial x_j'}\left(\frac{\partial\psi'}{\delta x_i'}\right) \\ &= \sum_i\sum_j \delta_{ij}\frac{\partial}{\partial x_j'}\left(\frac{\partial\psi'}{\partial x_i'}\right) = \sum_i \frac{\partial^2\psi'}{\partial x_i'^2} = \nabla^{2'}\psi' \\ &= \nabla'^2\,\{R\phi\}\,(x_1'x_2'x_3') \end{aligned} \tag{I.6}$$

Relation (I-5) can now be moulded into

$$\{R\nabla^2\phi\}(x'_1\ x'_2\ x'_3) = \nabla^{2'}\{R\phi\}(x'_1\ x'_2\ x_3).$$

Removal of primes from both sides, which does not affect the generality of the foregoing relation, yields

$$\{R\nabla^2\phi\}(x_1\ x_2\ x_3) = \nabla^2\{\mathbf{R}\phi\}(x_1\ x_2\ x_3) \qquad \text{(I.7)}$$

This indicates R and ∇^2 (and hence R and $-\frac{\hbar^2}{2m}\nabla^2$) commute. If, however, ϕ is a function of $x_1^1 x_2^1 x_3^1 \cdots\cdots x_1^n x_2^n x_3^n$ R will obviously commute separately with each ∇_i^2 and hence with each $-\frac{\hbar^2}{2m}\nabla i^2$. Finally, therefore, R commutes with their sum, viz., $-\frac{\hbar^2}{2m}\sum_i \nabla i^2$, the kinetic energy operator part of the hamiltonian.

2. Vibrational Motions.

The hamiltonian for nuclear vibration

$$\mathrm{H}_n = -\frac{\hbar^2}{2}\sum_a \frac{1}{m_a}\nabla_a^2 + \mathrm{V}_{nn},$$

where V_{nn} is the potential energy arising from inter-nuclear repulsions and contains, additionally, the electronic energy E_e, as a part of the potential energy. This E_e depends on the immediately adjustible electronic charge distribution during each moment of nuclear motions. The configuration of the system is represented by X, i.e., $x_1^1 x_2^1 x_3^1, \ldots . x_1^n x_2^n x_3^n$ representing vibrational displacements of the nuclei 1 to n measured along separate cartesian coordinate frames set up at the equilibrium positions of each vibrating nucleus.

(i) Potential energy–A symmetry operation R on the vibrational motions causes a shift of the individual displacements from one nucleus to another specific equivalent nucleus with an accompanying rearrangement of the electrons. The nuclei, however, remain fixed at their original locations. It is obvious that the set of internuclear distances in the configuration X changes to another equivalent set of identical distances in the configuration X′. The electronic displacements commensurate with R permit the retention of the old value of E_e. Hence $\mathrm{V}_{nn}(\mathrm{X})=\mathrm{V}_{nn}(\mathrm{X}')$. The potential energy operator $\mathrm{V_{nn}}$ is invariant under R.

(ii) Kinetic energy–Following the artifice adopted in electronic motions, it can be shown that the operator denoting kinetic energy of the nucleus a, viz. $-\frac{\hbar^2}{2m}\nabla a^2$ changes under R into a corresponding operator of the nucleus b, viz., $-\frac{\hbar^2}{2m_b}\nabla b^2$ where the nuclei a and b are equivalent under **R**. The relation actually obeyed is

$$\mathrm{R}\left[-\frac{\hbar^2}{2m_a}\nabla_a^2\right]\phi = \left[-\frac{\hbar^2}{2m_b}\nabla_b^2\right] R\phi$$

Summing over all nuclei leads to a commutation of the kinetic energy operator with **R**, the symmetry operator.

---

1. Quantum Chemistry– Eyring, Walter and Kimball (Wiley International Edition).

2. Molecular Symmetry–D. S. Schotland (Van Nostrand Co. Ltd. London, 1965).

APPENDIX II

Character Tables of Molecular Symmetry Groups

1. The nonaxial Groups

C_1	E
A	1

C_s	E	σ_h	basis vectors and components	basis functions
A′	1	1	x, y. R_z, T_x, T'_y	x^2, y^2, z^2, xy
A″	1	−1	z, R_x, R_y, T_z	yz, xz

C_i	E	i	basis vectors and components	basis functions
A_g	1	1	R_x, R_y, R_z	x^2, y^2, z^2, xy, xz, yz
A_u	1	−1	x, y, z, T_x, T_y, T_z	

2. The C_n Groups

C_2	E	C_2	basis vectors and components	basis functions
A	1	1	z, R_z, T_z	x^2, y^2, z^2, xy
B	1	−1	x, y, R_z, R_y, T_x, T_y	yz, xz

C_3	E	C_3	C_3^2	basis vectors and components	$\epsilon = \exp(2\pi i/3)$ basis functions
A	1	1	1	z, R_z, T_z	$x^2 + y^2$, z^2
E	1	$\left\{\begin{matrix} \epsilon \\ \epsilon^* \end{matrix}\right.$	$\left.\begin{matrix} \epsilon^* \\ \epsilon \end{matrix}\right\}$	(x, y) (R_x, R_y) (T_x, T_y)	($x^2 - y^2$, xy)(yz, xz)

C_4	E	C_4	C_2	C_4^3	basis vectors and components	basis functions
A	1	1	1	1	z, R_z, T_z	$x^2+y^2,\ z^2$
B	1	-1	1	-1		$x^2-y^2,\ xy$
E	$\left\{\begin{matrix}1\\1\end{matrix}\right.$	$\begin{matrix}i\\-i\end{matrix}$	$\begin{matrix}-1\\-1\end{matrix}$	$\left.\begin{matrix}-i\\i\end{matrix}\right\}$	$(x,y)(R_x,\ R_y)$ $(T_x,\ T_y)$	$(yz,\ xz)$

C_5	E	C_5	C_5^2	C_5^3	C_5^4	basis vectors and components	$\epsilon=\exp(2\pi i/5)$ basis functions
A	1	1	1	1	1	z, R_z, (T_z)	$x^2+y^2,\ z^2$
E_1	$\left\{\begin{matrix}1\\1\end{matrix}\right.$	$\begin{matrix}\epsilon\\\epsilon^*\end{matrix}$	$\begin{matrix}\epsilon^2\\\epsilon^{2*}\end{matrix}$	$\begin{matrix}\epsilon^{2*}\\\epsilon^2\end{matrix}$	$\left.\begin{matrix}\epsilon^*\\\epsilon\end{matrix}\right\}$	(x,y) $(R_x,\ R_y)$	(yz xz)
E_2	$\left\{\begin{matrix}1\\1\end{matrix}\right.$	$\begin{matrix}\epsilon^2\\\epsilon^{2*}\end{matrix}$	$\begin{matrix}\epsilon^*\\\epsilon\end{matrix}$	$\begin{matrix}\epsilon\\\epsilon^*\end{matrix}$	$\left.\begin{matrix}\epsilon^{2*}\\\epsilon^2\end{matrix}\right\}$		$(x^2-y^2,\ xy)$

C_6	E	C_6	C_3	C_2	C_3^2	C_6^5	basis vectors and components	ϵ=exp$(2\pi i/6)$ basis functions
A	1	1	1	1	1	1	z, R_z, T_z	$x^2+y^2,\ z^2$
B	1	-1	1	-1	1	-1		
E_1	$\left\{\begin{matrix}1\\1\end{matrix}\right.$	$\begin{matrix}\epsilon\\\epsilon^*\end{matrix}$	$\begin{matrix}-\epsilon^*\\-\epsilon\end{matrix}$	$\begin{matrix}-1\\-1\end{matrix}$	$\begin{matrix}-\epsilon\\-\epsilon^*\end{matrix}$	$\left.\begin{matrix}\epsilon^*\\\epsilon\end{matrix}\right\}$	$(x,y),(T_x,T_y)$ (R_x,R_y)	(xz,yz)
E_2	$\left\{\begin{matrix}1\\1\end{matrix}\right.$	$\begin{matrix}-\epsilon^*\\-\epsilon\end{matrix}$	$\begin{matrix}-\epsilon\\-\epsilon^*\end{matrix}$	$\begin{matrix}1\\1\end{matrix}$	$\begin{matrix}-\epsilon^*\\-\epsilon\end{matrix}$	$\left.\begin{matrix}-\epsilon\\-\epsilon^*\end{matrix}\right\}$		(x^2-y^2,xy)

C_7	E	C_7	C_7^2	C_7^3	C_7^4	C_7^5	C_7^6	basis vectors and components	ϵ=exp$(2\pi i/7)$ basis functions
A	1	1	1	1	1	1	1	z, R_z, T_z	$x^2+y^2,\ z^2$
E_1	$\left\{\begin{matrix}1\\1\end{matrix}\right.$	$\begin{matrix}\epsilon\\\epsilon^*\end{matrix}$	$\begin{matrix}\epsilon^2\\\epsilon^{2*}\end{matrix}$	$\begin{matrix}\epsilon^3\\\epsilon^{3*}\end{matrix}$	$\begin{matrix}\epsilon^{3*}\\\epsilon^3\end{matrix}$	$\begin{matrix}\epsilon^{2*}\\\epsilon^2\end{matrix}$	$\left.\begin{matrix}\epsilon^*\\\epsilon\end{matrix}\right\}$	$(x,y),(T_x,T_y)$ (R_x,R_y)	(xz,yz)
E_2	$\left\{\begin{matrix}1\\1\end{matrix}\right.$	$\begin{matrix}\epsilon^2\\\epsilon^{2*}\end{matrix}$	$\begin{matrix}\epsilon^{3*}\\\epsilon^3\end{matrix}$	$\begin{matrix}\epsilon^*\\\epsilon\end{matrix}$	$\begin{matrix}\epsilon\\\epsilon^*\end{matrix}$	$\begin{matrix}\epsilon^3\\\epsilon^{3*}\end{matrix}$	$\left.\begin{matrix}\epsilon^{2*}\\\epsilon^2\end{matrix}\right\}$		(x^2-y^2,xy)
E_3	$\left\{\begin{matrix}1\\1\end{matrix}\right.$	$\begin{matrix}\epsilon^3\\\epsilon^{3*}\end{matrix}$	$\begin{matrix}\epsilon^*\\\epsilon\end{matrix}$	$\begin{matrix}\epsilon^2\\\epsilon^{2*}\end{matrix}$	$\begin{matrix}\epsilon^{2*}\\\epsilon^2\end{matrix}$	$\begin{matrix}\epsilon\\\epsilon^*\end{matrix}$	$\left.\begin{matrix}\epsilon^{3*}\\\epsilon^3\end{matrix}\right\}$		

C_8	E	C_8	C_4	C_2	C_4^3	C_8^3	C_8^5	C_8^7	basis vectors and components	$\in = \exp(2\pi i/8)$ basis functions
A	1	1	1	1	1	1	1	1	z, R_z, T_z	$x^2 + y^2$, z^2
B	1	-1	1	1	1	-1	-1	-1		
E_1	1	$\in$	i	-1	$-i$	$-\in^*$	$-\in$	$\in^*$	(x, y), (T_x, T_y) (R_x, R_y)	(xz, yz)
	1	$\in^*$	$-i$	-1	i	$-\in$	$-\in^*$	$\in$		
E_2	1	i	-1	1	-1	$-i$	i	$-i$		($x^2 - y^2$, xy)
	1	$-i$	-1	1	-1	i	$-i$	i		
E_3	1	$-\in$	i	-1	$-i$	$\in^*$	$\in$	$-\in^*$		
	1	$-\in^*$	$-i$	-1	i	$\in$	$\in^*$	$-\in$		

3. The D_n Groups

D_2	E	$C_2(z)$	$C_2(y)$	$C_2(x)$	basis vectors and components	basis functions
A	1	1	1	1		x^2, y^2, z^2
B_1	1	1	-1	-1	z, T_z, R_z	xy
B_2	1	-1	1	-1	y, T_y, R_y	xz
B_3	1	-1	-1	1	x, T_x, R_x	yz

D_3	E	$2C_3$	$3C_2$	basin vectors and components	basis functions
A_1	1	1	1		$x^2 + y^2$, z^2
A_2	1	1	-1	z, R_z, T_z	
E	2	-1	0	(x, y)(T_x, T_y) (R_x, R_y)	($x^2 - y^2$, xy)(xz, yz)

D_4	E	$2C_4$	C_2 ($=C_4^2$)	$2C_2'$	$2C_2''$	basis vectors and components	basis functions
A_1	1	1	1	1	1		$x^2 + y^2$, z^2
A_2	1	1	1	-1	-1	z, R_z, T_z	
B_1	1	-1	1	1	-1		$x^2 - y^2$
B_2	1	-1	1	-1	1		xy
E	2	0	-2	0	0	(x, y)(R_x, R_y) (T_x, T_y)	(xz, yz)

D_5	E	$2C_5$	$2C_5^2$	$5C_2$	basis vectors and components	basis functions
A_1	1	1	1	1		(x^2+y^2), z^2
A_2	1	1	1	−1	z, R_z, T_z	
E_1	2	2cos 72°	2cos 144°	0	(x, y)(R_x, R_y) (T_x, T_y)	(xz, yz)
E_2	2	2cos 144°	2cos 72°	0		$(x^2-y^2,\ xy)$

D_6	E	$2C_6$	$2C_3$	C_2	$3C_2'$	$3C_2''$	basis vectors and components	basis functions
A_1	1	1	1	1	1	1		(x^2+y^2), z^2
A_2	1	1	1	1	−1	−1	z, R_z, T_z	
B_1	1	−1	1	−1	1	−1		
B_2	1	−1	1	−1	−1	1		
E_1	2	1	−1	−2	0	0	(x, y)(R_x, R_y)	(xz, yz)
E_2	2	−1	−1	2	0	0		(x^2-y^2)(xy)

4. The C_{nv} Groups

C_{2v}	E	C_2	$\sigma_v(xz)$	$\sigma_v(yz)$	basis vectors and components	basis functions
A_1	1	1	1	1	z, T_z	x^2, y^2, z^2
A_2	1	1	−1	−1	R_z	xy
B_1	1	−1	1	−1	x, R_y, T_x	xz
B_2	1	−1	−1	1	y, R_x, T_y	yz

C_{3v}	E	$2C_3$	$3\sigma_v$	basis vectors and components	basis functions
A_1	1	1	1	z, T_z	x^2+y^2, z^2
A_2	1	1	−1	R_z	
E	2	−1	0	(x, y)(R_x, R_y) (T_x, T_y)	((x^2-y^2), xy), (xz, yz)

C_{4v}	E	$2C_4$	C_2	$2\sigma_v$	$2\sigma_d$	basis vectors and components	basis functions
A_1	1	1	1	1	1	z, T_z	$x^2 + y^2,\ z^2$
A_2	1	1	1	−1	−1	R_z	
B_1	1	−1	1	1	−1		$x^2 - y^2$
B_2	1	−1	1	−1	1		xy
E	2	0	−2	0	0	(x, y)(R_x, R_y) (T_x, T_y)	(xz, yz)

C_{5v}	E	$2C_5$	$2C_5^2$	$5\sigma_v$	basis vectors and components	basis function
A_1	1	1	1	1	z, T_z	$x^2 + y^2,\ z^2$
A_2	1	1	1	−1	R_z	
E_1	2	2cos 72°	2cos 144°	0	(x, y) (R_x, R_y) (T_x, T_y)	(xz, yz)
E_2	2	2cos 144°	2cos 72°	0		$(x^2 - y^2,\ z^2)$

C_{6v}	E	$2C_6$	$2C_3$	C_2	$3\sigma_v$	$3\sigma_d$	basis vectors and components	basis functions
A_1	1	1	1	1	1	1	z, T_z	$x^2 + y^2,\ z^2$
A_2	1	1	1	1	−1	−1	R_z,	
B_1	1	−1	1	−1	1	−1		
B_2	1	−1	1	−1	−1	1		
E_1	2	1	−1	−2	0	0	(x, y)(R_x, R_y)	(xz, yz)
E_2	2	−1	−1	2	0	0		$(x^2 - y^2)$, (xy)

5. The Cnh Groups

C_{2h}	E	C_2	i	σ_h	basis vectors and components	basis functions
A_g	1	1	1	1	R_z	$x^2,\ y^2,\ z^2$, xy
B_g	1	−1	1	−1	R_x, R_y	xz, yz
A_u	1	1	−1	−1	z, T_z,	
B_u	1	−1	−1	1	x, y, T_x, T_y	

C_{3h}	E	C_3	C_3^2	σ_h	S_3	S_3^5	basis vectors and components	ϵ =exp$(2\pi i/3)$ basis functions
A′	1	1	1	1	1	1	R_z	$x^2+y^2,\ z^2$
E′	{ 1	ϵ	ϵ^*	1	ϵ	ϵ^* }	(x, y)(T_x, T_y)	$(x^2-y^2,\ xy)$
	{ 1	ϵ^*	ϵ	1	ϵ^*	ϵ }		
A″	1	1	1	−1	−1	−1	z, T_z	
E″	{ 1	ϵ	ϵ^*	−1	$-\epsilon$	$-\epsilon^*$ }	(R_x, R_y)	(xz, yz)
	{ 1	ϵ^*	ϵ	−1	$-\epsilon^*$	$-\epsilon$ }		

C_{4h}	E	C_4	C_2	C_4^3	i	S_4^3	σ_h	S_4	basis vectors and components	basis functions
A_g	1	1	1	1	1	1	1	1	R_z	$x^2+y^2,\ z^2$
b_g	1	−1	1	−1	1	−1	1	−1		x^2-y^2, xy
E_g	1	i	−1	$-i$	1	i	−1	$-i$	(R_x, R_y)	(xz, yz)
	1	$-i$	−1	i	1	$-i$	−1	i		
A_u	1	1	1	1	−1	−1	−1	−1	z, T_z	
B_u	1	−1	1	−1	−1	1	−1	1		
E_u	1	i	−1	$-i$	−1	$-i$	1	i	(x, y)(T_x, T_y)	
	1	$-i$	−1	i	−1	i	1	$-i$		

C_{5h}	E	C_5	C_5^2	C_5^3	C_5^4	σ_h	S_5	S_5^7	S_5^3	S_5^9	basis vectors and components	$\epsilon = exp(2\pi i/5)$ basis function
A′	1	1	1	1	1	1	1	1	1	1	R_z	$x^2 + y^2,\ z^2$
E_1'	1	ϵ	ϵ^2	ϵ^{2*}	ϵ^*	1	ϵ	ϵ^2	ϵ^{2*}	ϵ^*	(x, y)(T_x, T_y)	
	1	ϵ^*	ϵ^{2*}	ϵ^2	ϵ	1	ϵ^*	ϵ^{2*}	ϵ^2	ϵ		
E_2'	1	ϵ^2	ϵ^*	ϵ	ϵ^{2*}	1	ϵ^2	ϵ^*	ϵ	ϵ^{2*}		$(x^2 - y^2,\ xy)$
	1	ϵ^{2*}	ϵ	ϵ^*	ϵ^2	1	ϵ^{2*}	ϵ	ϵ^*	ϵ^2		
A″	1	1	1	1	1	−1	−1	−1	−1	−1	z, T_z	
E_1''	1	ϵ	ϵ^2	ϵ^{2*}	ϵ^*	−1	$-\epsilon$	$-\epsilon^2$	$-\epsilon^{2*}$	$-\epsilon^*$	(R_x, R_y)	(xz, yz)
	1	ϵ^*	ϵ^{2*}	ϵ^2	ϵ	−1	$-\epsilon^*$	$-\epsilon^{2*}$	$-\epsilon^2$	$-\epsilon$		
E_2''	1	ϵ^2	ϵ^*	ϵ	ϵ^{2*}	−1	$-\epsilon^2$	$-\epsilon^*$	$-\epsilon$	$-\epsilon^{2*}$		
	1	ϵ^{2*}	ϵ	ϵ^*	ϵ^2	−1	$-\epsilon^{2*}$	$-\epsilon$	$-\epsilon^*$	$-\epsilon^2$		

C_{6h}	E	C_6	C_3	C_2	C_3^2	C_6^5	i	S_3^5	S_6^5	σ_h	S_6	S_3	basis vectors and components	$\epsilon = exp(2\pi i/6)$ Basis functions
A_g	1	1	1	1	1	1	1	1	1	1	1	1	R_z	$x^2+y^2,\ z^2$
B_g	1	−1	1	−1	1	−1	1	−1	1	−1	1	−1		
E_{1g}	1	ϵ	$-\epsilon^*$	−1	$-\epsilon$	ϵ^*	1	ϵ	$-\epsilon^*$	−1	$-\epsilon$	ϵ^*	(R_x, R_y)	(xz, yz)
	1	ϵ^*	$-\epsilon$	−1	$-\epsilon^*$	ϵ	1	ϵ^*	$-\epsilon$	−1	$-\epsilon^*$	ϵ		
E_{2g}	1	$-\epsilon^*$	$-\epsilon$	1	$-\epsilon^*$	$-\epsilon$	1	$-\epsilon^*$	$-\epsilon$	1	$-\epsilon^*$	$-\epsilon$		$(x^2-y^2,\ xy)$
	1	$-\epsilon$	$-\epsilon^*$	1	$-\epsilon$	$-\epsilon^*$	1	$-\epsilon$	$-\epsilon^*$	1	$-\epsilon$	$-\epsilon^*$		
A_u	1	1	1	1	1	1	−1	−1	−1	−1	−1	−1	z, T_z	
B_u	1	−1	1	−1	1	−1	−1	1	−1	1	−1	1		
E_{1u}	1	ϵ	$-\epsilon^*$	−1	$-\epsilon$	ϵ^*	−1	$-\epsilon$	ϵ^*	1	ϵ	$-\epsilon^*$	(x, y)(T_x, T_y)	
	1	ϵ^*	$-\epsilon$	−1	$-\epsilon^*$	ϵ	−1	$-\epsilon^*$	ϵ	1	ϵ^*	$-\epsilon$		
E_{2u}	1	$-\epsilon^*$	$-\epsilon$	1	$-\epsilon^*$	$-\epsilon$	−1	ϵ^*	ϵ	−1	ϵ^*	ϵ		
	1	$-\epsilon$	$-\epsilon^*$	1	$-\epsilon$	$-\epsilon^*$	−1	ϵ	ϵ^*	−1	ϵ	ϵ^*		

6. The D_{nh} Groups

D_{2h}	E	C_2(z)	C_2(y)	C_2(x)	i	σ(xy)	σ(xz)	σ(yz)	basis vectors and components	basis functions
A_g	1	1	1	1	1	1	1	1		x^2, y^2, z^2
B_{1g}	1	1	−1	−1	1	1	−1	−1	R_z	xy
B_{2g}	1	−1	1	−1	1	−1	1	−1	R_y	xz
B_{3g}	1	−1	−1	1	1	−1	−1	1	R_x	yz
A_u	1	1	1	1	−1	−1	−1	−1		
B_{1u}	1	1	−1	−1	−1	−1	1	1	z, T_z	
B_{2u}	1	−1	1	−1	−1	1	−1	1	y, T_y	
B_{3u}	1	−1	−1	1	−1	1	1	−1	x, T_x	

D_{3h}	E	$2C_3$	$3C_2$	σ_h	$2S_3$	$3\sigma_v$	basis vectors and components	basis functions
A'_1	1	1	1	1	1	1		$x^2+y^2,\ z^2$
A'_2	1	1	−1	1	1	−1	R_z	
E'	2	−1	0	2	−1	0	(x,y)(T_x,T_y)	$(x^2-y^2,\ xy)$
A''_1	1	1	1	−1	−1	−1		
A''_2	1	1	−1	−1	−1	1	z, T_z	
E''	2	−1	0	−2	1	0	(R_x,R_y)	(xz,yz)

D_{4h}	E	$2C_4$	C_2	$2C_2'$	$2C_2''$	i	$2S_4$	σ_h	$2\sigma_v$	$2\sigma_d$	basis vectors and components	basis functions
A_{1g}	1	1	1	1	1	1	1	1	1	1		$x^2+y^2,\ z^2$
A_{2g}	1	1	1	−1	−1	1	1	1	−1	−1	R_z	
B_{1g}	1	−1	1	1	−1	1	−1	1	1	−1		x^2-y^2
B_{2g}	1	−1	1	−1	1	1	−1	1	−1	1		xy
E_g	2	0	−2	0	0	2	0	−2	0	0	(R_x,R_y)	(xz,yz)
A_{1u}	1	1	1	1	1	−1	−1	−1	−1	−1		
A_{2u}	1	1	1	−1	−1	−1	−1	−1	1	1	z, T_z	
B_{1u}	1	−1	1	1	−1	−1	1	−1	−1	1		
B_{2u}	1	−1	1	−1	1	−1	1	−1	1	−1		
E_u	2	0	−2	0	0	−2	0	2	0	0	(x,y)(T_x,T_y)	

D_{5h}	E	$2C_5$	$2C_5^2$	$5C_2$	σ_h	$2S_5$	$2S_5^3$	$5\sigma_v$	basis vectors and components	basis functions
A_1'	1	1	1	1	1	1	1	1		$x^2 + y^2, z^2$
A_2'	1	1	1	−1	1	1	1	−1	R_z	
E_1'	2	2cos 72°	2cos 144°	0	2	2cos 72°	2cos 144°	0	$(x, y)(T_x,\ T_y)$	
E_2'	2	2cos 144°	2cos 72°	0	2	2cos 144°	2cos 72°	0		$(x^2 - y^2, xy)$
A_1''	1	1	1	1	−1	−1	−1	−1		
A_2''	1	1	1	−1	−1	−1	−1	1	z, T_z	
E_1''	2	2cos 72°	2cos 144°	0	−2	−2cos 72°	−2cos 144°	0	$(R_x,\ R_y)$	(xy, yz)
E_2''	2	2cos 144°	2cos 72°	0	−2	−2cos 144°	−2cos 72°	0		

D_{6h}	E	$2C_6$	$2C_3$	C_2	$3C_2'$	$3C_2''$	i	$2S_3$	$2S_6$	σ_h	$3\sigma_d$	$3\sigma_v$	basis vectors and components	basis functions
A_{1g}	1	1	1	1	1	1	1	1	1	1	1	1		x^2+y^2, z^2
A_{2g}	1	1	1	1	–1	–1	1	1	1	1	–1	–1	R_z	
B_{1g}	1	–1	1	–1	1	–1	1	–1	1	–1	1	–1		
B_{2g}	1	–1	1	–1	–1	1	1	–1	1	–1	–1	1		
E_{1g}	2	1	–1	–2	0	0	2	1	–1	–2	0	0	(R_x, R_y)	(xy, yz)
E_{2g}	2	–1	–1	2	0	0	2	–1	–1	2	0	0		(x^2-y^2, xy)
A_{1u}	1	1	1	1	1	1	–1	–1	–1	–1	–1	–1		
A_{2u}	1	1	1	1	–1	–1	–1	–1	–1	–1	1	1	z, T_z	
B_{1u}	1	–1	1	–1	1	–1	–1	1	–1	1	–1	1		
B_{2u}	1	–1	1	–1	–1	1	–1	1	–1	1	1	–1		
E_{1u}	2	1	–1	–2	0	0	–2	–1	1	2	0	0	(x,y) (T_x, T_y)	
E_{2u}	2	–1	–1	2	0	0	–2	1	1	–2	0	0		

7. The D_{nd} Groups

D_{2d}	E	$2S_4$	C_2	$2C'_2$	$2\sigma_d$	basis vectors and components	basis functions
A_1	1	1	1	1	1		$x^2+y^2,\ z^2$
A_2	1	1	1	−1	−1	R_z	
B_1	1	−1	1	1	−1		x^2-y^2
B_2	1	−1	1	−1	1	z, T_z	xy
E	2	0	−2	0	0	(x,y)(T_x,T_y)(R_x,R_y)	(xz, yz)

D_{3d}	E	$2C_3$	$3C_2$	i	$2S_6$	$3\sigma_d$	basis vectors and components	basis functions
A_{1g}	1	1	1	1	1	1		$x^2 + y^2,\ z^2$
A_{2g}	1	1	−1	1	1	−1	R_z	
Eg	2	−1	0	2	−1	0	(R_x, R_y)	$(x^2–y^2, xy), (xz, yz)$
A_{1u}	1	1	1	−1	−1	−1		
A_{2u}	1	1	−1	−1	−1	1	z, T_z	
E_u	2	−1	0	−2	1	0	$(x,y)(T_x,T_y)$	

D_{4d}	E	$2S_8$	$2C_4$	$2S_8^3$	C_2	$4C_2'$	$4\sigma_d$	basis vectors and components	basis functions
A_1	1	1	1	1	1	1	1		$x^2+y^2,\ z^2$
A_2	1	1	1	1	1	−1	−1	R_z	
B_1	1	−1	1	−1	1	1	−1		
B_2	1	−1	1	−1	1	−1	1	z, T_z	
E_1	2	$\sqrt{2}$	0	$-\sqrt{2}$	−2	0	0	(x,y)(T_x,T_y)	
E_2	2	0	−2	0	2	0	0		(x^2-y^2, xy)
E_3	2	$-\sqrt{2}$	0	$\sqrt{2}$	−2	0	0	(R_x, R_y)	(xz, yz)

D_{5d}	E	$2C_5$	$2C_5^2$	$5C_2$	i	$2S_{10}^3$	$2S_{10}$	$5\sigma_d$	basis vectors and components	basis functions
A_{1g}	1	1	1	1	1	1	1	1		$x^2+y^2,\ z^2$
A_{2g}	1	1	1	−1	1	1	1	−1	R_z	
E_{1g}	2	2cos 72°	2cos 144°	0	2	2cos 72°	2cos 144°	0	(R_x,R_y)	(xz,yz)
E_{2g}	2	2cos 144°	2cos 72°	0	2	2cos 144°	2cos 72°	0		(x^2-y^2,xy)
A_{1u}	1	1	1	1	−1	−1	−1	−1		
A_{2u}	1	1	1	−1	−1	−1	−1	1	z,T_z	
E_{1u}	2	2cos 72°	2cos 144°	0	−2	−2cos 72°	−2cos 144°	0	$(x,y)(T_x,T_y)$	
E_{2u}	2	2cos 144°	2cos 72°	0	−2	−2cos 144°	−2cos 72°	0		

D_{6d}	E	$2S_{12}$	$2C_6$	$2S_4$	$2C_3$	$2S_{12}^5$	C_2	$6C_2'$	$6\sigma_d$	basis vectors and components	basis functions
A_1	1	1	1	1	1	1	1	1	1		x^2+y^2, z^2
A_2	1	1	1	1	1	1	1	−1	−1	R_z	
B_1	1	−1	1	−1	1	−1	1	1	−1		
B_2	1	−1	1	−1	1	−1	1	−1	1	z, T_z	
E_1	2	$\sqrt{3}$	1	0	−1	$-\sqrt{3}$	−2	0	0	(x,y)(T_x,T_y)	
E_2	2	1	−1	−2	−1	1	2	0	0		(x^2-y^2, xy)
E_3	2	0	−2	0	2	0	−2	0	0		
E_4	2	−1	−1	2	−1	−1	2	0	0		
E_5	2	$-\sqrt{3}$	1	0	−1	$\sqrt{3}$	−2	0	0	(R_x,R_y)	(xz,yz)

8. The S_n Groups

S_4	E	S_4	C_2	S_4^3	basis vectors and components	basis functions
A	1	1	1	1	R_z	x^2+y^2, z^2
B	1	−1	1	−1	z, T_z	x^2-y^2, xy
E	1	i	−1	$-i$	$(x,y),(T_x,T_y)$	(xz, yz)
	1	$-i$	−1	i	$(R_x\ R_y)$	

S_6	E	C_3	C_3^2	i	S_6^5	S_6	basis vectors and components	$\epsilon = exp(2\pi i/3)$ basis functions
A_g	1	1	1	1	1	1	R_z	$x^2 + y^2,\ z^2$
E_g	1	ϵ	ϵ^*	1	ϵ	ϵ^*	(R_x,R_y)	$(x^2 - y^2,\ xy)$
	1	ϵ^*	ϵ	1	ϵ^*	ϵ		(xz, yz)
A_u	1	1	1	-1	-1	-1	z,T_z	
E_u	1	ϵ	ϵ^*	-1	$-\epsilon$	$-\epsilon^*$	(x,y)(T_x,T_y)	
	1	ϵ^*	ϵ	-1	$-\epsilon^*$	$-\epsilon$		

S_8	E	S_8	C_4	S_8^3	C_2	S_8^5	C_4^3	S_8^7	basis vectors and components	$\epsilon = exp(2\pi i/8)$ basis functions
A	1	1	1	1	1	1	1	1	R_z	$x^2+y^2,\ z^2$
B	1	-1	1	-1	1	-1	1	-1	z, T_z	
E_1	1	ϵ	i	$-\epsilon^*$	-1	$-\epsilon$	$-i$	ϵ^*	$(x,y)(T_x,T_y)$; (R_x,R_y)	
	1	ϵ^*	$-i$	$-\epsilon$	-1	$-\epsilon^*$	i	ϵ		
E_2	1	i	-1	$-i$	1	i	-1	$-i$		$(x^2-y^2$,xy)
	1	$-i$	-1	i	1	$-i$	-1	i		
E_3	1	$-\epsilon^*$	$-i$	ϵ	-1	ϵ^*	i	$-\epsilon$		(xz,yz)
	1	$-\epsilon$	i	ϵ^*	-1	ϵ	$-i$	$-\epsilon^*$		

9. The Cubic Groups

T_d	E	$8C_3$	$3C_2$	$6S_4$	$6\sigma_d$	basis vectors and components	basis functions
A_1	1	1	1	1	1		$x^2 + y^2 + z^2$
A_2	1	1	1	−1	−1		
E	2	−1	2	0	0		$(2z^2 - x^2 - y^2, x^2 - y^2)$
T_1	3	0	−1	1	−1	(R_z,R_y,R_z)	
T_2	3	0	−1	−1	1	$(x,y,z)(T_x,T_y,T_z)$	(xy,xz,yz)

O_h	E	$8C_3$	$6C_2$	$6C_4$	$3C_2$ $(=C_4^2)$	i	$6S_4$	$8S_6$	$3\sigma_h$	$6\sigma_d$	basis vectors and components	basis functions
A_{1g}	1	1	1	1	1	1	1	1	1	1		$x^2+y^2+z^2$
A_{2g}	1	1	−1	−1	1	1	−1	1	1	−1		
E_g	2	−1	0	0	2	2	0	−1	2	0		$(2z^2-x^2-y^2, x^2-y^2)$
T_{1g}	3	0	−1	1	−1	3	1	0	−1	−1	(R_x,R_y,R_z)	
T_{2g}	3	0	1	−1	−1	3	−1	0	−1	1		(xz, yz, xy)
A_{1u}	1	1	1	1	1	−1	−1	−1	−1	−1		
A_{2u}	1	1	−1	−1	1	−1	1	−1	−1	1		
E_u	2	−1	0	0	2	−2	0	1	−2	0		
T_{1u}	3	0	−1	1	−1	−3	−1	0	1	1	(x,y,z)(T_x,T_y,T_z)	
T_{2u}	3	0	1	−1	−1	−3	1	0	1	−1		

10. The Groups $C_{\propto v}$ and $D_{\propto h}$

$C_{\propto v}$	E	$2C_{\propto}(\phi)$	$\cdots$	$\propto \sigma_v$	basis vectors and components	basis functions
$A_1 \equiv \Sigma^+$	1	1	$\cdots$	1	z, T_z	x^2+y^2, z^2
$A_2 \equiv \Sigma^-$	1	1	$\cdots$	-1	R_z	
$E_1 \equiv \Pi$	2	$2\cos\phi$	$\cdots$	0	(x,y)(T_x,T_y),(R_x,R_y)	(xz,yz)
$E_2 \equiv \Delta$	2	$2\cos 2\phi$	$\cdots$	0		(x^2-y^2, xy)
$E_3 \equiv \Phi$	2	$2\cos 3\phi$	$\cdots$	0		
$\cdots$	$\cdots$	$\cdots$	$\cdots$	$\cdots$		

$D_{\infty h}$	E	$2C_{\infty}(\phi)$	$\cdots$	$\infty\,\sigma_v$	i	$2S_{\infty}(\phi)$	$\cdots$	∞C_2	basis vectors and components	basis functions
Σ_g^+	1	1	$\cdots$	1	1	1	$\cdots$	1		x^2+y^2, z^2
Σ_g^-	1	1	$\cdots$	−1	1	1	$\cdots$	−1	R_z	
Π_g	2	$2\cos\phi$	$\cdots$	0	2	$-2\cos\phi$	$\cdots$	0	(R_x,R_y)	(xz,yz)
Δ_g	2	$2\cos 2\phi$	$\cdots$	0	2	$2\cos 2\phi$	$\cdots$	0		$(x^2-y^2$,xy)
$\cdots$	$\cdots$	$\cdots$	$\cdots$	$\cdots$	$\cdots$	$\cdots$	$\cdots$	$\cdots$	$\cdots$	
Σ_u^+	1	1	$\cdots$	1	−1	−1	$\cdots$	−1	z,T_z	
Σ_u^-	1	1	$\cdots$	−1	−1	−1	$\cdots$	1		
Π_u	2	$2\cos\phi$	$\cdots$	0	−2	$2\cos\phi$	$\cdots$	0	(x,y)(T_x,T_y)	
Δ_u	2	$2\cos 2\phi$	$\cdots$	0	−2	$-2\cos 2\phi$	$\cdots$	0		
$\cdots$	$\cdots$	$\cdots$	$\cdots$	$\cdots$	$\cdots$	$\cdots$	$\cdots$	$\cdots$		

APPENDIX III

Character Tables of Two-valued Representations of Point Groups

	D_2'	E	R	$C_2(z)$ $RC_2(z)$	$C_2(y)$ $RC_2(y)$	$C_2(x)$ $RC_2(x)$
Γ'_1	A'	1	1	1	1	1
Γ'_2	B_1'	1	1	1	−1	−1
Γ'_3	B_2'	1	1	−1	1	−1
Γ'_4	B_3'	1	1	−1	−1	1
Γ'_5	E'	2	−2	0	0	0

	D'_3	E	R	C_3^2R C_3	C_3R C_3^2	$3C_2$	$3C_2R$
Γ'_1	A'_1	1	1	1	1	1	1
Γ'_2	A'_2	1	1	1	1	−1	−1
Γ'_3	E'_1	2	2	−1	−1	0	0
Γ'_4	E'_2	{ 1 1	−1 −1	−1 −1	1 1	i $-i$	$-i$ i
Γ'_5	E'_3	2	−2	1	−1	0	0

	D'_4	E	R	C_4^2R C_4	C_4R C_4^3	C_2R C_2	$2C_2'R$ $2C_2'$	$2C_2''R$ $2C_2''$
Γ'_1	A'_1	1	1	1	1	1	1	1
Γ'_2	A'_2	1	1	1	1	1	−1	−1
Γ'_3	B'_1	1	1	−1	−1	1	1	−1
Γ'_4	B'_2	1	1	−1	−1	1	−1	1
Γ'_5	E'_1	2	2	0	0	−2	0	0
Γ'_6	E'_2	2	−2	$\sqrt{2}$	$-\sqrt{2}$	0	0	0
Γ'_7	E'_3	2	−2	$-\sqrt{2}$	$\sqrt{2}$	0	0	0

	O′	E	R	$4RC_3^2$ $4C_3$	$4RC_3$ $4C_3^2$	$3RC_2$ $3C_2$	$3RC_4^3$ $3C_4$	$3RC_4$ $3C_4^3$	$6RC_2'$ $6C_2'$
Γ'_1	A'_1	1	1	1	1	1	1	1	1
Γ'_2	A'_2	1	1	1	1	1	−1	−1	−1
Γ'_3	E'_1	2	2	−1	−1	2	0	0	0
Γ'_4	T'_1	3	3	0	0	−1	1	1	−1
Γ'_5	T'_2	3	3	0	0	−1	−1	−1	1
Γ'_6	E'_2	2	−2	1	−1	0	$\sqrt{2}$	$-\sqrt{2}$	0
Γ'_7	E'_3	2	−2	1	−1	0	$-\sqrt{2}$	$\sqrt{2}$	0
Γ'_8	G'	4	−4	−1	1	0	0	0	0

Subject Index